W9-CNG-632

Quantum Mechanics

Quantum Mechanics

P.J.E. Peebles

Princeton University Press Princeton, New Jersey

Published by Princeton University Press, 41 William Street,
Princeton, New Jersey 18540
In the United Kingdom: Princeton University Press, Oxford

Library of Congress Cataloging-in-Publication Data
Peebles, P. J. E. (Phillip James Edwin)
Quantum Mechanics / P.J.E. Peebles.
p. cm.
Includes index.
ISBN 0-691-08755-5 (CL : alk. paper)
1. Quantum theory. I. Title.
QC174.12.P4 1992
530.1'2—dc20 91-39340
CIP

This book has been composed in Computer Modern
and Optima using TEX

Princeton University Press books are printed on acid-free paper,
and meet the guidelines for permanence and durability of the
Committee on Production Guidelines for Book Longevity of
the Council on Library Resources

Printed in the United States of America
by Princeton University Press,
Princeton, New Jersey

10 9 8 7 6 5 4 3 2 1

To Alison

Contents

Preface

To understand quantum mechanics one ought to see how this remarkable world picture was discovered, and one must work through the details of some nontrivial applications that show how the theory works. Most of us also require a good deal of practice working physically interesting problems. My approach in this book, that grew out of a heavy but I hope not impossibly hard one-year course, is to introduce the full details of the theory as needed, so one can see specific and interesting applications, and to make room for this by dropping things like the analysis of special functions that are not important for the chosen applications. My hope is that this book will find a niche between introductory surveys, that can give some idea of what is going on but may leave the reader with the feeling that there are mysterious corners to the theory, and the standard treatises, which contain more than most of us want to know about quantum mechanics.

My arrangement of material roughly follows standard precedents, but with some exceptions that should be explained. The first chapter presents the origins of quantum mechanics in the usual pseudohistorical style of physics. It is important to convince the reader that the theory was not derived from measurements, nor discovered by a single theoretical stroke, but instead grew by a complicated interplay of experimental hints, theoretical insights, and good luck, intermingled with many wrong turns. An adequate study of this development would take a whole course; the survey presented here is misleading because it ignores the wrong turns, and incorrect because it doesn't even present the main advances in the right order, but there is not time to do better. Instead, I attempt to give some flavor of what went on by presenting a set of examples of physics that are well worth knowing independent of their historical interest. Here, and throughout the book, sections marked by an asterisk contain material I do not always include in the course for lack of time. For example, I like the treatment of phonons in section 12c as an introduction to the eigenvalue problem, but usually conclude that it takes more time than I can afford.

I hope the introductory chapter shows how people could have hit on wave mechanics, if not how it really happened. Chapter 2 develops the wave mechanics formalism. The emphasis here is on symmetries and conservation laws: parity, linear and angular momentum, and the electromagnetic interaction. The only specific physical application is the completion of the study of an isolated hydrogen atom, with some discussion of the motion of a particle in a magnetic field. This is a little dry, but of course it is needed if one is to do grown-up quantum mechanics, and I think the symmetry methods are clarified by presenting the standard cases all in the same chapter.

The formal development could end here, but I find it more satisfying, and not a lot more time consuming, to redevelop the theory in the abstract Dirac bracket formalism. It is fairly easy, and I think fascinating, to see how wave mechanics follows from the position representation and the canonical commutation relations. The main new application in this chapter is the treatment of spin.

At this point, which is about midway in the course, one is ready to practice quantum mechanics by applying it to real physical problems, but I think it is good to pause and consider measurement theory. Since this undoubtedly is part of the physics, it is striking to see how little space is devoted to it in many of the standard books (with notable exceptions, including David Bohm's *Quantum Theory*, and some more recent books such as Hans Ohanian's *Principles of Quantum Mechanics*). One reason is easy to see: a physicist can spend a career working on quantum mechanics without thinking about measurement theory beyond the bare prescription that can be written down in a few paragraphs. Another likely reason is that the attempt to decide what the measurement prescription really is telling us about the deeper nature of physical reality is a slippery business that so far has led to no fully satisfying conclusion. But much the same is true of any open research problem, and a discussion of open questions in the measurement puzzle may be a useful antidote to our tendency in physics textbooks to gloss over complexities. There is considerably more material on measurement theory in chapter 4 than I ever got through in the lectures, but I hope the notes may grab the imagination.

The next three chapters present a selection of the usual applications drawn from perturbation theory, atomic and molecular structure, and scattering theory. The particular choice of applications is less important than that one has a chance to see in detail how the computations go. The main topic in perturbation theory is the energy and spontaneous decay rate of the 21 cm hyperfine line in atomic hydrogen. I like this problem because it requires grown-up computations that can be presented in a reasonably short space. Chapter 6 presents some examples from atomic and molecular structure. I go somewhat further into atomic structure than is standard in books at this level, because the rich physics of atoms offers interesting and not overwhelmingly complicated computations. Chapter 7 presents some basic methods in scattering theory. I simplify the partial wave analysis by concentrating on s-wave scattering; this allows an easy treatment of interesting effects such as resonances and absorption.

I have placed more emphasis than is usual in a textbook on numerical quantities and order of magnitude estimates. People looking into physics for the first time tend not to like this part of the subject, but since it is an essential part of the game one does well to get used to it as early as possible.

I have not followed the movement toward the adoption of a standardized set of scientific units. The purpose is laudable: if we all used the same units it surely would aid communication with colleagues in different disciplines and subdisciplines. But the first imperative is to communicate with people working in the same field, and since progress toward adoption of fully rationalized units has not been rapid I have felt free to adopt the units I suspect most of us feel comfortable with in this subject. These include Gaussian cgs units for electromagnetism and the usual heterogeneous set of units for length, frequency, and energy.

I should also warn the reader that this book is not the place to find mathematically complete discussions. Some theorems are hinted at, but none of any substance is proved, and I have not even been careful to state the conditions under which theorems apply. This is a shortcoming, but I hope not one that will interfere with the communication of the physical ideas.

I have associated a few ideas with prominent physicists, but have made no attempt to identify these people or otherwise systematically to trace the history of ideas in quantum physics. However, if this book is successful it gives the basic physics needed to understand articles on the history of quantum mechanics, as well as a good start on the enormous variety of modern applications.

This book is the final form of the lecture notes I have distributed to the students in the quantum mechanics course I taught off and on for the last twenty years at Princeton University. Early generations of students suffered through handwritten and usually hard to read Ditto copies of the equations. The notes evolved into a printed version because I learned the word processing language TeX (and probably also because I did not anticipate how long it would take to add the words to the equations). I am grateful to the students and graders for their input; they made a considerable contribution to whatever merits this book may have. I particularly thank David Reiley for a thorough reading of the semifinal draft, and Jim Hartle and David Eichler for discussions of measurement theory.

It goes without saying, but nonetheless ought to be said, that the provenance of many of the problems in this book is only vaguely known to me. I made up some, others were adapted from the Princeton graduate general exams, and the rest were suggested by colleagues. Many problems, as well as my understanding of this subject, trace back to the people who taught me quantum mechanics: Sam Neamton at the University of Manitoba; Murph Goldberger at Princeton; and my professor of continuing education, Bob Dicke.

Quantum Mechanics

CHAPTER 1

HISTORICAL DEVELOPMENT

The story of how people hit on the highly nonintuitive world picture of quantum mechanics, in which the physical state of a system is represented by an element in an abstract linear space and its observable properties by operators in the space, is fascinating and exceedingly complicated. The theory could not have been deduced from experiment, for the elements of the linear space are in principle not observable. It is also true that the theory did not arise from one person's great insight, as happened in Einstein's discovery of general relativity theory. The much greater change from the classical world picture of Newtonian mechanics and general relativity to the quantum world picture came in many steps taken by many people, often against the better judgment of participants.

The goal of this chapter is to show how classical physicists could have hit on wave mechanics. The strategy is to select topics that still are (or ought to be) part of the fundamental lore of any modern physicist. There are three major elements in the story. The first is the experimental evidence that the energy of an isolated system can only assume special discrete or quantized values. The second is the idea that the energy is proportional to the frequency of a wave function associated with the system. (This is the famous de Broglie relation $E = h\nu$, for energy E and frequency ν). The third is the connection between the de Broglie relation and energy quantization through the mathematical result that a wave equation with fixed boundary conditions allows only discrete quantized values of the frequency of oscillation of the wave function (as in the fundamental and harmonics of the vibration of a violin string). Some substantial computations are presented in this chapter, but the physics is introduced piecemeal, as needed. The principles of wave mechanics are collected in the next chapter, and are generalized to an abstract linear space in chapter 3.

1 Energy Quantization and Heat Capacities

The Boltzmann Distribution

Consider an object—an atom, molecule, rock—in a mechanically stable state and well isolated from its surroundings. In classical or quantum theory the object has a definite energy, E, that is conserved. Also, if the object consists of several weakly coupled parts E is the sum of the energies of the parts. In classical mechanics, E can assume any value from some minimum to the maximum allowed by stability. In quantum theory the possible values of the energy are discrete, or quantized,

$$E = E_i, \quad i = 0, 1, 2, \ldots, \tag{1.1}$$

with E_0 the ground state energy, E_1 the energy in the first excited state, and so on. This remarkable quantization concept first appeared in 1900, in Planck's derivation of the blackbody radiation spectrum, as described in section 2. We will consider first the relevance of energy quantization to heat capacities of material objects, because the analysis is a little less lengthy.

To describe what happens when an object is heated to a given temperature T, let us imagine we have a statistical ensemble of $M \gg 1$ mechanically identical copies of the object, each of which has been placed in contact with a heat reservoir at temperature T, allowed to come to equilibrium, and then isolated. The reservoir is a macroscopic body much larger than the object. The ensemble might literally be a collection of objects, such as a large number of nearly free atoms, or we can think of the ensemble as representing one almost isolated object that is sampled at widely separated times.

The accidents of interaction of each object with the enormous number of atoms in the reservoir determine the probability distribution of final energies of the objects in the ensemble. Let N_i be the number of the M objects that are found to be in the ith energy level. Then in the limit $M \to \infty$ the probability of finding that a randomly chosen object from the ensemble is in level i is defined to

$$P_i = N_i/M. \tag{1.2}$$

The value of M is required to suppress sampling fluctuations. If the ensemble represents one object sampled at many different times, P_i is the probability that the object observed at a randomly chosen time is found to be in level i.

It will be assumed that the probability P_i in equation (1.2) depends only on the temperature T of the reservoir and on the energy E_i of the object (or more generally on the conserved quantities, which could include particle number), so at fixed temperature T the probability P_i is some function of energy,

$$P_i = F(E_i). \tag{1.3}$$

This assumption is justified below, in section 26 on measurement theory. For now the problem is to find the function $F(E_i)$.

Suppose the object consists of two weakly interacting parts, 1 and 2, so the allowed values of the energy of the object are of the form

$$E_i = E_a^1 + E_b^2, \tag{1.4}$$

for all combinations a, b of energy levels E_a^1 of part 1 and E_b^2 of part 2. The probability that part 1 is found to have energy E_a^1 is $P_a^1 = F(E_a^1)$, and part 2 has energy E_b^2 with probability $P_b^2 = F(E_b^2)$. Since the two parts are not interacting, the probability that one part has a given energy cannot depend on what the energy of the other part happens to be, that is, the parts are statistically independent. Since probabilities for independent events multiply, the probability that the object that consists of the two parts is in the energy level E_i in equation (1.4) is

$$P_i = P_a^1 P_b^2. \tag{1.5}$$

By equation (1.3) this is

$$F(E_a^1 + E_a^2) = F(E_a^1)F(E_b^2). \tag{1.6}$$

Since this equation is supposed to hold whatever the energies, we can write it as

$$F(E_a + E_b) = F(E_a)F(E_a), \tag{1.7}$$

for any values of E_a and E_b.

If it is not obvious that the solution to the functional equation (1.7) is an exponential, take the logarithm and differentiate with respect to E_a or E_b:

$$\frac{d}{dE_a} \log F(E) = \frac{1}{F(E)}\frac{dF}{dE} = \frac{1}{F(E_a)}\frac{dF}{dE_a} = \frac{1}{F(E_b)}\frac{dF}{dE_b} = -\beta, \tag{1.8}$$

where $E = E_a + E_b$. Here and always the natural logarithm is written as $\log F$. The first step follows from the chain rule in calculus. The second step follows from equation (1.7), and the third step follows by differentiating with respect to E_b instead of E_a. The final step is to define the expression to be equal to the negative of β, the minus sign being chosen to make the solution (eq. [1.9]) look sensible.

We see from the third part of equation (1.8) that β cannot depend on E_b, and from the fourth part that it cannot depend on E_a. That means β is a constant. The solution to equation (1.8) therefore is

$$P = F(E) \propto e^{-\beta E}. \tag{1.9}$$

With β positive, the probability P in equations (1.3) and (1.9) approaches zero at $E \to \infty$, as is reasonable.

We will define the temperature T of the reservoir by the equation

$$\beta = \frac{1}{kT}, \tag{1.10}$$

where Boltzmann's constant is

$$k = 1.38 \times 10^{-16}\,\mathrm{erg\,deg}^{-1}, \tag{1.11}$$

and the temperature is measured in degrees Kelvin. (Recall that zero degrees Centigrade is 273° K.) This has the reasonable feature that the higher the temperature the higher the probable values of the energy. Equation (1.10) is equivalent to the more formal definition of temperature in statistical mechanics.

The definition of temperature in equation (1.10) brings equation (1.9) to

$$P \propto e^{-E/kT}. \tag{1.12}$$

This is the Boltzmann probability distribution for the energy E of an object prepared by allowing it to relax to thermal equilibrium with a heat reservoir at temperature T.

The normalization condition on the Boltzmann distribution is that the probabilities summed over all possible values of the energy have to add to unity:

$$\sum_i P_i = 1. \tag{1.13}$$

A useful symbol for expressing the normalization is the partition function,

$$Z = \sum_i e^{-\beta E_i} = \sum_i e^{-E_i/kT}, \tag{1.14}$$

in terms of which the form of the normalized Boltzmann distribution is

$$P_i = \frac{e^{-E_i/kT}}{Z}. \tag{1.15}$$

The mean thermal energy of the object is the result of averaging the energies across the ensemble. Thus, if N_i of the M objects in the ensemble have energy E_i, the arithmetic mean value or expectation value of the energies of the objects is

$$\langle E \rangle = U = \frac{\sum_i E_i N_i}{M}. \tag{1.16}$$

With $P_i = N_i/M$ (eq. [1.2]), the average is

$$U = \sum_i E_i P_i. \tag{1.17}$$

For the Boltzmann distribution (1.15), this is

$$U = \frac{\sum E_i e^{-\beta E_i}}{\sum e^{-\beta E_i}}, \tag{1.18}$$

where β always means $1/kT$ (eq. [1.10]). Finally, using the partition function (1.14), we can write this expression for the average energy in the handy form

$$U = -\frac{d}{d\beta} \log Z. \tag{1.19}$$

Here and always log means the natural logarithm.

The Thermal Energy of a Simple Harmonic Oscillator

The Hamiltonian (the expression for the total energy) of a one-dimensional simple harmonic oscillator with displacement variable $x(t)$ is

$$\begin{aligned} H &= \frac{p^2}{2m} + \frac{Kx^2}{2}, \\ \omega &= 2\pi\nu = (K/m)^{1/2}. \end{aligned} \tag{1.20}$$

The momentum is $p = m\dot{x} = m\,dx/dt$, with m the mass, so the kinetic energy is $p^2/2m$. The spring constant is K, and the potential energy is $Kx^2/2$. As is readily checked, the natural frequency of the oscillator is $\omega = (K/m)^{1/2}$ (units of radians per second) or $\nu = \omega/(2\pi)$ (units of cycles per second = Hertz).

As discussed in the next section and in section 37, an electromagnetic radiation field can be described as a set of simple harmonic oscillators, one for each mode of oscillation. In working through the theory of thermal blackbody radiation, Planck introduced the constraint, as an intermediate step in the calculation, that the energy of each oscillator is only allowed to assume the discrete values

$$E_n = nh\nu = n\hbar\omega, \qquad n = 0, 1, 2, \ldots. \tag{1.21}$$

(We will use h and $\hbar \equiv h/2\pi$, as convenient.) Planck's sensible plan was to take the limit $h \to 0$ at the end of the calculation, but he noticed that the predicted blackbody spectrum would agree with the measurements if instead he took h to be a nonzero constant,

$$\hbar = \frac{h}{2\pi} = 1.05457 \times 10^{-27}\,\mathrm{erg\,s}. \tag{1.22}$$

The value quoted here is the modern result. The only other improvement to the energy spectrum (allowed values of the energy) of a simple harmonic oscillator is to replace the integers n with $n + 1/2$. The additive constant of course does not affect a heat capacity (which is the rate of change of mean energy with temperature).

Einstein proposed that Planck's quantization rule might apply to a material oscillator such as an atom oscillating about its equilibrium position in a solid. Let us see how that would affect the heat capacity.

With Planck's quantization rule (1.21), the partition function (eq. [1.14]) for a one-dimensional simple harmonic oscillator is

$$Z = \sum_0^\infty e^{-nh\nu/kT} = \sum A^n, \tag{1.23}$$

with $A = e^{-h\nu/kT} = e^{-\beta h\nu}$. The trick for evaluating this sum is to note that we can write it

$$\begin{aligned} Z &= 1 + A + A^2 + A^3 + \ldots \\ &= 1 + A[1 + A + A^2 + \ldots] \\ &= 1 + AZ. \end{aligned} \tag{1.24}$$

Thus we see that the sum is

$$Z = \frac{1}{1 - A} = \frac{1}{1 - e^{-\beta h\nu}} = \frac{1}{1 - e^{-h\nu/kT}}. \tag{1.25}$$

Equation (1.19) gives the mean thermal energy,

$$U = \frac{d}{d\beta} \log(1 - e^{-\beta h\nu}). \tag{1.26}$$

On differentiating this expression out we get

$$U = \frac{h\nu}{e^{h\nu/kT} - 1}. \tag{1.27}$$

This is the wanted expression for the mean thermal energy of a one-dimensional simple harmonic oscillator with natural frequency ν at temperature T.

The classical limit is obtained at high temperature, $kT \gg h\nu$. When $h\nu/kT$ is small, the Taylor series expansion of the exponential in equation (1.27), keeping only the first nontrivial term, is

$$e^{h\nu/kT} \sim 1 + \frac{h\nu}{kT}. \tag{1.28}$$

This brings equation (1.27) to

$$U = kT. \tag{1.29}$$

The heat capacity in this limit is $C = dU/dT = k$. This is a special case of the classical energy equipartition theorem. The theorem says that for every quadratic term in position or momentum in the Hamiltonian there is a contribution $kT/2$ to the mean thermal energy of the system. There are two quadratic terms in equation (1.20), giving a net value of kT, which checks equation (1.29). Of course, a reasonable quantum theory must agree with classical physics in the high energy limit where we know classical physics works.

In the opposite low temperature limit, $kT \ll h\nu$, the mean energy in equation (1.27) is suppressed by the exponential in the denominator, as is the heat capacity. That is, Planck's energy quantization assumption in equation (1.21) leads to a characteristic temperature $T_c = h\nu/k$ for an oscillator with natural frequency ν. If the temperature is much larger than T_c the energy quantization is scarcely noticeable, and we see classical behavior. If the temperature is well below T_c the situation is decidedly nonclassical: the oscillator is forced to the ground state that has the minimum allowed value of the energy. As discussed next, a similar effect applies to the kinetic energy of tumbling of a molecule in a gas.

Heat Capacity of Molecular Hydrogen

From the energy equipartition theorem of classical statistical mechanics we would have expected that the mean thermal energy of a gas of N hydrogen molecules is

$$\begin{aligned} U =& \frac{1}{2}NkT[3 \text{ (for the kinetic energy of translation in 3 dimensions)} \\ &+ 2 \text{ (for rotation of the axis in two directions)} \\ &+ 2 \text{ (for vibration along the axis)} \\ &+ 1 \text{ (for rotation about the axis)}], \end{aligned} \tag{1.30}$$

plus maybe more for vibrations of the internal structures of the individual atoms.

At $T \lesssim 100\,\mathrm{K}$ the measured heat capacity is $dU/dT \sim 3Nk/2$, so the hydrogen molecules act like a gas of pointlike particles, the only energy being the kinetic energy of translation. Following the discussion of the simple harmonic oscillator, we conclude that the energy levels corresponding to the kinetic energy of translation are close together compared to kT at $T \sim 100\,\mathrm{K}$, so classical energy equipartion applies to the motions of the molecules, and that the energy levels corresponding to the other modes of motion in equation (1.30) are more broadly separated, so these modes are not appreciably excited at $T \sim 100\,\mathrm{K}$.

At $T \sim 200$ to $400\,\mathrm{K}$ the heat capacity of molecular hydrogen gas is close to $dU/dT \sim 5Nk/2$, which is that of a classical gas of rigid dumbbells (the first two lines of eq. [1.30]). This means the energy of the first rotationally excited state of the molecule exceeds that of the ground state by the amount

$$E_1 - E_0 \sim kT_R, \qquad T_R \sim 200\,\mathrm{K}. \tag{1.31}$$

The allowed values of angular momentum in quantum mechanics will be computed in section 17. A useful order of magnitude approximation is Bohr's assumption, that the rotationally excited states are spaced at increments of angular momentum equal to $\hbar$. (This is discussed in section 4 below.) Let us check that these numbers make sense.

If the hydrogen molecule has angular momentum $\hbar$ in the first rotationally excited state, and the moment of inertia of the molecule is I, then the kinetic energy of rotation in this state is

$$U_R = \frac{\hbar^2}{2I} \sim kT_R, \tag{1.32}$$

with $T_R \sim 200K$. The first equation is the classical expression for kinetic energy of rotation. The second equation with equation (1.11) for k and (1.22) for $\hbar$ gives $I \sim 2 \times 10^{-41}\,\mathrm{g\,cm}^2$. We are only interested in checking the orders of magnitude, so let us approximate the moment of inertia of the molecule as $I \sim m_p r^2$, where

$$m_p = 1.67 \times 10^{-24}\,\mathrm{g} \tag{1.33}$$

is the proton mass and r is the separation of the two protons in the molecule. That gives $r \sim 3 \times 10^{-9}\,\mathrm{cm} = 0.3\,\text{Å}$. The size of a hydrogen atom is set by the Bohr radius (eq. [4.9] below). Our result is about half a Bohr radius, reasonably close considering the rough approximations.

At $T \sim 2000\,\mathrm{K}$ the heat capacity approaches that of a classical gas of dumbbells each of which can vibrate in length. This means the first vibrationally excited state of the molecule has energy roughly an order of magnitude above the first rotationally excited state. At $T \sim 3000\,\mathrm{K}$ the gas dissociates into atomic hydrogen.

Einstein and Debye Solids

A solid stores energy in the vibrations of the atoms about their equilibrium positions. In the simplest approximation, which Einstein considered, each atom vibrates with the same frequency, ν, in each of three dimensions, so a solid containing N atoms can be thought of as $3N$ one-dimensional simple harmonic oscillators. The thermal energy of the solid is then, by equation (1.27),

$$U = \frac{3Nh\nu}{e^{h\nu/kT} - 1}. \tag{1.34}$$

The high temperature limit is $U = 3NkT$, as in equations (1.28) and (1.29), so the heat capacity at high temperature is the classical energy equipartition expression

$$C = \frac{\partial U}{\partial T} = 3Nk. \tag{1.35}$$

By 1900 it was known that equation (1.35) is a good approximation to the heat capacities of solids at room temperature (this is the empirical law of Dulong and Petit), but Nernst had found that the heat capacity drops well below this value at low temperature, approaching zero at $T \to 0$. Einstein (1907) showed how the energy quantization assumption allows us to understand the decrease of heat capacity at low temperature: the heat capacity in equation (1.34) is strongly suppressed at $T \ll h\nu/k$.

Though the Einstein model gives the right qualitative picture, it says the heat capacity goes to zero at low temperature much faster than the measurements. It is easy to see why. When an atom moves it can bring its neighbors with it. This lowers the restoring force, which greatly lowers the frequency. That is, a solid acts like a collection of oscillators with a wide range of different frequencies. The lower frequency modes of oscillation are thermally excited at lower temperatures, so the heat capacity varies more slowly with temperature than it would if all the frequencies were the same. The Debye model to be discussed next approximates the low frequency modes of vibration of the solid as sound or pressure waves. The computation is lengthy but worth knowing, because it is used not only here but in the theory of blackbody radiation (section 2) and radiative transitions (section 37).

The low frequency modes that can be excited at low temperatures have long wavelengths and so are not much affected by the fact that the mass is in discrete lumps, in the atoms. For these long wavelength modes it is a good approximation to treat the solid as a continuous fluid, with smoothly varying mass density $\rho(\mathbf{r}, t)$ and velocity $\mathbf{v}(\mathbf{r}, t)$.

The mass and velocity functions obey two equations that express mass conservation and momentum conservation. The former is

$$\frac{\partial \rho}{\partial t} + \nabla \cdot \rho \mathbf{v} = 0, \tag{1.36}$$

while Newton's law $\mathbf{F} = m\mathbf{a}$ generalizes for a fluid to

$$\frac{\partial \mathbf{v}}{\partial t} + (\mathbf{v} \cdot \nabla)\mathbf{v} = -\nabla P/\rho. \tag{1.37}$$

The pressure is P, and $-\nabla P$ is the pressure force per unit volume. This force per unit volume divided by the mass per unit volume is the acceleration of a given fluid element. On the left-hand side of equation (1.37), $\partial \mathbf{v}/\partial t$ is the rate of change of the fluid velocity at a fixed point, and the second term converts this to the rate of change of velocity of a fixed fluid element. This combination is called the convective derivative. We see the same combination in the mass conservation equation (1.36) if we rewrite it as

$$\frac{\partial \rho}{\partial t} + \mathbf{v} \cdot \nabla \rho = -\rho \nabla \cdot \mathbf{v}. \tag{1.38}$$

Gauss's law can be used to rewrite the mass conservation equation as

$$\begin{aligned} \frac{dM}{dt} &= \frac{d}{dt} \int dV \rho = \int dV \frac{\partial \rho}{\partial t} = -\int dV \, \nabla \cdot \rho \mathbf{v} \\ &= -\oint dA \, \mathbf{n} \cdot \rho \mathbf{v}. \end{aligned} \tag{1.39}$$

The last integral is over a fixed surface that contains mass M. The last line says the time rate of change of M is fixed by the surface integral of the mass flux,

$$\mathbf{F} = \rho \mathbf{v}, \tag{1.40}$$

which is the rate at which mass is flowing through the surface.

We are interested in low amplitude vibrations, for which $\mathbf{v}$ is small and ρ close to homogeneous, so we will write the mass density as

$$\rho(\mathbf{r}, t) = \rho_o(1 + \delta(\mathbf{r}, t)), \tag{1.41}$$

where ρ_o is the constant mean value, and keep only terms of first order in the perturbations δ or $\mathbf{v}$. In this approximation, we can write the pressure as

$$P = P(\rho) = P_o + c_s^2 \rho_o \delta, \qquad c_s^2 = \frac{dP}{d\rho}. \tag{1.42}$$

As indicated, we are assuming the pressure P is a single valued function of density alone. The function has been expanded in a Taylor series, keeping only the constant part at $\delta = 0$ and the first order correction. In this order in perturbation theory, where we drop terms of order δ^2, $\mathbf{v}\delta$, and v^2, equations (1.37) and (1.38) become

$$\begin{aligned} \frac{\partial \delta}{\partial t} &= -\nabla \cdot \mathbf{v}, \\ \frac{\partial \mathbf{v}}{\partial t} &= -c_s^2 \nabla \delta. \end{aligned} \tag{1.43}$$

We can eliminate the velocity by taking the time derivative of the first equation and the divergence of the second, and exchanging order of differentiation. The result is

$$\frac{\partial^2 \delta}{\partial t^2} = c_s^2 \nabla^2 \delta. \tag{1.44}$$

To see why equation (1.44) is called a wave equation note that, as is readily checked, a solution is

$$\delta = F(x - c_s t), \tag{1.45}$$

where F is a differentiable function of the single variable $w = x - c_s t$, and x is the position along the x axis in a cartesian coordinate system. This solution represents a pressure wave moving without change of shape at the speed of sound, c_s, in the x direction.

The allowed frequencies of sound waves in an isolated solid depend on its shape. To simplify things, let us consider a cube of the solid with volume V, side $L = V^{1/3}$. The surface will be assumed to be free, meaning that the pressure at the surface vanishes. Therefore the constant P_o in equation (1.42) has to vanish, and δ has to vanish at the surface. A set of solutions of the wave equation that satisfy this boundary condition is

$$\delta = A \cos(\omega t - \phi) \sin(k_x x) \sin(k_y y) \sin(k_z z), \tag{1.46}$$

where A and ϕ are constants. This function satisfies the wave equation (1.44) if the frequency ω satisfies the relation

$$\omega = k c_s, \qquad k^2 = k_x^2 + k_y^2 + k_z^2. \tag{1.47}$$

To assign the boundary condition $\delta = 0$, equation (1.46) places the sides of the cube at $x = 0$ and $x = L$; $y = 0$ and $y = L$; and $z = 0$ and $z = L$. Then we satisfy the boundary conditions if the constants k_x, k_y, and k_z are chosen so $k_x L$, $k_y L$, and $k_z L$ are integer multiples of π. We will write these conditions as

$$k_\alpha L = n_\alpha \pi, \qquad n_\alpha = 0, 1, 2 \ldots, \tag{1.48}$$

where the index $\alpha = 1, 2, 3$ refers to the x, y, and z components. (A word about notation: a vector may be specified by a boldface symbol, tably averaged over pressure and shear modes. This gives the Debye ıation,

$$\frac{U}{V} = \frac{\pi^2}{10} \frac{(kT)^4}{(\hbar c_o)^3}. \tag{1.63}$$

will be recalled that this equation applies at low temperature where y long wavelength modes are excited. It gives a good approximation the low temperature heat capacity of many solids. In others there significant additional contributions, such as from thermal motions of trons.

Blackbody Radiation

at Was Known in 1900

sider a black cavity with walls at temperature T and a small hole t us sample the radiation it contains. "Black" means that any light enters the hole from outside is absorbed; any radiation coming out emitted by the walls. At thermal equilibrium, the radiation energy frequency in the range ω to $\omega + d\omega$ found in the volume element n the cavity is

$$du = u_\omega dV d\omega. \tag{2.1}$$

ndicated, the energy has to be proportional to the size dV of the me element and to the bandwidth $d\omega$. The constant of proportion-, u_ω, is the spectral energy density, the energy per unit volume and bandwidth.

The second law of thermodynamics says u_ω can only depend on ω on the wall temperature, T, independent of the nature of the wall. we can imagine connecting two cavities made of different materials e same temperature by a light pipe that passes only frequencies in ange ω to $\omega + d\omega$. If the radiation energy densities were different e two cavities, we would find that heat is moving spontaneously one reservoir to another at the same temperature, which alas is lden by the second law.

he net energy density is

$$u = \int_0^\infty u_\omega d\omega = aT^4. \tag{2.2}$$

as $\mathbf{r}$, or by its components, as $\mathbf{r} = (x, y, z)$, or by the index notation r_α with $r_1 = x$, $r_2 = y$, and $r_3 = z$.)

Equations (1.46) to (1.48) describe the normal modes of pressure oscillations of the solid in the fluid model, which we have noted is a good approximation at low temperatures where only the low frequency long wavelength modes are excited. (In a normal mode each mass element vibrates with the same frequency, as in eq. [1.46]. The word normal refers to the orthogonality relations discussed in section 12.) Since each mode behaves as a simple harmonic oscillator, we will follow Planck and Einstein in assuming that the allowed values of the energy of each mode are quantized, $E = h\nu = \hbar\omega$, where ω is the classical frequency of vibration of the mode (eq. [1.21]). (This assumption is justified in section 12 below.) Then at temperature T the mean thermal energy of the solid is given by equation (1.27):

$$U = \sum_{n_\alpha \geq 0} \frac{\hbar\omega_{\mathbf{n}}}{e^{\hbar\omega_{\mathbf{n}}/kT} - 1}. \tag{1.49}$$

The sum is over the triplets of nonnegative integers, with $\omega_{\mathbf{n}}$ given by equations (1.47) and (1.48),

$$\omega_{\mathbf{n}} = \frac{\pi c_s}{L}(n_x^2 + n_y^2 + n_z^2)^{1/2}. \tag{1.50}$$

The sum in equation (1.49) can be approximated by an integral, as follows. Let us write the change in k_α in equation (1.48) when n_α is incremented by unity, to $n_\alpha + 1$, as

$$\Delta k_\alpha = \frac{\pi}{L}. \tag{1.51}$$

Then we can write the sum over n_α as

$$\sum_{n_\alpha} = \frac{L}{\pi} \sum \Delta k_\alpha \sim \frac{L}{\pi} \int_0^\infty dk_\alpha. \tag{1.52}$$

The last step is a good approximation if the temperature is not exceedingly low, so that the sum extends to large n_α before the exponential in the denominator in equation (1.49) becomes large. In this case the fractional increment in k_α on each increment of n_α is small, so the sum is well approximated as an integral.

In three dimensions, equation (1.52) generalizes to

$$\sum = \frac{L^3}{\pi^3}\int_{k_\alpha>0} d^3k. \tag{1.53}$$

At this point it is convenient to introduce new and even simpler boundary conditions. If the thermal energy is dominated by modes with wavelengths much shorter than the size of the solid, the heat capacity cannot depend on the shape of the object—we just have to specify some shape in order to fix definite boundary conditions for the wave equation. Mathematically convenient boundary conditions are that the solid fills a space periodic in a cube of width L, volume $V = L^3$, so the point (x, y, z) is the same as the point $(x + L, y, z)$ and so on for the other three directions. We can write solutions to the wave equation that satisfy these periodic boundary conditions as the real part of

$$\delta \propto e^{i\mathbf{k}\cdot\mathbf{r}-\omega t}, \qquad \omega = kc_s. \tag{1.54}$$

The periodic boundary condition is that δ cannot change if x is shifted to $x + L$, so the propagation vector $\mathbf{k}$ has to satisfy

$$k_\alpha = \frac{2\pi n_\alpha}{L} \quad \text{or} \quad \mathbf{k} = \frac{2\pi\mathbf{n}}{L}. \tag{1.55}$$

Here $\mathbf{n}$ means the triplet of integers n_α of either sign,

$$n_\alpha = 0, \pm 1, \pm 2, \ldots. \tag{1.56}$$

Note that in the standing wave solution in equation (1.46) negative and positive integers (which means negative and positive k_α) are physically equivalent, the only difference being a change of sign which can be absorbed in the phase ϕ. Equation (1.54) represents a running wave, so a change of sign of n_α means a change in the direction the wave is running, which is a physical difference. Thus here we must sum over all eight octants of $\mathbf{n}$, while the sum in equation (1.53) is over the first octant only. A second difference is that here the increment in k_α for a unit increment of n_α is, by equation (1.55),

$$\Delta k_\alpha = \frac{2\pi}{L}, \tag{1.57}$$

twice the value in equation (1.51). Thus with periodic
tions the sum over modes is approximated as the integ

$$\sum_{\mathbf{n}} = \frac{V}{(2\pi)^3}\int d^3k.$$

The integral is over all octants, 8 times the volume of
the first octant in equation (1.53). This cancels the
the denominator in equation (1.58).

Collecting equations (1.49) and (1.58), we see tha
ergy of the solid is

$$U = \frac{V}{(2\pi)^3}\int d^3k \frac{\hbar\omega}{e^{\hbar\omega/kT} - 1}.$$

Because of the appearance of the factor V from the
sum to an integral, this equation says the energy per
is independent of the volume of the solid, which make

Since the integrand in equation (1.59) depends o
tude of $\mathbf{k}$, we can write the volume element as $d^3k = 4$
$\omega = kc_s$ (eq. [1.54]) and the change of variables

$$y = \frac{\hbar\omega}{kT},$$

(and taking care not to confuse Boltzmann's constant
of the propagation vector) we get

$$U = \frac{V(kT)^4}{2\pi^2(\hbar c_s)^3}\int_0^\infty \frac{y^3 dy}{e^y - 1}.$$

The dimensionless integral is

$$\int_0^\infty \frac{y^3 dy}{e^y - 1} = \frac{\pi^4}{15}.$$

The final step is to note that energy can be st
waves, of which there are two for every pressure w
two orthogonal directions perpendicular to the prop
so we should multiply U by three and replace c_s with

This T^4 law was found empirically by Stefan (1879) and derived (apart from the value of Stefan's constant, a) by Boltzmann (1884) from thermodynamics.

A Quick Review of Electromagnetism

As a first step in the derivation of u_ω, let us write down Maxwell's equations. We will use Gaussian cgs units, where the electric and magnetic fields satisfy

$$\nabla \cdot \mathbf{E} = 4\pi\rho, \qquad \nabla \cdot \mathbf{B} = 0,$$
$$\nabla \times \mathbf{E} + \frac{1}{c}\frac{\partial \mathbf{B}}{\partial t} = 0, \tag{2.3}$$
$$\nabla \times \mathbf{B} - \frac{1}{c}\frac{\partial \mathbf{E}}{\partial t} = \frac{4\pi}{c}\mathbf{j}.$$

The charge density is ρ, and the charge conservation equation is

$$\frac{\partial \rho}{\partial t} + \nabla \cdot \mathbf{j} = 0, \tag{2.4}$$

where $\mathbf{j}$ is the current density. This can be compared to equation (1.36) for mass conservation. The force on a charge q moving at velocity $\mathbf{v}$ is

$$\mathbf{F} = q\,(\mathbf{E} + \mathbf{v} \times \mathbf{B}/c). \tag{2.5}$$

The charge is measured in electrostatic units, where the static electric field at position $\mathbf{r}$ relative to a point charge q is

$$\mathbf{E} = \frac{q\mathbf{r}}{r^3}. \tag{2.6}$$

The electric and magnetic fields have the same units; for $\mathbf{B}$ the unit is called a Gauss. The velocity of light is c.

To get the electromagnetic wave equation we need the identity

$$\mathbf{A} \times (\mathbf{B} \times \mathbf{C}) = \mathbf{B}(\mathbf{A} \cdot \mathbf{C}) - \mathbf{C}(\mathbf{A} \cdot \mathbf{B}). \tag{2.7}$$

This also applies to the gradient operator and a vector function of position, as long as we are careful not to change the order of differentiation. Thus we find from equation (2.7)

$$\nabla \times (\nabla \times \mathbf{E}) = \nabla(\nabla \cdot \mathbf{E}) - \nabla^2 \mathbf{E}. \tag{2.8}$$

In a pure radiation field there are no charges or currents: $\rho = 0$ and $\mathbf{j} = 0$. In this case the result of taking the curl of the third of Maxwell's equations (2.3), applying the identity (2.8), and then simplifying with the help of the other Maxwell equations is

$$\frac{\partial^2 \mathbf{E}}{\partial t^2} = c^2 \nabla^2 \mathbf{E}. \tag{2.9}$$

This is a wave equation for each component of the electric field, as in equation (1.44).

Just as for sound waves in a solid (eq. [1.54]), we can write a complete set of solutions to the vector wave equation (2.9) as the real part of

$$\mathbf{E} = \mathbf{E}_o \, e^{i\mathbf{k}\cdot\mathbf{r} - \omega t}, \tag{2.10}$$

where $\mathbf{E}_o$ is a complex constant vector. This expression in the wave equation (2.9) gives the relation

$$\omega = kc. \tag{2.11}$$

To describe radiation in a cavity, let us adopt the periodic boundary conditions from the last section. Then the propagation vector $\mathbf{k}$ has to satisfy (eq. [1.55])

$$k_\alpha = 2\pi n_\alpha / L, \qquad n_\alpha = 0, \pm 1, \pm 2, \ldots. \tag{2.12}$$

There also is a transversality condition: on substituting equation (2.10) into the condition $\nabla \cdot \mathbf{E} = 0$ in the absence of charges we get

$$\mathbf{k} \cdot \mathbf{E_o} = 0. \tag{2.13}$$

This says the electric field $\mathbf{E}$ has to be perpendicular to the direction $\mathbf{k}$ of propagation of the wave.

It is left as an exercise to get the magnetic field $\mathbf{B}$ in terms of $\mathbf{E}_o$ and $\mathbf{k}$.

The Planck Spectrum

Planck's blackbody spectrum follows by the same procedure used in the last section to find the low temperature heat capacity of a solid. The increment of k_α per unit increment of the integer n_α is $dk_\alpha = 2\pi/L$

(eqs. [1.57], [2.12]), so the number of independent modes of oscillation of the electromagnetic field with wave number $\mathbf{k}$ in the range d^3k is

$$dN = 2\frac{Vd^3k}{(2\pi)^3}. \tag{2.14}$$

This is a factor of two larger than in equation (1.58) because, by the transversality condition (2.13), there are two independent (orthogonal) directions for the electric field $\mathbf{E}$ for given $\mathbf{k}$, so there are two independent modes of oscillation for given $\mathbf{k}$. Summing over directions gives $d^3k = 4\pi k^2 dk$. Using $k = \omega/c$ (eq. [2.11]), we find that the number of modes with frequency between ω and $\omega + d\omega$ is

$$dN = \frac{V\omega^2 d\omega}{\pi^2 c^3}. \tag{2.15}$$

On multiplying this by the mean energy per mode (eq. [1.27]), and dividing by the volume V, we arrive at the thermal energy per unit volume and per unit frequency interval,

$$u_\omega = \frac{du}{d\omega} = \frac{1}{\pi^2 c^3}\frac{\hbar\omega^3}{e^{\hbar\omega/kT} - 1}. \tag{2.16}$$

This is the Planck blackbody radiation spectrum.

In the classical limit, $\hbar\omega \ll kT$, we have as before $e^{\hbar\omega/kT} \sim 1 + \hbar\omega/kT$ (eq. [1.28]), so equation (2.16) becomes

$$u_\omega = \frac{kT\omega^2}{\pi^2 c^3}. \tag{2.17}$$

Planck's constant does not appear in this Rayleigh-Jeans law, as expected because the equation can be derived from classical statistical mechanics.

The net energy density is obtained by integrating equation (2.16) over all frequencies. On changing variables to $x = \hbar\omega/kT$ and using equation (1.62) we get

$$u = \int_0^\infty u_\omega d\omega = \frac{\pi^2}{15}\frac{(kT)^4}{(\hbar c)^3}. \tag{2.18}$$

This is the Stefan-Boltzmann law. Using the measured values of Stefan's constant a, the frequency at the peak of the spectrum (2.16) at a given temperature, and the velocity of light c, Planck could solve for Boltzmann's constant k and $\hbar$; both were within 2 percent of the modern values.

Planck emphasized that, if his approach has any validity, $\hbar$ ought to show up somewhere else in physics. Einstein gave the first two examples: heat capacities, as discussed in the last section, and the photoelectric effect to be discussed next.

3 Photons

Light shining on a metal knocks out electrons. Einstein (1905) proposed an interpretation of this effect based on Planck's prescription $E = n\hbar\omega$ (eq. [1.21]) for the energy of an oscillator. Planck's prescription indicates that light can only transfer energy in discrete units—photons, or quanta of the electromagnetic field—of amount $\hbar\omega$. If one of these units of energy is given to an electron in a metal, then the electron ought to leave the metal with energy

$$E \leq \hbar\omega - \Phi, \tag{3.1}$$

where Φ is the binding energy (the work required to pull an electron out of the metal). The inequality takes account of the fact that the electron may lose energy before reaching the surface. By 1917, Millikan had found that there is a linear relation between the maximum energy of the electrons released and the frequency of the incident light, consistent with equation (3.1), and had found that the slope $\hbar$ of the relation agrees with Planck's value within the errors, again about 1 percent.

As discussed in chapter 8, the relativistic relation between the energy E of a particle, its momentum p, and its rest mass m is

$$E^2 = p^2c^2 + m^2c^4. \tag{3.2}$$

If the energy in light acts as discrete units, photons, perhaps the photons move as particles. Because these particles would have to move at the velocity of light, their rest mass would have to vanish, $m = 0$. The relativistic relation (3.2) indicates $E = pc$ for massless particles. Thus a photon with energy $E = \hbar\omega$ would be expected to have momentum

$$p = E/c = \hbar\omega/c. \tag{3.3}$$

Einstein was cautious about referring to the momentum of a photon; that it really has momentum in agreement with this equation was made clear by the Compton effect, that refers to the recoil of an electron that scatters a photon, as follows.

Imagine a photon of energy pc and momentum $\mathbf{p}$ incident on an electron of mass m that initially is at rest. The net energy and momentum of the system are therefore $E = pc + mc^2$ and $\mathbf{p}$. If the photon scatters off the electron and leaves with momentum $\mathbf{p}'$, then to conserve momentum the electron must end up with momentum $\mathbf{P} = \mathbf{p} - \mathbf{p}'$. The final energy, which must be the same as the initial, is

$$E = pc + mc^2 = p'c + [(\mathbf{p} - \mathbf{p}')^2c^2 + m^2c^4]^{1/2}. \tag{3.4}$$

On subtracting $p'c$, squaring, and simplifying we get

$$p' = \frac{mcp}{mc + p - p\cos\theta}, \tag{3.5}$$

where θ is the angle between $\mathbf{p}$ and $\mathbf{p}'$.

Using equation (3.3), we can write equation (3.5) as a relation between the initial and final frequencies ω and ω' of the photon,

$$\frac{1}{\omega'} = \frac{1}{\omega} + \frac{\hbar}{mc^2}(1 - \cos\theta). \tag{3.6}$$

Compton experimentally found this relation between the initial and final frequencies and the scattering angle, θ. This shows that light does scatter like a gas of massless particles, photons, with the usual relation (3.2) between energy and momentum.

4 Spectra and Energy Quantization of Atoms

The Combination Principle

A hot dilute gas of atoms or molecules emits light at sharply defined frequencies, ν_i. The set of values of these frequencies for a given material can be written as differences among a list of quantities called terms:

$$\nu_i = T_\alpha - T_\beta. \tag{4.1}$$

This provides a handy way to record the frequencies, because a list of terms gives a much longer list of term differences. But this combination principle also is telling us something about physics. If we multiply equation (4.1) by Planck's constant, we get

$$E_i = h\nu_i = h(T_\alpha - T_\beta). \tag{4.2}$$

Since $h\nu_i$ is the energy of the photon produced by the atom, this equation suggests the possible values of the energy of the atom are the discrete, or quantized, quantities $E_\alpha = hT_\alpha$. The atom would produce a photon with frequency ν_i when it makes a transition from the energy level α to the level β in equation (4.2). This is yet another example of energy quantization.

In atomic hydrogen, the terms have the particularly simple form

$$T_n = Rc/n^2, \qquad n = 1, 2, 3, \ldots, \tag{4.3}$$

where R is called the Rydberg constant. The frequencies corresponding to transitions from term $n > 2$ to term $n' = 2$ are the Balmer series,

$$\nu = Rc\left[\frac{1}{4} - \frac{1}{n^2}\right], \tag{4.4}$$

with $n = 3$ giving the prominent red line seen in the spectrum of atomic hydrogen. The next series to be discovered in atomic hydrogen were the Paschen series $n \to n' = 3$, in the infrared, and the Lyman series, $n \to n' = 1$, in the ultraviolet.

The Bohr Model

Equation (4.3) is so simple it ought to be understandable. Bohr found the first successful model. He started with the assumption that the angular momentum, L, of the electron in orbit around the proton in a hydrogen atom is quantized,

$$L = n\hbar, \qquad n = 1, 2, 3, \ldots. \tag{4.5}$$

The original reasons for this assumption, adduced by Bohr and Ehrenfest, are not worth going into here.

Following Bohr, let us imagine the electron with charge $-e$ is moving in a circular orbit of radius a at speed v around a fixed proton in a hydrogen atom. (The proton is fixed because it is much more massive than the electron.) The proton has charge e. The balance of the electrostatic force of attraction of the electron to the proton and the acceleration of the electron moving in a circle is

$$\frac{e^2}{a^2} = \frac{mv^2}{a}, \tag{4.6}$$

where m is the mass of the electron. Bohr's condition (4.5) says the angular momentum of the electron

$$L = mav = n\hbar, \tag{4.7}$$

where n is a positive integer. The result of eliminating the velocity from these equations is

$$a = n^2 a_o, \tag{4.8}$$

where the Bohr radius is

$$a_o = \frac{\hbar^2}{me^2} = 5.3 \times 10^{-9}\ \mathrm{cm}. \tag{4.9}$$

The energy of the electron is the sum of its kinetic and potential energies,

$$E_n = \frac{mv^2}{2} - \frac{e^2}{a} = -\frac{e^2}{2a} = -\frac{e^4 m}{2\hbar^2}\frac{1}{n^2}, \tag{4.10}$$

where the last steps follow from equations (4.6), (4.8), and (4.9).

We have argued that the energy $\Delta E = E_n - E_{n'}$ released in the transition of an atom from level n to level n' is given to a photon with frequency $\omega = \Delta E/\hbar$, so the frequencies of the radiation emitted by the atom are predicted from equation (4.10) to be

$$\omega = \frac{e^4 m}{2\hbar^3}\left(\frac{1}{(n')^2} - \frac{1}{n^2}\right). \tag{4.11}$$

With Planck's value for $\hbar$, Bohr found excellent agreement with the measured frequencies in the spectrum of atomic hydrogen. This was considered a great but certainly mysterious triumph.

5 Matter Waves

The de Broglie Relations

Since light, which clearly exhibits the properties of a wave (interference and all that), can also act like a gas of particles, as discussed in section 3, one might guess that material particles exhibit wave properties. The connection between the energy and momentum of a particle like an electron and the frequency and wavelength of the wave associated with the particle was introduced by de Broglie.

Let us begin with some definitions. Consider the plane wave function

$$\psi \propto e^{i\phi}, \qquad \phi = \mathbf{k} \cdot \mathbf{r} - \omega t. \tag{5.1}$$

The propagation vector in this function is $\mathbf{k}$, the magnitude of $\mathbf{k}$ is the wavenumber, and the angular frequency of the wave is ω. An example is the pressure wave discussed in section 1 (eq. [1.54]). We will assume here that ω depends on the magnitude of $\mathbf{k}$ alone; the function $\omega = \omega(k)$ is called the dispersion relation.

As indicated in figure 5.1, the positions $\mathbf{r}$ at which the phase ϕ at a fixed time t has a fixed value define a plane perpendicular to the propagation vector $\mathbf{k}$. (This is because a displacement of $\mathbf{r}$ in a direction perpendicular to $\mathbf{k}$ does not change the value of $\mathbf{k} \cdot \mathbf{r}$ in eq. [5.1]). The distance between planes with ϕ differing by 2π at a fixed instant of time is the wavelength,

$$\lambda = 2\pi/k. \tag{5.2}$$

A surface of fixed phase advances along the direction of $\mathbf{k}$ at the phase velocity,

$$v_p = \omega/k, \tag{5.3}$$

as one sees by considering the displacement δr that would balance the time shift δt to hold ϕ in equation (5.1) constant.

We saw in section 2 that the dispersion relation for a photon is $\omega = kc$ (eq. [2.11]). According to the Planck relation (1.21), the photon has energy $E = \hbar\omega$, and equation (3.3) says the photon has momentum $p = E/c = \hbar\omega/c = \hbar k$. De Broglie proposed that electrons may exhibit the same wave-particle duality with the same de Broglie relations between

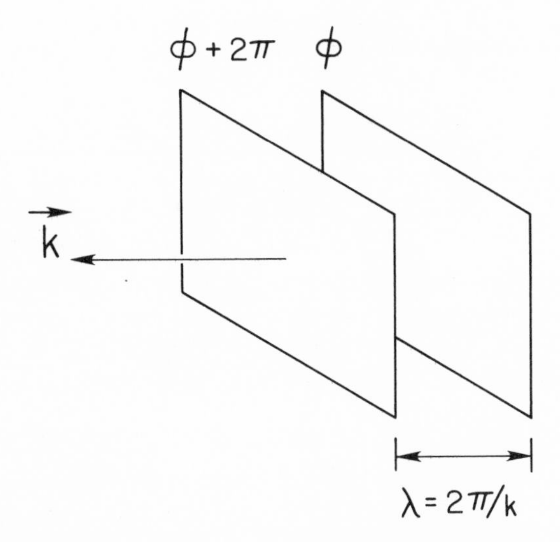

Fig. 5.1 The propagation vector for a plane wave is $\mathbf{k}$. The positions $\mathbf{r}$ that lie in a plane of constant phase at a given instant of time satisfy $\mathbf{r} \cdot \mathbf{k} = \text{constant}$. The plane of constant phase is normal to the propagation vector $\mathbf{k}$. Two planes of constant phase with phase difference 2π are separated by the wavelength $\lambda = 2\pi/k$.

energy and frequency, and momentum and propagation vector,

$$E = \hbar\omega,$$
$$\mathbf{p} = \hbar\mathbf{k}. \tag{5.4}$$

The wavelength (5.2) is then

$$\lambda = 2\pi/k = 2\pi\hbar/p = h/p. \tag{5.5}$$

This is equivalent to the second of the de Broglie relations (5.4).

The de Broglie momentum relation offers a way to understand the Bohr-Ehrenfest angular momentum condition in equation (4.5). Equation (4.6) refers to an electron moving around a circle of radius a. Imagine the electron is represented by a wave with wavelength $\lambda = h/p$ (eq. [5.5]) that runs around the circle. We want the wave to be continuous, so there has to be an integral number, n, of wavelengths around the circumference of the circle:

$$\text{circumference} = 2\pi a = n\lambda. \tag{5.6}$$

The de Broglie relation (5.5) thus says the angular momentum of the particle is

$$L = ap = ah/\lambda = nh/2\pi = n\hbar. \tag{5.7}$$

This is equation (4.5).

A nonrelativistic particle of mass m moving with momentum p has kinetic energy $E = p^2/2m$. Thus the dispersion relation derived from the de Broglie relations (5.4) is

$$\omega = \frac{E}{\hbar} = \frac{\hbar k^2}{2m}. \tag{5.8}$$

The phase velocity (eq. [5.3]) in this case is $v_p = \omega/k = \hbar k/2m = p/2m$. This differs from the usual relation between momentum and velocity by a factor of two. However, as we will now discuss, that is because we need another measure for the velocity.

Group Velocity of a Wave Packet

De Broglie assumed that the plane waves (5.1) can be superimposed (added together) to get more general wave functions, of the form

$$\psi(\mathbf{r}, t) = \int d^3k \; f(\mathbf{k}) e^{i\mathbf{k}\cdot\mathbf{r} - \omega(k)t}. \tag{5.9}$$

The factor $f(\mathbf{k})$ is the weight assigned to each plane wave. With periodic boundary conditions, the allowed values of $\mathbf{k}$ are discrete, as discussed in section 1 (eq. [1.55]), so with periodic boundary conditions the integral (5.9) would be replaced with a sum.

Now let us choose the function $f(\mathbf{k})$ so ψ is a wave packet, which is to say ψ is fairly sharply peaked at one position. This is done by taking $f(\mathbf{k})$ to be bell-shaped, with its maximum at $\mathbf{k} = \mathbf{k}_o$, having a width Δk in all three directions, and with f rapidly approaching zero at $|\mathbf{k} - \mathbf{k}_0| > \Delta k$. To find the values of $\mathbf{r}$ and t for which ψ is large, consider the phase

$$\phi = \mathbf{k}\cdot\mathbf{r} - \omega(k)t, \tag{5.10}$$

in the exponential in the integral in equation (5.9). For most choices of $\mathbf{r}$ and t, the phase ϕ varies rapidly as $\mathbf{k}$ varies over the range Δk around $\mathbf{k}_o$ where f is appreciably large. This variation of ϕ makes $e^{i\phi}$ oscillate, and the oscillation makes the integral ψ small. But if ϕ as a function

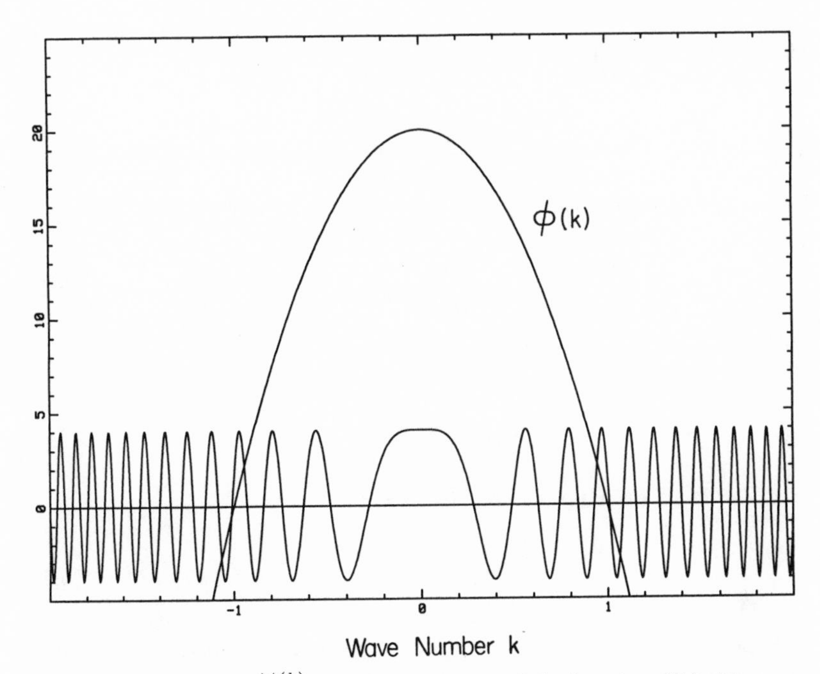

Fig. 5.2 Behavior of $e^{i\phi(k)}$ near an extremum of the function $\phi(k)$. The upper curve shows the parabola $\phi(k)$, in radians, as a function of wave number, k. The bottom curve shows the real part of $e^{i\phi(k)}$, in arbitrary units. This function oscillates rapidly except near the extremum of $\phi(k)$.

of $\mathbf{k}$ happens to have an extremum at $\mathbf{k} = \mathbf{k}_o$, the oscillation of $e^{i\phi}$ near $\mathbf{k} = \mathbf{k}_o$ is suppressed, because ϕ varies only slowly with k near the extremum. This is illustrated in figure 5.2. When the oscillation of $e^{i\phi}$ is suppresed where $f(\mathbf{k})$ is large, the integral ψ is large.

The condition for a large value of $\psi(\mathbf{r}, t)$ is then that ϕ have an extremum at $\mathbf{k} \sim \mathbf{k}_o$. This means the first derivative of ϕ vanishes at $\mathbf{k} \sim \mathbf{k}_o$. Since we want ϕ to be an extremum with respect to variations of each component of $\mathbf{k}$, large ψ requires

$$\frac{\partial}{\partial k_\alpha}[\mathbf{k} \cdot \mathbf{r} - \omega(k)t] = 0 \quad \text{at} \quad \mathbf{k} = \mathbf{k}_o. \tag{5.11}$$

This is three equations, $\alpha = 1, 2, 3$ representing the three orthogonal components of $\mathbf{k}$. On differentiating out equation (5.11), we see that the peak of ψ is at

$$\mathbf{r} = \mathbf{v}_g t, \tag{5.12}$$

where the group velocity is

$$v_\alpha = \frac{\partial \omega}{\partial k_\alpha} = \frac{d\omega}{dk}\frac{\partial k}{\partial k_\alpha}. \tag{5.13}$$

The result of differentiating the expression $k^2 = \sum k_\beta^2$ with respect to the component k_α is

$$\frac{\partial k}{\partial k_\alpha} = \frac{k_\alpha}{k}. \tag{5.14}$$

This brings the group velocity (5.13) to

$$\mathbf{v}_g = \frac{d\omega}{dk}\frac{\mathbf{k}}{k}, \tag{5.15}$$

with $\mathbf{k} = \mathbf{k}_o$. The magnitude of this expression is

$$v_g = \frac{d\omega}{dk}. \tag{5.16}$$

The speed of motion of a wave packet thus is determined by the dispersion relation $\omega(k)$.

The group velocity (5.3) agrees with the phase velocity (5.16) if $\omega \propto k$, as in a pressure or electromagnetic wave (eq. [1.54]). For the nonrelativistic de Broglie dispersion relation in equation (5.8), the magnitude of the group velocity is

$$v_g = \frac{d\omega}{dk} = \frac{\hbar k}{m} = \frac{p}{m}. \tag{5.17}$$

This is the usual relation between velocity and momentum.

6 Schrödinger's Equation

Single-Particle Wave Equation

Schrödinger in 1926 wrote down the differential equation satisfied by the matter wave associated with a nonrelatistic particle moving in a potential well. He was guided by an analogy Hamilton had noted between the motion of a particle in classical mechanics and the short wavelength limit of the motion of a light wave in a medium with index of refraction that is a function of position. There is another clue. As we have seen in sections 1 and 4, the allowed values of the energy of an isolated atom are quantized. According to de Broglie's relation $E = \hbar\omega$ (eq. [5.4]), that means the matter wave associated with the atom can only oscillate at discrete frequencies. One gets discrete frequencies from the solutions to a wave equation with fixed boundary conditions, as we saw in the normal mode analyses in sections 1 and 2 (eq. [1.50]). Thus a theory for the discrete energies might be based on a wave equation.

By de Broglie's relations, a wave with energy E oscillates with frequency $\omega = E/\hbar$,

$$\psi \propto e^{-i\omega t}. \tag{6.1}$$

The wave function therefore satisfies the differential equation

$$i\hbar\frac{\partial\psi}{\partial t} = E\psi. \tag{6.2}$$

If the wave has momentum $\mathbf{p}$, de Broglie's relations say it has wave number $\mathbf{k} = \mathbf{p}/\hbar$. That means the wave function varies with position as

$$\psi \propto e^{i\mathbf{k}\cdot\mathbf{r}}, \tag{6.3}$$

as in equation (5.1). Equation (6.3) satisfies the relation

$$-i\hbar\nabla\psi = \hbar\mathbf{k}\,\psi = \mathbf{p}\psi. \tag{6.4}$$

Applying this equation twice, and recalling that the kinetic energy is $E = p^2/2m$, one might guess that the wave equation for a free particle with energy E would be

$$E\psi = \frac{p^2}{2m}\psi = -\frac{\hbar^2}{2m}\nabla^2\psi. \tag{6.5}$$

A particle in a potential well, such as an electron in a hydrogen atom, has potential energy, $V(\mathbf{r})$, as well as kinetic energy. Schrödinger generalized the wave equation for the case of a particle with energy E and potential energy $V(\mathbf{r})$ by adding the potential energy to the kinetic energy in equation (6.5), to get

$$E\psi = -\frac{\hbar^2}{2m}\nabla^2\psi + V(\mathbf{r})\psi, \tag{6.6}$$

This is Schrödinger's equation for a single particle with definite energy E.

The more general time-dependent Schrödinger equation is obtained by eliminating E from equations (6.2) and (6.6), to get

$$i\hbar\frac{\partial\psi}{\partial t} = \left[-\frac{\hbar^2}{2m}\nabla^2 + V(\mathbf{r})\right]\psi. \tag{6.7}$$

We arrived at equation (6.7) from equations (6.2) and (6.5), in which the system has a definite energy, E. We will assume that the more general solutions to this equation, where ψ does not vary with time as $e^{-iEt/\hbar}$, also are allowed. (Thus has the interesting consequence, that Schrödinger was reluctant to accept, that in states represented by such solutions the system does not have a definite energy.)

It is customary to write equation (6.7) in the form

$$i\hbar\frac{\partial\psi}{\partial t} = H\psi, \tag{6.8}$$

for the time-dependent case, and, when the energy is known to be E,

$$H\psi = E\psi. \tag{6.9}$$

Here H is the derivative operator

$$H = -\frac{\hbar^2}{2m}\nabla^2 + V(\mathbf{r}). \tag{6.10}$$

In equation (6.9) it is understood that the wave function for the system is of the form $\psi(\mathbf{r})e^{-iEt/\hbar}$, where $\psi(\mathbf{r})$ is a time-independent solution to equation (6.9). This requires that V be a function of position alone; when V varies with time energy is not conserved, and one must use the more general equation (6.8).

The Schrödinger differential equation (6.8) has to be supplemented with boundary conditions. For a bounded system like a hydrogen atom it is reasonable to require that the ψ wave be bounded too, that is, that ψ go to zero at large distance from the proton. This will be formalized in section 8 in the condition that the wave function, ψ, for a physical system be square integrable, that is, that the integral of $|\psi|^2$ over all space be finite.

The expression in equation (6.10) is called an operator, because it yields a new function when applied to a given function, as in equation (6.6). The operator in equation (6.10) is called the Hamiltonian, for reasons that will become clear in sections 14 and 19. Some other jargon might be mentioned: equation (6.9) is called an eigenvalue equation; the solutions ψ that are square integrable are the eigenfunctions of the operator H, and the constant E is the eigenvalue belonging to the eigenfunction ψ. The prefix "eigen" refers to the result that the energy eigenvalues E for a bounded system are discrete.

Schrödinger's guess has proved to be remarkably successful: equation (6.8) still is the basic dynamical equation of quantum physics. Before developing the formalism any further let us look at two applications: the ground state energy of a hydrogen atom, to see that Schrödinger can do as well as Bohr, and the energy levels of a one-dimensional simple harmonic oscillator, to see that Schrödinger can do as well as Planck.

Ground State of a Hydrogen Atom

Techniques for solving (or approximating solutions to) the Schrödinger energy eigenvalue problem $H\psi = E\psi$ will be developed below at some length. Here for the purpose of a preliminary exploration of what Schrödinger's equation has to offer we will suppose that the potential energy is spherically symmetric, V a function only of distance $r = |\mathbf{r}|$ from the origin, and we will consider only those eigenfunctions that happen to be spherically symmetric.

We need the form of $\nabla^2\psi$ when the wave function ψ is spherically symmetric, $\psi = \psi(r)$. A useful relation is obtained by differentiating the expression $r^2 = x^2 + y^2 + z^2$ with respect to x and rearranging:

$$\frac{\partial r}{\partial x} = \frac{x}{r}. \tag{6.11}$$

The same of course applies to y and z. When ψ is a function of the radius r alone, we have from this relation

$$\frac{\partial \psi}{\partial x} = \frac{d\psi}{dr}\frac{\partial r}{\partial x} = \frac{x}{r}\frac{d\psi}{dr}. \tag{6.12}$$

The result of applying this operation again is

$$\frac{\partial^2 \psi}{\partial x^2} = \frac{x^2}{r^2}\frac{d^2\psi}{dr^2} + \frac{1}{r}\frac{d\psi}{dr} - \frac{x^2}{r^3}\frac{d\psi}{dr}. \tag{6.13}$$

Similar expressions of course apply to the second derivatives with respect to y and z. Since

$$\nabla^2\psi = \frac{\partial^2\psi}{\partial x^2} + \frac{\partial^2\psi}{\partial y^2} + \frac{\partial^2\psi}{\partial z^2}, \tag{6.14}$$

equation (6.13) gives

$$\nabla^2\psi = \frac{d^2\psi}{dr^2} + \frac{2}{r}\frac{d\psi}{dr}. \tag{6.15}$$

An equivalent way to write equation (6.15), as may be checked by differentiating the expression out, is

$$\nabla^2\psi = \frac{1}{r}\frac{d^2}{dr^2}r\psi. \tag{6.16}$$

This is the wanted expression for $\nabla^2\psi$, under the assumption that ψ is spherically symmetric, a function of radius r alone. The general expression for $\nabla^2\psi$ is given in equation (18.13) below.

On substituting equation (6.16) into Schrödinger's equation (6.6) and multiplying through by r, we get

$$-\frac{\hbar^2}{2m}\frac{d^2}{dr^2}u(r) + V(r)u(r) = Eu(r), \tag{6.17}$$

where the product $r\psi(r)$ has been written as the radial wave function,

$$u(r) = r\psi(r). \tag{6.18}$$

Equation (6.17) is just the Schrödinger equation for motion in one dimension, with the added constraint that the radial wave function $u(r)$ has to vanish at $r = 0$, because we want ψ to be well behaved at the origin. It will be seen in section 18 below that this very convenient relation between the three-dimensional problem and a simpler one-dimensional case applies more generally: when the potential energy is spherically symmetric and the wave function is not, the potential energy in the one-dimensional problem has an added centrifugal term, in close analogy to what is done in classical mechanics.

Let us consider now the wave function for an electron in a hydrogen atom. The radius of a proton is much smaller than an atom, so it is a good approximation to write the potential energy of the electron in the electric field of the proton as

$$V = -e^2/r, \tag{6.19}$$

where r is the distance from the proton approximated as a point charge, the electron having charge $-e$ and the proton charge e. Because the proton is much more massive than the electron, we will imagine that the electron is moving in the fixed electric potential well produced by the proton.

Equation (6.17) with (6.19) for the potential is

$$-\frac{\hbar^2}{2m}\frac{d^2}{dr^2}u(r) - \frac{e^2}{r}u(r) = Eu(r) \equiv -Bu(r). \tag{6.20}$$

Because the energy E is negative for bound states, the binding energy (the energy that must be added to the system to pull the electron away from the proton and leave it at rest at large separation) has been written as $B = -E$.

A solution to equation (6.20), that we will see represents the ground state, has the functional form

$$u = re^{-\alpha r}. \tag{6.21}$$

The result of substituting this function into equation (6.20) and differentiating out is

$$\frac{\hbar^2}{2m}(\alpha^2 re^{-\alpha r} - 2\alpha e^{-\alpha r}) + e^2 e^{-\alpha r} = Bre^{-\alpha r}. \tag{6.22}$$

This equation is consistent if the coefficients of $e^{-\alpha r}$ and of $re^{-\alpha r}$ agree on both sides of the equation. This requires that

$$\alpha = me^2/\hbar^2 \equiv 1/a_o, \qquad \alpha^2 = 2mB/\hbar^2. \tag{6.23}$$

On eliminating α from these two equations, we get

$$B = \frac{me^4}{2\hbar^2}. \tag{6.24}$$

Equation (6.24) agrees with Bohr's equation for the ground state energy of a hydrogen atom, $n = 1$ in equation (4.10). Since Bohr's result agrees with the measurements this is a Good Thing. It will be noted also that the reciprocal of α in the first of equations (6.23) is the Bohr radius (eq. [4.9]). In the Bohr model, a_o is the radius of the orbit of the electron in the ground state. In the Schrödinger equation solution, the ground state wave function (6.21) spreads over a region on the order of this radius.

One-Dimensional Simple Harmonic Oscillator

This is a one-dimensional problem, with the wave function a function of one space variable, x. The potential energy is the quadratic form

$$V = Kx^2/2, \tag{6.25}$$

as in equation (1.20). In classical mechanics, K is the spring constant.

Schrödinger's energy eigenvalue equation (6.9) with this potential can be solved by a direct analysis, that shows that the eigenfunctions are products of a Gaussian and a polynomial function g of x, $\psi = g(x)e^{-\beta x^2}$, with β a constant. We will instead find the energy eigenvalues and eigenfunctions by an operator technique that figures heavily in quantum physics.

The Schrödinger equation (6.9) belonging to a simple harmonic oscillator state with definite energy E is

$$H\psi = E\psi, \qquad H = -\frac{\hbar^2}{2m}\partial^2 + \frac{1}{2}Kx^2. \tag{6.26}$$

The symbol ∂ means the derivative operator:

$$\begin{aligned} \partial \cdot \psi &= \frac{d\psi}{dx}, \\ \partial^2 \cdot \psi = \partial \cdot \partial \cdot \psi &= \partial \cdot \frac{d\psi}{dx} = \frac{d^2\psi}{dx^2}. \end{aligned} \tag{6.27}$$

That is, ∂ operating on any given (and differentiable) function ψ maps ψ into a new function ϕ:

$$\phi(x) = \partial \cdot \psi = \frac{d\psi}{dx}. \tag{6.28}$$

This is a linear operation: if $\partial\psi = \phi$ and $\partial\chi = \theta$ then $\partial(c\psi + d\chi) = c\phi + d\theta$, where c and d are constants. Another linear operator is x, which maps the function $\psi(x)$ into the new function, $x\psi(x)$.

The usual rules for adding and multiplying apply to linear operators such as ∂ and x, with the exception that the operators need not commute. For example, we have

$$\partial \cdot x \cdot \psi = \partial \cdot (x\psi) = \frac{d}{dx} x\psi = \psi + x\partial\psi. \tag{6.29}$$

This can be written as the relation

$$(\partial x - x\partial)\psi = \psi. \tag{6.30}$$

Since this is true for any ψ, the operator in parenthesis is the identity:

$$\partial x - x\partial = 1. \tag{6.31}$$

This combination appears often enough to deserve a symbol. The commutator of the operators A and B is

$$[A, B] = AB - BA. \tag{6.32}$$

Equation (6.31) is the commutation relation

$$[\partial, x] = 1. \tag{6.33}$$

This particular commutation relation will play an important part in the discussion of momentum in quantum mechanics. Here we are interested in using it to find an operator solution to the Schrödinger eigenvalue equation for the energy of an oscillator.

Consider the operators

$$\begin{aligned} a &= \left(\frac{K}{2}\right)^{1/2} x + \frac{\hbar}{(2m)^{1/2}} \partial, \\ a^\dagger &= \left(\frac{K}{2}\right)^{1/2} x - \frac{\hbar}{(2m)^{1/2}} \partial, \end{aligned} \tag{6.34}$$

Their product is

$$a^\dagger a = \frac{K}{2} x^2 - \frac{\hbar^2}{2m} \partial^2 + \frac{\hbar K^{1/2}}{2m^{1/2}} [x, \partial]. \tag{6.35}$$

In the cross terms of the product $a^\dagger a$ the operators x and ∂ appear in the combinations $x\partial$ and ∂x with opposite signs; that gives the commutator.

The first two terms on the right-hand side of equation (6.35) are the Hamiltonian operator in equation (6.26). In the last term we recognize the classical natural frequency of the oscillator (eq. [1.20]),

$$\omega = (K/m)^{1/2}. \tag{6.36}$$

Since the commutator is $[x, \partial] = -1$ (eq. [6.33]), the product in equation (6.35) is

$$a^\dagger a = H - \hbar\omega/2. \tag{6.37}$$

The same calculation in the other order gives

$$aa^\dagger = H + \hbar\omega/2. \tag{6.38}$$

The difference of equations (6.37) and (6.38) is the commutation relation

$$[a^\dagger, a] = -\hbar\omega. \tag{6.39}$$

Now let us find the commutator of $a^\dagger$ with H. Equation (6.37) says

$$[a^\dagger, H] = [a^\dagger, a^\dagger a + \hbar\omega/2]. \tag{6.40}$$

This is a sum of two commutators, the second of which vanishes because $\hbar\omega/2$ is a constant, and the first of which is

$$\begin{aligned} [a^\dagger, a^\dagger a] &= a^\dagger a^\dagger a - a^\dagger a a^\dagger \\ &= a^\dagger [a^\dagger, a] \\ &= -\hbar\omega a^\dagger, \end{aligned} \tag{6.41}$$

by equation (6.39). Thus the wanted commutator is

$$[a^\dagger, H] = -\hbar\omega a^\dagger. \tag{6.42}$$

The same calculation yields

$$[a, H] = \hbar\omega a. \tag{6.43}$$

We can use these operator relations to find the allowed values of the energy, as follows. Suppose ψ is an eigenfunction of the oscillator Hamiltonian H with eigenvalue E,

$$H\psi = E\psi. \tag{6.44}$$

Operate on both sides of this equation with $a^\dagger$:

$$\begin{aligned} a^\dagger H\psi &= a^\dagger E\psi, \\ ([a^\dagger, H] + Ha^\dagger)\psi &= Ea^\dagger\psi, \\ (-\hbar\omega a^\dagger + Ha^\dagger)\psi &= Ea^\dagger\psi. \end{aligned} \tag{6.45}$$

The second line uses the definition (6.32) of a commutator. The third line uses equation (6.42) to eliminate the commutator. On rearranging the third line, we get

$$H(a^\dagger\psi) = (E + \hbar\omega)(a^\dagger\psi). \tag{6.46}$$

Equation (6.46) indicates that $E + \hbar\omega$ is an eigenvalue of the Hamiltonian H belonging to the eigenfunction $a^\dagger\psi$. For this reason, $a^\dagger$ is called a raising operator (or ladder operator): it maps the eigenfunction ψ with energy E to the eigenfunction $a^\dagger\psi$ with energy $E + \hbar\omega$.

A similar calculation, which you are invited to check, yields

$$H(a\psi) = (E - \hbar\omega)(a\psi). \tag{6.47}$$

Thus a is another ladder operator, a lowering operator for energy.

Equation (6.47) says that the result of applying a to $a\psi$ is a new function $a^2\psi$ with eigenvalue $E - 2\hbar\omega$. And this procedure can of course be iterated. Can E be lowered indefinitely? Not in classical physics, because the energy is the sum of two nonnegative numbers, $p^2/2m$ and $Kx^2/2$ (eq. [1.20]). As will be seen in the next section, the same applies in quantum mechanics: the energy has to be larger than the minimum value of the potential energy, which for this Hamiltonian means the energy eigenvalues must satisfy $E > 0$. Let us see what this implies.

The function $\psi = 0$ clearly solves the eigenvalue equation $H\psi = E\psi$, but trivially, so that cannot be an acceptable solution. The chain of lower energies obtained by iteratively applying the lowering operator a thus stops if at some point the eigenfunction satisfies

$$a\psi_0 = 0. \tag{6.48}$$

The energy belonging to ψ_0 follows from equation (6.37):

$$H\psi_0 = (a^\dagger a + \hbar\omega/2)\psi_0 = \frac{\hbar\omega}{2}\psi_0. \tag{6.49}$$

The state represented by ψ_0 thus has energy $E_0 = \hbar\omega/2$. We know this is the lowest possible value for the energy, for if it were assumed that ψ_- has lower energy, then $a\psi_-$ could not vanish, so $a\psi_-$ would represent a state with negative energy, which we are asserting (and will check in the next section) is not possible.

The ground state wave function thus is ψ_0, and the ground state energy is $\hbar\omega/2$. The result of applying the raising operator $a^\dagger$ to ψ_0 n times is the nth energy level,

$$E_n = (n + 1/2)\hbar\omega, \quad n = 0, 1, 2 \ldots. \tag{6.50}$$

This sequence has to include all energy eigenvalues, because the lowering operator a applied enough times to any eigenfunction has to lead back to $a\psi_0 = 0$.

Equation (6.50) agrees with Planck's assumption in equation (1.21) up to the zero point energy term $\hbar\omega/2$. This constant term does not affect heat capacities because it is a fixed addition to the energy, so all the preceding discussion is unaffected. As we shall see, however, the zero point energy certainly is part of the measurable energy of the system.

Equation (6.48) is a differential equation for the wave function ψ_0 (because a is the differential operator defined in eq. [6.34]). You are invited to check that the solution to this differential equation is a Gaussian, which is a square integrable function. The eigenfunction for the nth energy level is found by applying the raising operator $a^\dagger$ to ψ_0 n times. Because $(a^\dagger)^n$ contains n powers of x or the derivative with respect to x, $(a^\dagger)^n\psi_0$ is a Gaussian multiplied by an nth order polynomial, which is square integrable.

The functions $(a^\dagger)^n\psi_0$ we have generated by the algebra of the operators are not the only solutions to the differential equation $H\psi = E\psi$ with H given by equation (6.26); they are the special subset of square integrable functions. As will be discussed in section 8, this condition is required so the wave functions can be properly normalized. The condition entered the operator algebra through the bound $E > 0$ to be discussed next.

7 Remarks on Motion in One Dimension

Curvature and Quantization

To get some feeling for the origin of energy quantization and for the behavior of the wave function ψ it is useful to consider the general features of the shape of $\psi(x)$ for a particle that moves in one dimension as a bound state in the potential well $V(x)$.

In a one-dimensional problem with definite energy E, Schrödinger's equation (6.6) is

$$-\frac{\hbar^2}{2m}\frac{d^2\psi}{dx^2} + V(x)\psi = E\psi. \tag{7.1}$$

This is a real differential equation, so we can demand that the wave function ψ be real. Let us rewrite the equation as

$$C(x) \equiv \frac{1}{\psi}\frac{d^2\psi}{dx^2} = \frac{2m}{\hbar^2}(V(x) - E). \tag{7.2}$$

The function $C(x)$ measures the curvature of $\psi(x)$, in the sense that the larger the magnitude of C the more rapidly the slope of ψ varies with x. Where C is positive, ψ is convex toward the axis. That is, if $C > 0$ and $\psi > 0$ then the second derivative of ψ is positive, so ψ bends upward away from the axis $\psi = 0$. If $C > 0$ and $\psi < 0$, the second derivative is negative, and ψ bends downward, again away from the axis. If, on the other hand, C is less than zero, then ψ is concave toward the axis: where $\psi > 0$, the second derivative of ψ is negative, so the function bends toward the axis, and where $\psi < 0$ the second derivative is positive, so ψ again bends toward the axis.

A potential well and its ground state wave function, $\psi_0(x)$, are sketched in figure 7.1. In the classically allowed range of values of x, the energy E_0 of the system is greater than the potential energy. (In classical mechanics, E_0 is the sum of the potential and kinetic energies, and since the kinetic energy cannot be negative the total energy E_0 has to be greater than the potential energy $V(x)$.) In this classically allowed

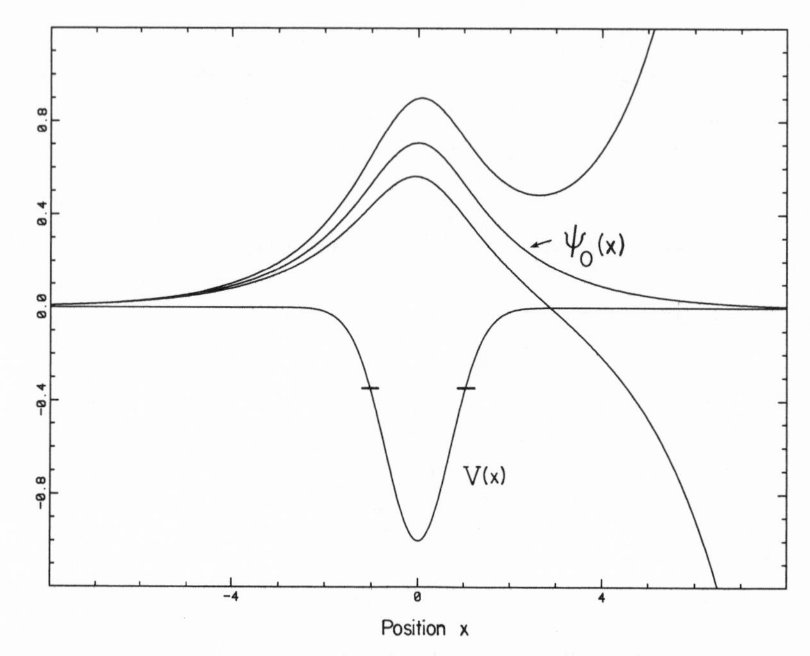

Fig. 7.1 Solutions to Schrödinger's one-dimensional time-independent wave equation. The bottom curve shows the potential well $V(x)$, with the short horizontal dashes marking the ground state energy, $E_0 = -0.35$ in the units of the vertical axis. The middle of the top three curves is the ground state energy eigenfunction. The lower of the top three curves shows the solution to Schrödinger's differential equation with the boundary conditions on the left side of the graph that are the same as for the ground state wave function, but with the energy E 10% higher than the ground state energy E_0. The top curve shows what happens to the solution when the energy is 10% lower than E_0. Neither of these latter two functions is acceptable because neither is square integrable.

region, the curvature C in equation (7.2) is negative. This means the wave function always bends toward the axis in the classically allowed region. This curvature toward the axis is seen in the central parts of the functions shown in figure 7.1. In the classically forbidden regions, at large and small x in this example, the total energy E_0 is less than the potential energy $V(x)$, so the curvature C is positive. This means that in the classically forbidden regions the wave function bends away from the axis. This curvature away from the axis allows the ground state wave function $\psi_0(x)$ to approach zero in the limits of large and small x, so as to make ψ_0 square integrable. The curvature toward the axis in the classically allowed region allows ψ_0 to bend over to join the decaying parts of the function at large and small x.

The top curve in the figure shows what happens when the energy E is taken to be slightly less than the ground state energy eigenvalue E_0. The solution to the differential equation (7.1) has been started at $x \to -\infty$ with $\psi \to 0$ but positive. The function hooks away from the axis in the left-hand forbidden region, because here $C > 0$, and then the function bends back toward the axis in the classically allowed region. But in the classically forbidden region the curvature is stronger than for the ground state wave function, because we have increased C by reducing E. Also, in the classically allowed region we have reduced the magnitude of C by bringing E closer to V, so $\psi(x)$ bends back toward the axis less strongly than does $\psi_0(x)$. The net result is that where $\psi(x)$ reenters the classically forbidden region on the right-hand side of the graph it is not sloping down steeply enough to allow the function to approach zero as x goes to positive infinity. Instead, the function bends too sharply away from the axis, so $\psi(x) \to \infty$ as $x \to \infty$. The function is not square integrable, so it is not allowed for a bound state.

The lower of the top three curves in figure 7.1 shows what happens to the solution to equation (7.1) if E is taken to be a little larger than the ground state energy eigenvalue. Here $\psi(x)$ curves away from the axis less strongly than $\psi_0(x)$ in the left-hand forbidden region, because we have reduced the curvature C in equation (7.1) by increasing the energy. Then in the classically allowed region the function curves back toward the axis more strongly than $\psi_0(x)$, because here the magnitude of the curvature has been increased. The result is that $\psi(x)$ for this larger energy passes through zero in the right-hand forbidden region and then curves off to negative infinity, an equally Bad Thing.

If the potential well were deep or broad enough, there would be a first excited energy eigenstate $\psi_1(x)$. With increasing x, this function would curve through $\psi_1(x) = 0$ at a point of inflection within the allowed region, and then hook back toward zero, approaching the axis at the right-hand boundary of the allowed region with slope just steep enough that $\psi_1(x)$ asymptotically approaches zero at $x \to \infty$. Evidently, the first excited state wave function ψ_1 has one zero, or node. The second excited state, if it exists, has two nodes, the third three, and so on.

It will be noted that to have an acceptable wave function $\psi(x)$, that approaches zero at $|x| \to \infty$, there must be an interval of x where $E > V$ so $C(x)$ is negative and $\psi(x)$ can bend toward the axis. That is why it is clear that $E > 0$ for the simple harmonic oscillator in the last section, where the minimum value of the potential is $V = 0$. A one-dimensional potential well like the one in figure 7.1, which vanishes outside an isolated minimum where $V(x) < 0$, always has a bound state, with energy between zero and the minimum value of $V(x)$. There need not be a bound excited state. As it happens, the potential well in figure 7.1 ($V(x) \propto -e^{-x^2}$) has no bound excited state: at $E \to 0$ the curvature of the wave function in the classically allowed region is too weak to make the function complete a full oscillation.

A standard and useful example of bounded one-dimensional motion uses the square potential well,

$$\begin{aligned} V(x) &= 0, \quad \text{for} \quad 0 < x < L, \\ &= V_o \quad \text{for} \quad x \leq 0 \text{ or } x \geq L. \end{aligned} \tag{7.3}$$

Here V_o and L are real positive constants. The bound state energy eigenfunctions in this potential, with energy $E < V_o$, curve away from the axis at $x < 0$ and $x > L$, and in between, in the classically allowed region, the functions curve toward the axis. In the limit of arbitrarily large V_o, the curvature away from the axis at $x < 0$ and $x > L$ is so strong that the eigenfunctions are squeezed to negligibly small values in the classically forbidden regions. In this limit, Schrödinger's equation becomes

$$-\frac{\hbar^2}{2m}\frac{d^2\psi}{dx^2} = E\psi, \tag{7.4}$$

for $0 < x < L$, with the boundary conditions $\psi = 0$ at $x = 0$ and $x = L$. The solutions are sine waves. The two lowest energy solutions to this equation consistent with the boundary conditions are the ground state and first excited state eigenfunctions,

$$\begin{aligned} \psi_0(x) &\propto \sin(\pi x/L), \\ \psi_1(x) &\propto \sin(2\pi x/L). \end{aligned} \tag{7.5}$$

On substituting $\psi_0(x)$ into equation (7.4), we see that the ground state energy is $E_0 = \pi^2\hbar^2/(2mL^2)$. Because Schrödinger's equation is linear, these eigenfunctions can be multiplied by any nonzero constant. The normalization is discussed in the next section.

WKB Approximation

In the high energy limit of one-dimensional motion, the number of nodes in the wave function is large because the large value of the magnitude of the curvature C in equation (7.2) makes $\psi(x)$ oscillate rapidly in the allowed region. The analysis of this limiting case of large energy is variously called the WKB (for Wentzel, Kramers, and Brillouin), adiabatic or semiclassical approximation.

To separate the rapid oscillation of the wave function $\psi(x)$ from the slower variation with x of the amplitude of the oscillation, let us seek an approximate solution to Schrödinger's equation (7.1) of the form

$$\psi = A(x)e^{i\phi(x)}, \tag{7.6}$$

where A and ϕ are real. It will be assumed that the amplitude A varies relatively slowly with x compared to the oscillation of $\psi(x)$ due to the advance of the phase $\phi(x)$. Since the phase factor $e^{i\phi}$ completes one oscillation when $\phi(x)$ advances by 2π, the order of magnitude of the distance between zeros of the real or imaginary part of ψ is

$$\delta x \sim 1/|\phi'|, \tag{7.7}$$

where the prime means derivative with respect to x (and 2π is of order unity). If the potential energy were constant the wave function would be a plane wave, with constant amplitude. The characteristic distance Δx over which the amplitude $A(x)$ changes (by a factor ~ 2, say) thus is fixed by the distance x over which the potential changes, so we have

$$\Delta x \sim |A/A''|^{1/2}. \tag{7.8}$$

In the case to be considered here, $\Delta x \gg \delta x$, so $A(x)$ varies with x slowly compared to the oscillations of $\psi(x)$ due to the variation of $\phi(x)$.

The first derivative of equation (7.6) is

$$\psi' = (A' + iA\phi')e^{i\phi}. \tag{7.9}$$

On differentiating this again and substituting in Schrödinger's equation (7.2), we get

$$A''/A + i\phi'' + 2iA'\phi'/A - (\phi')^2 = -\frac{2m}{\hbar^2}(E - V(x)). \tag{7.10}$$

Since A and ϕ are real, we can separate the real and imaginary parts of this equation as

$$\begin{aligned} A''/A - (\phi')^2 &= -\frac{2m}{\hbar^2}(E - V(x)), \\ \phi'' + 2A'\phi'/A &= 0. \end{aligned} \tag{7.11}$$

According to equations (7.7) and (7.8), the second term on the left side of the first of equations (7.11) dominates when $\delta x \ll \Delta x$, so we have

$$(\phi')^2 = \frac{2m}{\hbar^2}(E - V(x)), \tag{7.12}$$

with the solution

$$\phi(x) = \int^x dx\, [2m(E - V(x))]^{1/2}/\hbar. \tag{7.13}$$

This fixes the x dependence of the phase factor, up to an additive constant.

The solution to the second of equations (7.11) is

$$A^2 \propto 1/\phi' \propto (E - V)^{-1/2}, \tag{7.14}$$

by equation (7.12).

In this adiabatic limit, where the variation of the amplitude with position is slow compared to the oscillation of the wave function, the wave function in the allowed region is, with equations (7.13) and (7.14),

$$\psi \propto \frac{1}{(E - V(x))^{1/4}} \exp\left(i \int^x dx \frac{[2m(E - V(x))]^{1/2}}{\hbar}\right). \tag{7.15}$$

We can use equation (7.15) to make wave packets, as was done in section 5. Here the parameter we can vary is the energy E rather than the momentum, so we will make a wave packet by adding waves of the form of equation (7.15) with amplitudes that are some smooth function $g(E)$ of the energy. The function $g(E)$ will have a single maximum, with $g(E)$ small outside some range of values of the energy around this maximum. This gives the wave function

$$\begin{aligned} \chi(x,t) = \int dE\, g(E) \frac{1}{(E - V(x))^{1/4}} \\ \times \exp\left(i \int^x dx \frac{[2m(E - V(x))]^{1/2}}{\hbar} - i\frac{Et}{\hbar}\right). \end{aligned} \tag{7.16}$$

The second term in the exponential is the time dependence of the function ψ in equation (7.15), that has definite energy $E = \hbar\omega$, as in equation (6.1).

Now let us consider the values of position x and time t for which the function $\chi(x,t)$ can be appreciably different from zero. The argument is the same as for equation (5.11). For most values of x and t, the exponential in the integral in equation (7.16) oscillates rapidly as E varies over the range of values where $g(E)$ is large, and the oscillation makes χ negligibly small. This does not happen if the phase of the exponential function has an extremum as a function of E at the maximum value of $g(E)$. On writing the phase of the exponential as

$$\theta = \int^x dx \frac{[2m(E - V(x))]^{1/2}}{\hbar} - \frac{Et}{\hbar}, \tag{7.17}$$

we can write the condition that $\chi(x,t)$ is appreciable as

$$\partial\theta/\partial E = 0, \tag{7.18}$$

at E equal to the maximum of $g(E)$. On differentiating the expression in equation (7.17) with respect to E, and using the extremum condition (7.18), we get

$$t = \int^x dx \left(\frac{m}{2(E - V(x))}\right)^{1/2}. \tag{7.19}$$

We have taken the constant of integration from equation (7.13) to be independent of E.

Because the classical kinetic energy is $E - V(r)$, we see that the denominator of equation (7.19) is the classical expression for the particle velocity,

$$v_{cl} = [2(E - V(x))/m]^{1/2}, \tag{7.20}$$

so equation (7.19) is

$$t = \int^x \frac{dx}{v_{cl}}. \tag{7.21}$$

This is the usual classical expression for the time required to move a given distance. That is, in WKB approximation the wave packet $\chi(x,t)$ moves through the potential well $V(x)$ at the same rate a classical particle moves through the well.

A second notable feature of equation (7.15) is that the amplitude varies with position as $A \propto 1/(E - V)^{1/4} \propto 1/v_{cl}^{1/2}$. The significance of this result will be discussed next.

8 Probability Interpretation

Probability Distribution in Position

An electron undoubtedly acts as a particle whose position can be measured. What is the meaning of its wave function? Born introduced the successful assumption: if the position of a particle with wave function $\psi(\mathbf{r}, t)$ is measured at time t, the probability that the particle is found to be at position $\mathbf{r}$ in the range d^3r is

$$dP \propto |\psi(\mathbf{r}, t)|^2 \, d^3r. \tag{8.1}$$

Here $|\psi|^2 = \psi^*\psi$ is the square of the absolute value of the wave function. This expression is real and not negative, which is required for a probability.

If the particle were constrained to move in one dimension, we would rewrite equation (8.1) as

$$dP \propto |\psi(x)|^2 dx. \tag{8.2}$$

This is the probability for finding that the particle is found to be at position in the range x to $x + dx$. In the semiclassical approximation, where the wave function for a particle with definite energy E is given by equation (7.15), we have from equation (8.2) the probability

$$dP \propto \frac{dx}{(E - V(x))^{1/2}} \propto \frac{dx}{v_{cl}}, \tag{8.3}$$

where v_{cl} is the classical velocity (eq. [7.20]). This makes sense: in the classical limit, the time the particle spends in the interval x to $x + dx$ is inversely proportional to the speed of the particle at x, so the probability of finding the particle in dx is inversely proportional to v_{cl} at x, consistent with equation (8.3).

The probability in equation (8.1) must integrate up to unity, for the particle must be found somewhere, so the wave function must be normalized by the condition

$$\int |\psi(\mathbf{r}, t)|^2 d^3r = 1. \tag{8.4}$$

The integral is over all allowed values of the position. The square integrability condition just means that the integral of $|\psi|^2$ has to be nonzero and finite so we can normalize ψ to equation (8.4) by multiplying it by a suitable constant. Because Schrödinger's equation (6.8) is linear this multiplication by a constant does not affect the solution to Schrödinger's equation.

For motion in one dimension, the three-dimensional integral in the normalization equation (8.4) is replaced by an integral over one dimension. For example, the normalized form of the ground state wave function in equation (7.5) is

$$\psi_0(x) = \left(\frac{2}{L}\right)^{1/2} \sin(\pi x/L). \tag{8.5}$$

This function can be multiplied by a phase factor, $e^{i\phi}$, where ϕ is real and constant, without affecting the normalization or the probability distribution in position. For the wave function in equation (8.5), equation (8.2) says the probability that the particle is found in the interval x to $x + dx$ is

$$dP = \frac{2}{L} \sin^2(\pi x/L)\, dx. \tag{8.6}$$

The particle is not likely to be found near the walls at $x = 0$ and $x = L$, where $|\psi_0|^2$ is small.

Because Schrödinger's equation is linear in ψ, we can always satisfy the normalization condition (8.4) at a fixed instant of time (assuming the integral of $|\psi|^2$ is neither zero nor infinity) by multiplying ψ by a suitably chosen constant. However, we must check that the normalization does not change with time. Also, we have to decide what happens to ψ when the position of the particle is measured.

Conservation of Probability

Schrödinger's equation (6.7) for a single particle moving in three dimensions is

$$i\hbar\frac{\partial\psi}{\partial t} = -\frac{\hbar^2}{2m}\nabla^2\psi + V(\mathbf{r},t)\psi. \tag{8.7}$$

The complex conjugate of this equation is

$$-i\hbar\frac{\partial\psi^*}{\partial t} = -\frac{\hbar^2}{2m}\nabla^2\psi^* + V(\mathbf{r},t)\psi^*. \tag{8.8}$$

We have then

$$\begin{aligned}\frac{\partial|\psi|^2}{\partial t} &= \psi^*\frac{\partial\psi}{\partial t} + \psi\frac{\partial\psi^*}{\partial t} \\ &= \frac{i\hbar}{2m}(\psi^*\nabla^2\psi - \psi\nabla^2\psi^*).\end{aligned} \tag{8.9}$$

It is important that the potential energy $V(\mathbf{r})$ is real, so the potential energy term cancels out of this expression. With the identity

$$\psi^*\nabla^2\psi - \psi\nabla^2\psi^* = \nabla\cdot(\psi^*\nabla\psi - \psi\nabla\psi^*), \tag{8.10}$$

we get from equation (8.9) a conservation equation like that for mass or charge (eqs. [1.36] and [2.4]),

$$\frac{\partial\rho}{\partial t} + \nabla\cdot\mathbf{j} = 0. \tag{8.11}$$

Here the probability density in the particle position is

$$\rho = |\psi|^2, \tag{8.12}$$

as in equation (8.1), and the probability flux density is

$$\mathbf{j} = -\frac{i\hbar}{2m}(\psi^*\nabla\psi - \psi\nabla\psi^*). \tag{8.13}$$

For motion in one dimension, the conservation equation (8.11) becomes

$$\frac{\partial\rho}{\partial t} + \frac{\partial f}{\partial x} = 0. \tag{8.14}$$

The probability that the particle is found between x and $x + dx$ is $dP = |\psi(x)|^2 dx = \rho(x)dx$, and the flux of the probability in the positive x direction is

$$f = -\frac{i\hbar}{2m}(\psi^*\frac{\partial\psi}{\partial x} - \psi\frac{\partial\psi^*}{\partial x}). \tag{8.15}$$

For the wave $\psi = Ae^{ikx}$, with A and k real constants, the probability density is $\rho = |A|^2$, and the flux density is

$$f = |A|^2\hbar k/m. \tag{8.16}$$

Since de Broglie's relations (5.4) say $\hbar k$ is the momentum p, this equation shows that the probability flux for the plane wave $\psi = Ae^{ikx}$ is the probability density ρ multiplied by the classical velocity p/m, which is reasonable.

The integral of the conservation law (8.11) over a volume bounded by a fixed surface is, by Gauss's law,

$$\frac{d}{dt}\int|\psi|d^3r = -\int \mathbf{j}\cdot\mathbf{n}\,dA, \tag{8.17}$$

where dA is a surface element with outward pointing unit normal $\mathbf{n}$. The left side of this equation is the time rate of change of the probability of finding the particle within the surface. The right-hand side is the flux of probability out of the surface. If $|\psi|$ goes to zero at $r \rightarrow \infty$ faster than $|\psi| \sim r^{-3/2}$, the volume integral converges and the surface integral approaches zero as the surface is moved to infinity, so in this limit we get

$$\frac{d}{dt}\int_{\infty}|\psi|^2 d^3r = 0. \tag{8.18}$$

This means the normalization is independent of time, as required.

The condition quoted several times above, that the wave function must be square integrable, is just the condition that the wave function be normalizable. As we have also noted, a normalized wave function still can be multiplied by the factor $e^{i\phi}$, with ϕ a real constant. The significance of this freedom to adjust the phase of the wave function will be discussed in section 19.

Collapse of the Wave Function

We come now to perhaps the most curious feature of quantum mechanics. What happens to the wave function when the position of the particle is measured?

In quantum physics, the wave function ψ (or its generalization to an element of a linear space, as discussed in chapter 3) is the fullest possible description of a physical system. The wave function predicts among other things the probability distribution of where a particle can be found if its position is measured (eq. [8.1]).

Suppose the particle position is observed at time t and found to be in the small volume element δV at position $\mathbf{r}$. Suppose we look for the particle again at time $t+\delta t$, with δt arbitrarily small. We might expect, and experimental evidence shows, that the particle will be found to be arbitrarily close to the same spot, δV, at $t+\delta t$; the particle has not had time to move away. It follows that the wave function after the first measurement and before the second has to be

$$\begin{aligned} \psi'(\mathbf{r}, t+\delta t) &= 0, \ \text{ for } \mathbf{r} \text{ not in } \delta V, \\ &= \text{constant}, \ \text{for } \mathbf{r} \text{ in } \delta V. \end{aligned} \tag{8.19}$$

That is, we are forced to the conclusion that the act of observing the particle has changed the wave function from the original one, ψ, to a new one, ψ', that is consistent with the result of the first observation. This situation will be discussed at length in sections 26 to 29 on measurement theory.

9 Cold Fusion*

A deuteron is the bound state of a neutron and proton, the typical distance between the two particles being on the order of the characteristic size r_n of a light atomic nucleus,

$$r_n \sim 1 \times 10^{-13}\,\text{cm}. \tag{9.1}$$

When two deuterons are separated by a distance $\sim r_n$ the strong nuclear interaction can rapidly exchange particles, perhaps changing the two deuterons into a triton (the bound state of two neutrons and a proton) and a free proton, or into a ^{3}He nucleus (two protons and a neutron) along with a free neutron. Either of these fusion reactions yields energy. If the reactions could be controlled in an efficient way it would provide a considerable source of energy, because deuterium is quite abundant.

Two deuterons can be bound by one or two electrons in a molecule, just like the hydrogen molecule discussed in section 44 below (but with the slight difference that, because the deuterons are heavier than protons, the zero point contribution $\hbar\omega/2$ to the energy of vibration of the molecule is smaller). The wave functions of the deuterons in such a molecule overlap slightly, so there is a small probability that the deuterons find themselves within separation $\sim r_n$ and suffer one of the above reactions. As an illustration of the techniques developed above, in an interesting grown-up computation, let us find an order of magnitude estimate of the rate of this reaction in a deuterium molecule.

We can imagine that the two deuterons in the molecule move in the electrostatic potential produced by the electric charges of the deuterons and the mean charge distribution of the electrons. (This is discussed in section 44.) Treatment of the two-body problem for the two deuterons requires one more result, that is obtained in section 12 below: just as in classical mechanics, the analysis of the relative motion of the two deuterons is equivalent to a one-body problem, with reduced mass m. (Here m is half the deuteron mass, or very close to the mass of a proton.)

The potential energy $V(r)$ as a function of the separation r of the two deuterons has the following features. Since the deuterons are bound in a hydrogen-like molecule, we know $V(r)$ has a minimum with depth $\sim 4\,\mathrm{eV}$ (the binding energy of the two atoms in a hydrogen molecule) at separation $r_a \sim 0.7 \times 10^{-8}$ cm. At smaller separations, the two deuterons see their bare charges, e, because the charge distribution in the electron wave function is spread over a distance $\sim r_a$. Thus the potential may be approximated as

$$V(r) \sim e^2/r, \quad \text{at} \quad r_n \lesssim r \lesssim r_a. \tag{9.2}$$

At a separation comparable to the characteristic nuclear radius $r_n \sim 10^{-13}$ cm, the strong nuclear interaction becomes dominant. At this radius, the potential rather abruptly drops from the large positive value in equation (9.2) to the deeply negative value characteristic of the nuclear interaction,

$$V(r) \sim V_n \sim -1\,\mathrm{MeV} = -10^6\,\mathrm{eV}, \quad \text{at} \quad r < r_n. \tag{9.3}$$

The potential energy function is sketched (in highly distorted scales) in figure 9.1.

The wave function for the relative motion of the deuterons is spherically symmetric, for this gives the lowest energy state (as will be shown in section 18). As discussed in section 6, the radial wave function $u(r) = r\psi(r)$ satisfies a one-dimensional Schrödinger equation, as in equations (6.17) and (6.18). (It is left as exercise, in problem I.17, to check that the probability density and flux one writes down in this one-dimensional problem, using the radial wave function $u(r) = r\psi(r)$ in eqs. [8.14] and [8.15], are consistent with the corresponding probability density and flux in three dimensions obtained from $\psi(r)$.)

At radii $r \sim r_a$, the radial wave function $u(r)$ for the relative motion of the deuterons has very nearly the shape of the ground state wave

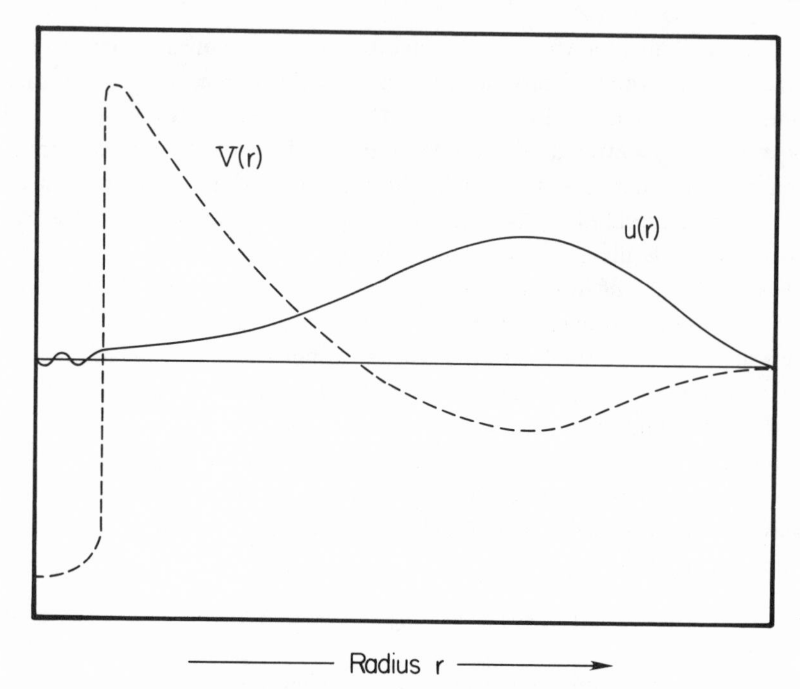

Fig. 9.1 The potential energy (dashed line) and real part of the radial wave function $u(r)$ (solid line) for the relative motion of two deuterons in a molecule. The horizontal and vertical axes have been strongly distorted so as to allow a sketch of the behavior of the functions over a broad range of scales. At separations smaller than the nuclear radius $r_n \sim 10^{-13}$ cm, the wave function is $u \propto e^{-ikr}$, representing motion to destruction at small radius. In the classically forbidden region between r_n and an atomic scale of separation, $r \sim r_a \sim 10^{-8}$ cm, the wave function bends sharply away from the axis. In the classically allowed region of motion, the wave function bends back toward the axis.

function of a particle bound in an ordinary potential well. However, $u(r)$ has a small tail extending to very small r, that produces a small flux of probability moving into the interaction region at $r < r_n$. This tunneling effect through the classically forbidden region is what causes the eventual destruction of the deuterons. The goal is to find an order of magnitude estimate of the probability flux into the interaction region, that is, the probability per unit time that the deuterons react and turn into something else, such as a triton and proton.

The first approximation in the computation is to ignore the effect of the flux of probability at $r \sim r_n$ on the behavior of the wave function at $r \sim r_a$. This is reasonable, because a deuteron molecule lasts a very long time. In this approximation, the radial wave function $u(r)$ satisfies

Schrödinger's time independent wave equation (7.1), with definite energy E, and with the boundary conditions that $u(r)$ is very small where r is very large and very small compared to r_a.

Next, we have to say what happens when the deuterons find themselves separated by a distance less than $\sim r_n$. For simplicity, it will be assumed that they inevitably react. Then the boundary condition at small r really is that pairs of deuterons never are found to be moving apart at $r \sim r_n$. This means the radial wave function at $r = r_n$ must match onto the function

$$u_n \propto e^{-ikr}, \quad \text{for} \quad r < r_n. \tag{9.4}$$

In the example in equation (8.16), it was seen that the wave function $\psi = Ae^{ikx}$ describes probability flux moving toward positive x. The flux in equation (9.4) is moving toward smaller r, to destruction at $r < r_n$.

The wave function just outside the nuclear potential has to match the value and first derivative of equation (9.4). (That is because Schrödinger's equation contains second derivatives with respect to r. If the value or slope of the wave function $u(r)$ were discontinuous at a point where the potential energy is finite, the wave function would not satisfy Schrödinger's equation at that point.) The convenient way to deal with the undetermined constant of proportionality in equation (9.4) is to express this matching condition as

$$\frac{u'}{u} = \frac{d}{dr}\log u = -ik, \quad \text{at} \quad r = r_n. \tag{9.5}$$

As in equation (7.7), the prime means a derivative with respect to r. The derivative of the logarithm eliminates the constant of proportionality. The value for the logarithmic derivative is given by equation (9.4). The solution for the wave function immediately outside the nuclear potential well has to have the same value of the logarithmic derivative. The wave function, again in a highly distorted scale, is sketched in figure 9.1.

It is convenient to write the radial wave function outside the nuclear interaction region in the form

$$u = A(r)e^{i\phi(r)}, \tag{9.6}$$

where A and ϕ are real, as in section 7. The boundary condition (9.5) is

$$A'/A + i\phi' = -ik, \tag{9.7}$$

at $r = r_n$. The real and imaginary parts of this expression are

$$(A'/A)_n = 0, \qquad \phi'_n = -k. \tag{9.8}$$

The subscript reminds us that these equations apply at the edge of the region of nuclear interaction, at $r \sim r_n$.

The probability flux (eq. [8.15]), with the wave function (9.6), is

$$f = -\frac{i\hbar}{2m}(u^* u' - u u'^*) = \frac{\hbar}{m}\phi' A^2. \tag{9.9}$$

This is the probability per unit time that the deuteron pair moves into the nuclear interaction region. With the boundary condition (9.8), the flux is

$$f = -\frac{\hbar k}{m} A_n^2. \tag{9.10}$$

This expression is negative because the flux of probability is moving toward smaller r. As in equation (8.16), the flux f is the product of the classical velocity $v_{cl} = p/m = \hbar k/m$ in the nuclear interaction region and the probability per unit length, $|u|^2 = A^2$, of finding the particle there.

The Schrödinger equation for the wave function in the form of equation (9.6) was written down in equation (7.11). The imaginary part is

$$\phi'' + 2A'\phi'/A = 0, \tag{9.11}$$

with the solution

$$\phi' \propto 1/A^2. \tag{9.12}$$

This says the flux f in equation (9.9), which is proportional to $A^2\phi'$, is independent of radius. This is reasonable because we are assuming a steady state situation, in which the wave function u is assigned a definite energy, E, so u varies with time as $e^{-iEt/\hbar}$. The probability for finding that the separation of the deuteron pair is in the interval r to $r + dr$, which is $dP \propto |u(r)|^2 dr$, thus is independent of time, so the flux $f(r+dr)$ of probability entering the interval at $r + dr$ must be equal to the flux $f(r)$ of probability leaving the interval at r. That is not quite realistic, for the probability of finding the deuteron at atomic separations $\sim r_a$ really is decreasing, but at such a slow rate that we can ignore it.

The real part of equation (7.11), with equations (9.8) and (9.12), is

$$\frac{A''}{A} - \frac{k^2 A_n^4}{A^4} = \frac{2m}{\hbar^2}(V(r) - E). \tag{9.13}$$

The second term on the left side is normalized to the value A_n of the amplitude $A(r)$ at the edge of the interaction region $r = r_n$, where $\phi' = -k$.

The next step is to solve equation (9.13), by numerical integration if an accurate result is wanted, subject to the conditions that the amplitude $A(r)$ is very small at large r and satisfies equation (9.8) at $r = r_n$. This solution must then be normalized in the usual way, so the probability distribution dP/dr integrates to unity at $r > r_n$. The normalized solution fixes the value of the amplitude A_n at $r = r_n$. That amplitude in equation (9.10) gives the probability flux into the interaction region. This probability flux is the decay rate of the deuterium molecule in the approximation that the deuterons inevitably destroy each other when they interact.

The first thing one does at this point in a computation is to establish orders of magnitude. For this purpose, we note that in the interesting range of values of the separation r the second term on the left side of equation (9.13) is small compared to the first, because $A(r)$ increases rapidly with increasing r. In the strongly classically forbidden region, where $V \gg |E|$, we can take the potential to be approximated by equation (9.2). This brings the Schrödinger equation (9.13) to

$$\frac{A''}{A} \sim \frac{2me^2}{\hbar^2 r}. \tag{9.14}$$

In the adiabatic approximation of section 7, the solution to this equation is

$$\begin{aligned} A(r) &\propto \exp\left[\left(\frac{2me^2}{\hbar^2}\right)^{1/2} \int^r \frac{dr}{r^{1/2}}\right] \\ &\propto \exp\left[\left(\frac{8me^2}{\hbar^2}\right)^{1/2} r^{1/2}\right]. \end{aligned} \tag{9.15}$$

This is a crude approximation, but remember that we are only attempting to fix orders of magnitude.

We can find the normalization of equation (9.15), to an approximation consistent with the accuracy of our solution, by noting that most of the probability is concentrated in an interval of width $\sim r_a$ centered on the minimum of the potential at $r \sim r_a$, so the amplitude at r_a is on the order

$$A_a \sim 1/r_a^{1/2}. \tag{9.16}$$

Because r_n is much smaller than r_a, we see from equation (9.15) that the amplitude at the edge of the interaction region is

$$A_n \sim r_a^{-1/2} \exp -(8me^2 r_a/\hbar^2)^{1/2}. \tag{9.17}$$

The factor $\hbar k$ that appears in equation (9.10) is the momentum in the interaction region. The kinetic energy is almost entirely due to the depth V_n of the nuclear potential well, so the momentum is

$$\hbar k = (2m|V_n|)^{1/2}. \tag{9.18}$$

Since our approximation (9.17) for A_n is crude, it makes no sense to keep the numerical factors multiplying the expression for the probability flux f. Dropping these prefactors, we get from equations (9.10) and (9.17)

$$f \sim \frac{1}{r_a}\left(\frac{|V_n|}{m}\right)^{1/2} \exp -(32me^2 r_a/\hbar^2)^{1/2}. \tag{9.19}$$

This is the wanted estimate of the flux of probability into the nuclear interaction region, that is, of the reaction probability per unit time per molecule.

It is traditional and convenient to express the argument of the exponential in terms of some standard physical constants. The magnitude of the electron charge, $e = 4.80 \times 10^{-10}$ esu, defines the dimensionless fine-structure constant,

$$\alpha = \frac{e^2}{\hbar c} \sim \frac{1}{137}. \tag{9.20}$$

The mass m is very nearly equal to the proton mass; it defines the proton Compton wavelength

$$\lambda_c = \frac{\hbar}{mc} \sim 2.1 \times 10^{-14}\,\text{cm}. \tag{9.21}$$

The potential $V(r)$ in a hydrogen molecule is minimum at $r \sim 0.7 \times 10^{-8}$ cm. Since we used the bare coulomb potential in equation (9.2) all the way to r_a, it is reasonable to take $r_a \sim 0.5 \times 10^{-8}$ cm, a little smaller than the radius at the minimum of $V(r)$, to take account of the rounding of the potential near the minimum. With these parameters, the exponential factor in equation (9.19) is

$$\exp -(32\alpha r_a/\lambda_c)^{1/2} \sim e^{-235}. \tag{9.22}$$

The argument of the exponential is large, and was only roughly estimated; that is why it makes no sense to be careful about the prefactor multiplying the exponential in equation (9.19). This prefactor is

$$\frac{c}{r_a}\left(\frac{|V_n|}{mc^2}\right)^{1/2} \sim e^{40}\,\mathrm{sec}^{-1}, \tag{9.23}$$

where the proton mass is $mc^2 \sim 1\,\mathrm{GeV} = 10^9\,\mathrm{eV}$, and $V_n \sim -10^6\,\mathrm{eV}$. The product is the reaction rate per molecule,

$$f \sim 10^{-85}\,\mathrm{sec}^{-1}. \tag{9.24}$$

A more accurate computation is of course possible. One would use a more realistic approximation to the potential $V(r)$, and a numerical integration of equation (9.13). An additional refinement takes account of the fact that some part of the flux entering the nuclear interaction region is reflected: deuterons do not inevitably destroy themselves. The result is not far from equation (9.24).

Could the deuteron reaction rate at room temperature be higher in some other molecular or condensed matter environment? Since superconductivity was not anticipated, but is consistent with quantum mechanics, could the same be true for cold fusion? The central point of the computation is that the wave function at the nuclear interaction region is strongly suppressed by the repulsive coulomb potential $\sim e^2/r$ of the charges of the deuterons. This repulsive potential, that is much greater than the energy of atoms in a molecule, causes the wave function to curve strongly away from the axis, as discussed in section 7. The curvature away from the axis makes the wave function very small at $r \sim r_n$. This makes the "tunneling rate" through the coulomb barrier very slow. The two ways to increase the tunneling rate are to increase the energy or reduce the width of the barrier through which the particles must pass. Deuterons react rapidly in stars because of the high thermal energy, but of course that option is not available at room temperature. If by mechanical compression the radius of the electron wave function around the deuteron were reduced to $\sim$ one fifth of r_a, it would raise the reaction rate to $\sim 10^{-28}\,\text{sec}^{-1}$, which would be detectable. However, that would make $\nabla^2\psi_e$ for the electron wave function, ψ_e, about 25 times its normal value. By Schrödinger's equation, that would mean the kinetic energy of the electron around each deuteron is $\sim$ 25 times its normal value, which is not a realistic possibility. The situation with superconductivity is very different because it involves low energy interactions among many particles. An appreciable cold fusion rate would require a large compression of the electron wave function that shields the charges of the deuterons, leading to energies that are known to be unrealistically large. That is why, despite the fact that there are many unresolved problems in condensed matter physics, people are confident that, if conventional quantum mechanics is valid, the deuterium fusion reaction rate in solids at room temperature is exceedingly slow.

10 Momentum

Short Review of Fourier Transforms

As one might expect from the de Broglie relation $\mathbf{p} = \hbar\mathbf{k}$, where the wavelength is $2\pi/k$, the momentum $\mathbf{p}$ of a particle is related to the Fourier transform of its wave function. To begin, let us recall some properties of Fourier transforms.

Consider a periodic function $f(x)$ of one variable, with period L. The periodicity means that, for any value of x,

$$f(x) = f(x + L). \tag{10.1}$$

If $f(x)$ is reasonably well behaved (all wave functions in quantum mechanics are reasonably well behaved) it can be expanded in a Fourier series in the periodic functions $e^{2\pi inx/L}$ with integer n:

$$f(x) = \sum_{n=-\infty}^{+\infty} f_n e^{2\pi inx/L}. \tag{10.2}$$

The sum is over all integers n, positive and negative. The Fourier expansion coefficients f_n are obtained from the function $f(x)$ by using the orthogonality relation, which you are invited to check,

$$\int_0^L dx\ e^{2\pi i(n-m)x/L} = 0 \text{ if } n \neq m, \\ = L \text{ if } n = m. \tag{10.3}$$

This is written for short as

$$\int_0^L dx\ e^{2\pi i(n-m)x/L} = L\delta_{nm}. \tag{10.4}$$

The Kronecker delta function is

$$\delta_{nm} = 0 \text{ if } m \neq n, \\ = 1 \text{ if } m = n. \tag{10.5}$$

The result of multiplying the Fourier series (10.2) by $e^{-2\pi imx/L}$, with m an integer, integrating the result from $x = 0$ to $x = L$, and using equation (10.4), is

$$f_m = \frac{1}{L}\int_0^L dx\ f(x)\ e^{-2\pi imx/L}. \tag{10.6}$$

A useful relation that follows from equations (10.2) and (10.4) is

$$\int_0^L dx|f(x)|^2 = \int_o^L dx \sum_{m,n} f_n f_m^* e^{2\pi i(n-m)x/L} \\ = L\sum_n |f_n|^2. \tag{10.7}$$

Another useful relation is obtained by substituting equation (10.6) for the expansion coefficients into the Fourier series (10.2), and then exchanging the order of the sum and integration:

$$f(x) = \int_0^L dy f(y) \sum_n e^{2\pi in(x-y)/L}/L. \tag{10.8}$$

The function multiplying $f(y)$ in the integrand has the property that it always returns the value of f at x, whatever $f(y)$. This is useful enough to deserve special notation. The Dirac delta function is

$$\delta(z) = \frac{1}{L}\sum_n e^{2\pi i n z/L}, \tag{10.9}$$

so equation (10.8) is

$$f(x) = \int_0^L dy\; f(y)\; \delta(x-y). \tag{10.10}$$

It is apparent that the function $\delta(x-y)$ can only be nonzero at $y = x$, because that is the only argument where the value of $f(y)$ matters in determining the value of the integral. And since $\delta(x-y) = 0$ at $x \neq y$, we can set $f(y) = f(x)$ in equation (10.10) to get

$$\int dz\; \delta(z) = 1. \tag{10.11}$$

One can (perhaps should) think of the Dirac delta function as the limit of a sequence of non-singular functions

$$\delta_m(x) = \frac{1}{L}\sum_{n=-m}^{+m} e^{2\pi i n x/L}. \tag{10.12}$$

The function $\delta_m(x)$ has unit integral,

$$\int_0^L \delta_m(x)dx = 1, \tag{10.13}$$

in agreement with equation (10.11). The peak value at $x = 0$ is $\delta_m = (2m+1)/L$, and the width of the peak is $\Delta x \sim L/m$. In the limit $m \rightarrow \infty$, the sequence of functions $\delta_m(x)$ approaches the Dirac function $\delta(x)$ that has unit integral and vanishes at $x \neq 0$.

Let us consider next the Fourier integral transform of a function that is not periodic but does go to zero at $|x| \rightarrow \infty$. We can take this to be the limit of a periodic function as the period $L \rightarrow \infty$, as was done in sections 1 and 2. Let

$$k_n = 2\pi n/L, \qquad \delta k = 2\pi/L, \qquad g(k_n) = Lf_n/2\pi. \tag{10.14}$$

Then equations (10.2) and (10.6) can be written as

$$f(x) = \sum_n \delta k \, g(k_n) e^{ik_n x}, \qquad g(k_m) = \int \frac{dx}{2\pi} f(x) e^{-ik_m x}. \tag{10.15}$$

In the limit $L \to \infty$, the value of δk goes to zero and we can replace the sum with an integral to get the Fourier integral relations,

$$f(x) = \int_{-\infty}^{+\infty} dk \; g(k) e^{ikx}, \qquad g(k) = \frac{1}{2\pi} \int_{-\infty}^{+\infty} dy \; f(y) e^{-iky}. \tag{10.16}$$

On substituting the second of the Fourier integral relations into the first and exchanging the order of integration, we get

$$f(x) = \int dy \; f(y) \frac{1}{2\pi} \int_{-\infty}^{\infty} dk \; e^{ik(x-y)}. \tag{10.17}$$

This is the same as equation (10.10) except that the integral has been expanded to all y. Thus another expression for Dirac's delta function is

$$\delta(x) = \frac{1}{2\pi} \int_{-\infty}^{\infty} dk \; e^{ikx}. \tag{10.18}$$

This expression is worth memorizing.

The discussion is readily generalized to three dimensions. The three-dimensional δ function is a product of delta functions of the three cartesian position components:

$$\delta(\mathbf{r}) = \delta(x)\delta(y)\delta(z). \tag{10.19}$$

This is normalized to

$$\int d^3r \; \delta(\mathbf{r}) = 1. \tag{10.20}$$

We have from equation (10.18)

$$\delta(\mathbf{r}) = \frac{1}{(2\pi)^3} \int d^3k \; e^{i\mathbf{k}\cdot\mathbf{r}}. \tag{10.21}$$

The integral is over all $\mathbf{k}$, positive and negative values of the three components.

The three-dimensional version of equation (10.16) is

$$\begin{aligned} f(\mathbf{r}) &= \int d^3k\ g(\mathbf{k})e^{i\mathbf{k}\cdot\mathbf{r}}, \\ g(\mathbf{k}) &= \frac{1}{(2\pi)^3}\int d^3y\ f(\mathbf{y})e^{-i\mathbf{k}\cdot\mathbf{y}}. \end{aligned} \tag{10.22}$$

This is consistent with equation (10.21) for the three-dimensional Dirac delta function, as one can check by substituting the expression for $g(\mathbf{k})$ into the integral for $f(\mathbf{r})$, and exchanging the order of integration.

Momentum Measurement by Time of Flight

The momentum of a single free particle (mass m, potential energy $V = 0$ so there is no classical force) can be measured by the time of flight method. The wave function, ψ, of the particle is arranged so that at time $t = 0$ the particle is localized to some bounded region around $\mathbf{r} = 0$. At some later time, t, the position of the particle is measured; if it is found at $\mathbf{r}$ then we can say that since the particle moved a distance $\sim \mathbf{r}$ in time t it had velocity $\mathbf{v} \sim \mathbf{r}/t$, and momentum $\mathbf{p} \sim m\mathbf{r}/t$. The measurement is uncertain because of the uncertainty in the initial position of the particle, but we can make that arbitrarily small by making t large enough.

For simplicity, we will consider the one-dimensional case. Schrödinger's equation for a free particle moving in one dimension is

$$i\hbar\frac{\partial\psi}{\partial t} = -\frac{\hbar^2}{2m}\frac{\partial^2\psi}{\partial x^2}. \tag{10.23}$$

We can write the solutions to this equation as Fourier sums or integrals over plane waves:

$$\psi \propto e^{i(px - p^2t/2m)/\hbar}. \tag{10.24}$$

This is a solution to equation (10.23), as may be checked by differentiating it out. Since the differential equation (10.23) is linear, we can write the general solution as a linear combination of the solutions (10.24),

$$\psi = \int \frac{dp}{(2\pi\hbar)^{1/2}} f(p)e^{i(px-p^2t/2m)/\hbar}, \tag{10.25}$$

where the function $f(p)$ is fixed by initial conditions. The factor $(2\pi\hbar)^{1/2}$ is chosen to give a convenient form for the normalization integral

$$\int_{-\infty}^{+\infty} dx|\psi|^2 = \int \frac{dpdp'}{2\pi\hbar} f(p)f(p')^* e^{i((p')^2-p^2)t/2m\hbar} \int dx\, e^{i(p-p')x/\hbar}. \tag{10.26}$$

On changing the variable of integration in the last integral to $y = x/\hbar$, we see that the last integral becomes $2\pi\hbar\delta(p-p')$ (eq. [10.18]). Then we can immediately do the integral over p', as in equation (10.10), to get

$$\int_{-\infty}^{+\infty} dx|\psi(x)|^2 = \int_{-\infty}^{+\infty} dp|f(p)|^2. \tag{10.27}$$

This can be compared to equation (10.7). As will be seen, the normalization in equation (10.27) is particularly useful.

We want the particle to be in the neighborhood of $x = 0$ at time $t = 0$. The way to do this was discussed in section 5: let the function $f(p)$ in equation (10.25) vary smoothly with p, with $f(p)$ tending to be positive at small p and approaching zero at large $|p|$. Then at $t = 0$ and large $|x|$, the rapid oscillation of the exponential in equation (10.25) as p varies makes the integral ψ small. At small enough x, the exponential is nearly constant over the range of values of p for which $f(p)$ is large, so the integral is large. That is, the wave function, $\psi(x, 0)$, is large only at $x \sim 0$. Since the probability of finding the particle between x and $x + dx$ is $|\psi(x,t)|^2 dx$, the particle certainly is near $x = 0$ at $t = 0$.

To see the behavior of ψ at large time, rewrite the argument of the exponential in equation (10.25) by completing squares:

$$\begin{aligned} px - \frac{p^2 t}{2m} &= -\frac{t}{2m}(p^2 - 2mxp/t) \\ &= -\frac{t}{2m}(p - mx/t)^2 + \frac{mx^2}{2t}. \end{aligned} \tag{10.28}$$

This in equation (10.25) gives

$$\psi = \phi e^{imx^2/2\hbar t}, \tag{10.29}$$

with

$$\phi = \int \frac{dp}{(2\pi\hbar)^{1/2}} f(p) \exp -\frac{it}{2m\hbar}(p - mx/t)^2. \tag{10.30}$$

It will be noted that the exponential multiplying ϕ in equation (10.29) does not affect the probability distribution of the particle position: $|\psi|^2 = |\phi|^2$.

Now if the time t is large, the exponential in the integral in equation (10.30) tends to oscillate very rapidly as a function of p, except at $p \sim mx/t$, where the argument of the exponential is small and does not vary rapidly because $(p - mx/t)$ appears as the square. The oscillation makes the integral over p very small everywhere save at $p \sim mx/t$, so that is the only place where the value of $f(p)$ can matter. Thus we can replace $f(p)$ with $f(mx/t)$ in the integral, to reduce equation (10.30) to

$$\phi(x,t) = \frac{f(mx/t)}{(2\pi\hbar)^{1/2}} \int dp \exp -\frac{it}{2m\hbar}(p - mx/t)^2. \qquad (10.31)$$

With the change of variables

$$p = \frac{mx}{t} + \frac{(2m\hbar)^{1/2}}{t^{1/2}} z, \qquad (10.32)$$

we get

$$\phi = \left(\frac{m}{\pi t}\right)^{1/2} f(mx/t) \int dz\, e^{-iz^2}. \qquad (10.33)$$

The integral is along the real axis of z. To evaluate it, let $z = (1 - i)y/2^{1/2}$, and note that, because the integrand is an analytic function, we can bend the contour back to the real axis of y, to get

$$\int dze^{-iz^2} = \frac{1-i}{2^{1/2}} \int_{-\infty}^{\infty} dye^{-y^2} = (1-i)(\pi/2)^{1/2}. \qquad (10.34)$$

The last step follows from

$$\int_{-\infty}^{\infty} dy\, e^{-y^2} = \pi^{1/2}. \qquad (10.35)$$

Collecting all this, we see that the wave function (10.33) approaches

$$|\psi(x,t)|^2 = |\phi(x,t)|^2 = \frac{m}{t}|f(mx/t)|^2, \qquad (10.36)$$

at large time t. This result indicates that the probability of finding the particle at x in the range dx approaches

$$dP = |\psi|^2 dx = |f(mx/t)|^2 \frac{m\,dx}{t}, \qquad (10.37)$$

at very large time.

If the particle is found at x then it has moved a distance $\sim x$ because it started at $x \sim 0$, so the velocity measured by the time of flight is $\sim x/t$. The measured momentum is then

$$p = mx/t. \tag{10.38}$$

Therefore, equation (10.37) says that the probability that the particle is found to have momentum p in the range dp is

$$dP = |f(p)|^2\, dp. \tag{10.39}$$

It will be noted from equation (10.27) that, if ψ is normalized, so is f:

$$\int dP = \int |f|^2\, dp = \int |\psi|^2\, dx. \tag{10.40}$$

Also, because $\psi(x, 0)$ in equation (10.25) at time $t = 0$ is the Fourier transform of $f(p)$ (up to a multiplicative factor), we can use the Fourier transform relations to write $f(p)$ as an integral over $\psi(x, 0)$. On setting $t = 0$ in equation (10.25), multiplying the equation by $e^{-ip'x}$, integrating over x, and using equation (10.18) to evaluate the integral over x, one finds

$$f(p) = \int \frac{dx}{(2\pi\hbar)^{1/2}} \psi(x, 0) e^{-ipx/\hbar}. \tag{10.41}$$

These equations refer to the result of a momentum measurement by time of flight for a free particle. As discussed next, it will be assumed that the momentum distribution in general is given by equations (10.39) and (10.41).

Momentum Wave Function

The discussion is generalized to motion in three dimensions, and to a particle that is not free (that is, may have potential energy $V(\mathbf{r})$), as follows. The particle is associated with a wave function, $\psi(\mathbf{r}, t)$. This function evaluated at time t can be expressed as a Fourier sum (under periodic boundary conditions) or integral over plane waves. In the latter case, we have

$$\psi(\mathbf{r}, t) = \int \frac{d^3p}{(2\pi\hbar)^{3/2}} f(\mathbf{p}, t) e^{i\mathbf{p}\cdot\mathbf{r}/\hbar}. \tag{10.42}$$

The factor $(2\pi\hbar)^{3/2}$ gives a convenient normalization in the inverse Fourier transform relation, which follows from equation (10.21):

$$f(\mathbf{p}, t) = \int \frac{d^3r}{(2\pi\hbar)^{3/2}} \psi(\mathbf{r}, t) e^{-i\mathbf{p}\cdot\mathbf{r}/\hbar}. \tag{10.43}$$

According to de Broglie's relation $\mathbf{p} = \hbar\mathbf{k}$, the plane wave component $\propto e^{i\mathbf{p}\cdot\mathbf{r}/\hbar}$ in equation (10.42) has momentum $\mathbf{p}$. For a free particle, this component has definite energy $E = p^2/2m$, so the amplitude of this component varies with time as $f(\mathbf{p}, t) = f(\mathbf{p}, 0)e^{-ip^2t/2m\hbar}$, as in equation (10.24). If the particle is not free, the time dependence of $f(\mathbf{p}, t)$ has to be computed from Schrödinger's equation.

The probability distribution in the result of a measurement of the momentum of the particle at time t is assumed to be

$$dP = |f(\mathbf{p}, t)|^2 \, d^3p. \tag{10.44}$$

If the particle is free, this follows from the time of flight argument, as in equation (10.39). We shall adopt equations (10.42) to (10.44) as the general definition in quantum mechanics of the momentum of a single particle moving in three dimensions.

The functions $\psi(\mathbf{r}, t)$ and $f(\mathbf{p}, t)$ play symmetrical roles in the measurement of position (eq. [8.1]) and the measurement of momentum (eq. [10.44]). These functions are related by the symmetrical equations (10.42) and (10.43). Just as $\psi(\mathbf{r}, t)$ is the position wave function, one can call $f(\mathbf{p}, t)$ the momentum wave function. This prescription is generalized to other observables in section 14.

Uncertainty Principle

For a specific example, suppose a particle that moves in one dimension has wave function

$$\begin{aligned} \psi &= 0, \quad \text{for} \quad |x| \geq x/2, \\ &= 1/L^{1/2}, \quad \text{for} \quad |x| < L/2, \end{aligned} \tag{10.45}$$

at time t. This function is properly normalized to $\int |\psi|^2 dx = 1$. If the position x of a particle with this wave function is measured at the given time, t, it may be found to be anywhere between $-L/2$ and $L/2$ with uniform probability: $dP = dx/L$ at $-L/2 < x < L/2$.

The momentum wave function follows on computing the integral in equation (10.41):

$$f(p) = \left(\frac{2\hbar}{\pi L}\right)^{1/2} \frac{\sin pL/(2\hbar)}{p}. \tag{10.46}$$

The probability distribution in the momentum is then

$$dP = |f(p)|^2 = \frac{2\hbar}{\pi L} \frac{\sin^2 pL/(2\hbar)}{p^2} dp. \tag{10.47}$$

Using

$$\int_{-\infty}^{\infty} \frac{\sin^2 x}{x^2} dx = \pi, \tag{10.48}$$

you can check that this is normalized to $\int dP = 1$, as we knew it had to be.

Equation (10.47) is maximum at $p = 0$, and is small at momenta significantly larger than the characteristic width

$$\delta_p \sim \hbar/L. \tag{10.49}$$

This means that in the state represented by equation (10.45) the momentum is predicted to be $p \sim 0$ with an uncertainty δ_p. This uncertainty can be made arbitrarily small by taking L to be large, but that means the position is predicted with the large uncertainty $\delta_x \sim L$. The result of measuring the position can be predicted with small uncertainty by taking L to be small, but that means the uncertainty in momentum is large. The product of the uncertainties in the prediction of position and momentum is

$$\delta_x \delta_p \sim \hbar. \tag{10.50}$$

A more general derivation of this Heisenberg uncertainty relation is given in problem III.3 below.

The uncertainty relation (10.50) is a useful guide to an intuitive understanding of the behavior of particles in quantum mechanics. For example, the electron in the ground state of a hydrogen atom is confined to a region of size $\sim a_o$, the Bohr radius (eq. [6.23]). Since the position of the electron is known to this accuracy, the momentum is uncertain by the amount $\delta_p \sim \hbar/a_o$. Since the mean value of the momentum vanishes if the atom is not moving, the magnitude of the momentum must typically be $p \sim \delta_p$, so the kinetic energy is $\sim p^2/2m \sim \hbar^2/(ma_o^2)$. The electron typically is at distance $\sim a_o$ from the proton, so the potential energy $\sim -e^2/a_o$. The net energy is the sum of kinetic and potential,

$$E \sim \frac{\hbar^2}{ma_o^2} - \frac{e^2}{a_o}. \tag{10.51}$$

In the ground state this energy is as small as possible. The results of minimizing the expression (10.51) are $a_o \sim \hbar^2/(me^2)$ and $E \sim me^4/\hbar^2$, which are the right order of magnitude (eqs. [6.23] and [6.24]).

11 Expectation Values and the Momentum Operator

Equations (8.1) and (10.44) give probability distributions for the result of measuring either the position or the momentum of a particle with wave function ψ. These distributions are usefully characterized by the expectation values of functions of position or of momentum, as follows.

Consider a random variable x that assumes discrete values x_i. Suppose the probability that the variable is found to have the value x_i is P_i. That means that if we had a statistical ensemble of $M \gg 1$ cases, the variable would be found to have the value x_i in $N_i = MP_i$ members of the ensemble (as in eq. [1.2]). The arithmetic mean value of the result of measuring the variable is

$$\langle x \rangle = \sum N_i x_i / M = \sum P_i x_i. \tag{11.1}$$

This is the expectation value or ensemble average value of the random variable x. The expectation value of the function $V(x)$ of x similarly is

$$\langle V \rangle = \sum P_i V(x_i). \tag{11.2}$$

If the random variable can assume a continuous range of values, then the probability of finding that the result of measuring the variable is between x and $x + dx$ is written as

$$dP = f(x)dx, \tag{11.3}$$

just as was done for position and momentum in equations (8.1) and (10.44). The expectation value of a function $V(x)$ of the continuous variable x is again the arithmetic mean value of the result of measuring the variable in many members of an ensemble. In this continuous limit equation (11.2) goes to

$$\langle V \rangle = \int V(x) f(x) dx. \tag{11.4}$$

This definition of the expectation value of a random variable can be applied to the probability distributions of quantum mechanics. For a particle that moves in three dimensions with wave function ψ, the probability that the particle will be found at $\mathbf{r}$ in the range d^3r is $dP = |\psi(\mathbf{r},t)|^2 d^3r$. Thus the ensemble average value, or expectation value, of the position of the particle if measured at time t is

$$\langle \mathbf{r} \rangle = \int d^3r \; \mathbf{r} \; |\psi(\mathbf{r},t)|^2. \tag{11.5}$$

Because integrals of this form are so common they are given a special notation: for functions ψ and ϕ, let

$$(\psi, \phi) = \int d^3r \, \psi^*(\mathbf{r}) \phi(\mathbf{r}). \tag{11.6}$$

The integral is over all space, which may be the volume V of space under periodic boundary conditions, or to infinity. The integral is called the inner product of the function ψ with the function ϕ, for reasons to become clear in sections 13 and 22. In this notation, the normalization condition is

$$(\psi, \psi) = 1, \tag{11.7}$$

and equation (11.5) is

$$\langle \mathbf{r} \rangle = (\psi, \mathbf{r}\psi). \tag{11.8}$$

The expectation value of the momentum follows from equation (10.44):

$$\langle \mathbf{p} \rangle = \int d^3p \, \mathbf{p} |f(\mathbf{p},t)|^2. \tag{11.9}$$

We can rewrite this in a simple form in terms of the wave function $\psi(\mathbf{r},t)$ by introducing the momentum operator,

$$\hat{\mathbf{p}} = -i\hbar\nabla. \tag{11.10}$$

The hat (which will be dropped shortly; confusion between operators and ordinary numbers is not likely) means this is an operator, as ∂ in equation (6.27). The result of applying this operator to the wave function in equation (10.42) is

$$\chi(\mathbf{r},t) = \hat{\mathbf{p}}\psi(\mathbf{r},t) = \int \frac{d^3p}{(2\pi\hbar)^{3/2}} f(\mathbf{p},t)\mathbf{p}e^{i\mathbf{p}\cdot\mathbf{r}/\hbar}. \tag{11.11}$$

The inner product of this function $\chi = \hat{\mathbf{p}}\psi$ with the wave function ψ is

$$\begin{aligned}(\psi,\chi) &= \int d^3r \int \frac{d^3p'}{(2\pi\hbar)^{3/2}} \int \frac{d^3p}{(2\pi\hbar)^{3/2}} f^*(\mathbf{p}',t)f(\mathbf{p},t)\mathbf{p}e^{i(\mathbf{p}-\mathbf{p}')\cdot\mathbf{r}/\hbar} \\ &= \int d^3p\,\mathbf{p}\,|f(\mathbf{p},t)|^2 = \langle\mathbf{p}\rangle. \end{aligned} \tag{11.12}$$

In the first line, the wave function ψ has been written as the integral over the momentum wave function in equation (10.42). This expression is simplified by using equation (10.21): on evaluating the integral over $\mathbf{r}$ first, we arrive at $(2\pi\hbar)^3\delta(\mathbf{p}-\mathbf{p}')$, and the integral over the delta function is immediately evaluated to get the second line.

Since $\chi = \hat{\mathbf{p}}\psi$, equation (11.12) is

$$\langle\mathbf{p}\rangle = (\psi, \hat{\mathbf{p}}\psi). \tag{11.13}$$

This bears a pleasant resemblance to equation (11.8). The same form is obtained in sections 14 and 21 for all observables (eqs. [14.14] and [21.11]).

The momentum operator has some other useful properties. As discussed in section 6, the allowed values of the energy of a system are the eigenvalues E of the equation $H\psi = E\psi$, and the system associated with the eigenfunction ψ with eigenvalue E has the definite energy E (eq. [6.9]). We have the same situation for the momentum operator: its eigenfunctions are plane waves:

$$\hat{\mathbf{p}}e^{i\mathbf{p}\cdot\mathbf{r}/\hbar} = -i\hbar\nabla e^{i\mathbf{p}\cdot\mathbf{r}/\hbar} = \mathbf{p}\,e^{i\mathbf{p}\cdot\mathbf{r}/\hbar}. \tag{11.14}$$

By de Broglie's relation (5.4), the particle associated with this plane wave has definite momentum $\mathbf{p}$. (To make this plane wave square integrable, one would have to go to periodic boundary conditions with volume V; the normalized wave function is then $\psi = e^{i\mathbf{p}\cdot\mathbf{r}/\hbar}/V^{1/2}$.) That is, just as for energy, if the wave function for a particle is an eigenfunction of the momentum operator, with eigenvalue $\mathbf{p}$, then the particle has definite momentum $\mathbf{p}$.

In section 14 this relation between observables and eigenfunctions is generalized to the prescription that the allowed values of any observable are the eigenvalues of an operator associated with the observable.

We can use the momentum operator to simplify the energy operator (eq. [6.10]) in Schrödinger's equation to

$$\hat{H} = \frac{\hat{\mathbf{p}}^2}{2m} + V(\mathbf{r}). \tag{11.15}$$

This is the usual form for the Hamiltonian for a single particle moving in the potential $V(\mathbf{r})$, with operators replacing ordinary numbers. Following Dirac, one sometimes calls operators q-numbers (for quantum), whereas ordinary numbers are called c-numbers (for classical). From now on hats usually will not be used to distinguish q-numbers; the context does that.

12 Many-Particle Systems

Schrödinger's Equation

The wave mechanics formalism generalizes to a many-body system in a straightforward way. The wave function for N particles moving in three dimensions is a function of time and of $3N$ variables for the $3N$ cartesian position coordinates,

$$\psi = \psi(\mathbf{r}_1, \ldots, \mathbf{r}_N, t). \tag{12.1}$$

The joint probability of finding that particle 1 is at position $\mathbf{r}_1$ in the volume element δV_1, that particle 2 is at $\mathbf{r}_2$ in δV_2, and so on, is

$$\delta P = |\psi(\mathbf{r}_1, \ldots, \mathbf{r}_N, t)|^2 \delta V_1 \ldots \delta V_N. \tag{12.2}$$

The inner product of wave functions ψ and ϕ in equation (11.6) generalizes to

$$(\psi, \phi) = \int d^3r_1 \ldots d^3r_N \; \psi^* \phi. \tag{12.3}$$

The normalization condition for equation (12.2) is then

$$(\psi, \psi) = 1, \tag{12.4}$$

and the expectation value of the position of the ith particle is

$$\langle \mathbf{r}_i \rangle = (\psi, \mathbf{r}_i \psi), \tag{12.5}$$

as can be checked by repeating the argument that led to equation (11.5)

The Hamiltonian for a system of N particles with potential energy V that is some given function of the $\mathbf{r}_j$ is the natural generalization of equation (11.15):

$$H = \sum_{1,N} \frac{\mathbf{p}_j^2}{2m_j} + V(\mathbf{r}_1, \ldots, \mathbf{r}_N). \tag{12.6}$$

The mass of particle j is m_j, and the momentum operator belonging to particle j is

$$\mathbf{p}_j = -i\hbar \nabla_j. \tag{12.7}$$

This means the x-component of the momentum operator for particle j is

$$p_{x,j} = -i\hbar \frac{\partial}{\partial x_j}. \tag{12.8}$$

Two examples are presented here, the two-body problem with a central force, because it is simple and important, and the description of phonons in a solid, because it is important and a good introduction to the linear vector spaces of quantum mechanics.

Two-Body Problem

Consider a system of two particles, such as the electron and proton in a hydrogen atom, with particle masses m_1 and m_2. The potential energy is assumed to be a function only of the relative position of the particles. The Hamiltonian (12.6) is then

$$H = \frac{p_1^2}{2m_1} + \frac{p_2^2}{2m_2} + V(\mathbf{r}_1 - \mathbf{r}_2). \tag{12.9}$$

Schrödinger's energy eigenvalue equation $H\psi = E\psi$ thus is a linear partial differential equation in six variables, the three components of $\mathbf{r}_1$ and of $\mathbf{r}_2$.

The differential equation is simplified by changing to center of mass and relative position variables, as in classical mechanics:

$$\begin{aligned} \mathbf{r} &= (x, y, z) = \mathbf{r}_1 - \mathbf{r}_2, \\ \mathbf{R} &= (X, Y, Z) = (m_1\mathbf{r}_1 + m_2\mathbf{r}_2)/M, \\ M &= m_1 + m_2. \end{aligned} \tag{12.10}$$

As an aid to rewriting the differential equation $H\psi = E\psi$ in terms of these new variables, let us define new momentum operators,

$$p_x = -i\hbar\frac{\partial}{\partial x}, \qquad P_X = -i\hbar\frac{\partial}{\partial X}, \tag{12.11}$$

and so on, or for short

$$\mathbf{p} = -i\hbar\nabla_r, \qquad \mathbf{P} = -i\hbar\nabla_R. \tag{12.12}$$

The wave function can be considered a function of the new position variables,

$$\psi = \psi(\mathbf{r}, \mathbf{R}, t), \tag{12.13}$$

where $\mathbf{r}$ and $\mathbf{R}$ are the functions of $\mathbf{r}_1$ and $\mathbf{r}_2$ in equation (12.10). According to the usual rules for a change of variables in a function, we have

$$\begin{aligned} \frac{\partial\psi}{\partial x_1} &= \frac{\partial\psi}{\partial x}\frac{\partial x}{\partial x_1} + \frac{\partial\psi}{\partial X}\frac{\partial X}{\partial x_1} = \frac{\partial\psi}{\partial x} + \frac{m_1}{M}\frac{\partial\psi}{\partial X}, \\ \frac{\partial\psi}{\partial x_2} &= \frac{\partial\psi}{\partial x}\frac{\partial x}{\partial x_2} + \frac{\partial\psi}{\partial X}\frac{\partial X}{\partial x_2} = -\frac{\partial\psi}{\partial x} + \frac{m_2}{M}\frac{\partial\psi}{\partial X}. \end{aligned} \tag{12.14}$$

Since this is true for any ψ, the relations between old and new momentum operators are

$$\begin{aligned} \mathbf{p}_1 &= \frac{m_1}{M}\mathbf{P} + \mathbf{p}, \\ \mathbf{p}_2 &= \frac{m_2}{M}\mathbf{P} - \mathbf{p}. \end{aligned} \tag{12.15}$$

Equation (12.15) is the same as the relation in classical physics between the momenta of the two particles, $\mathbf{p}_1$ and $\mathbf{p}_2$, and the center of mass and relative momenta, $\mathbf{P}$ and $\mathbf{p}$. The usual algebra in classical mechanics by which the total kinetic energy $\mathbf{p}_1^2/2m_1 + \mathbf{p}_2^2/2m_2$ is expressed in terms of $\mathbf{P}$ and $\mathbf{p}$ works equally well for these operators, because $\mathbf{p}$ and $\mathbf{P}$ commute with each other (because partial derivatives can be taken in either order). So on substituting equations (12.15) into the Hamiltonian (eq. [12.9]) we find by the usual algebra

$$H = \frac{\mathbf{P}^2}{2M} + \frac{\mathbf{p}^2}{2\mu} + V(\mathbf{r}) \equiv H_R + H_r, \tag{12.16}$$

where $M = m_1 + m_2$ is the total mass, and the reduced mass is defined as

$$\mu = m_1 m_2 / M. \tag{12.17}$$

The Hamiltonian in equation (12.16) is the sum of two terms, $H = H_R + H_r$, with $H_R = \mathbf{P}^2/2M$ and $H_r = \mathbf{p}^2/2\mu + V(\mathbf{r})$. The first Hamiltonian, H_R, involves only the center of mass variables $\mathbf{R}$ (through the derivative operator $\mathbf{P}$). The second, H_r, involves only the relative position variables $\mathbf{r}$. This means Schrödinger's equation can be simplified by separation of variables, as follows.

If the functions $\Psi(\mathbf{R})$ and $\phi(\mathbf{r})$ satisfy the separate eigenvalue equations

$$H_R\Psi = E_R\Psi, \qquad H_r\phi = E_r\phi, \tag{12.18}$$

then, as is readily checked, the product,

$$\psi(\mathbf{R}, \mathbf{r}) = \Psi(\mathbf{R})\phi(\mathbf{r}), \tag{12.19}$$

satisfies the energy eigenvalue equation, $H\psi = E\psi$, with $H = H_R + H_r$, and energy

$$E = E_R + E_r. \tag{12.20}$$

The first of equations (12.18) is the Schrödinger equation for a free particle with mass M. A complete set of solutions is the plane waves

$$\Psi \propto e^{i\mathbf{P}\cdot\mathbf{R}/\hbar}, \qquad E_R = P^2/2M. \tag{12.21}$$

The second eigenvalue equation (12.18) is a one-particle problem with reduced mass μ and potential energy $V(\mathbf{r})$.

The conclusion is that, as in classical mechanics, a two-body problem with potential energy V that depends only on the separation $\mathbf{r}_1 - \mathbf{r}_2$ of the two particles can be reduced to two single-body problems, one for the free motion of the center of mass, the other for the relative motion of the particles. The solutions we have presented assume the special product form in equation (12.19) for the two-body wave function. The general solution to Schrödinger's equation may be expressed as a linear combination of these products, as discussed in chapter 2 (section 13).

*Phonons**

The experimentally successful Debye theory in section 1 for the low energy lattice vibrations in a solid assumes the classical normal modes of vibration of the solid, with frequencies ω_α, have energies $E_\alpha = n_\alpha \hbar\omega_\alpha$, with $n_\alpha = 0, 1, 2, \ldots$. Let us see how this follows in quantum mechanics.

The dynamical variables will be taken to be the positions of the ions, ignoring the electrons, and the ions will be taken to have all the same mass, so the Hamiltonian is

$$H = \sum \mathbf{p}_j^2/2m + V(\mathbf{r}_1 \ldots \mathbf{r}_N). \tag{12.22}$$

The position variables may be set to $\mathbf{r}_j = 0$ at the classical equilibrium positions, where the gradient of V vanishes (so the classical force vanishes). We are interested in low energy excitations, for which displacements from equilibrium are small, so we will expand the potential energy V in a Taylor series in displacement from classical equilibrium and keep only the lowest nontrivial terms, which are quadratic in position. This brings the Hamiltonian (12.22) to

$$H = \sum_{1,3N} \frac{p_i^2}{2m} + \frac{1}{2} \sum_{1,3N} V_{ij} x_i x_j. \tag{12.23}$$

To simplify the expressions, the N position vectors for the N particles have been replaced with the set of $3N$ coordinates x_i. As usual, p_i means the derivative operator with respect to the ith one of these $3N$ coordinates, $p_i = -i\hbar\partial/\partial x_i$. The V_{ij} are the constant coefficients $\partial^2 V/\partial x_i \partial x_j$ in the Taylor series expansion. These coefficients are symmetric, $V_{ij} = V_{ji}$, because the partial derivatives can be taken in either order.

Now the game is to change variables from the x_i to normal mode coordinates, just as one does in the classical theory of small vibrations. Recall that one introduces a linear transformation to $3N$ new coordinates y_a,

$$x_i = \sum L_i^a y_a, \tag{12.24}$$

with the constants L_i^a chosen to simplify the potential energy term in the Hamiltonian. This is done by seeking the eigenvectors of the matrix V_{ij}.

Some elements of the matrix eigenvalue problem might be recalled. A symmetric M-by-M matrix, H_{ij}, can multiply a column vector, v_j, by the rule

$$\sum_j H_{ij} v_j = u_i, \tag{12.25}$$

yielding the column vector u_i. In the eigenvalue problem one seeks column vectors such that u_i is proportional to v_i:

$$\sum_j H_{ij} v_j = q v_i \ ; \tag{12.26}$$

the solutions v_i are the eigenvectors of the matrix H_{ij}, with eigenvalues q. We have used the same terminology for Schrödinger's energy equation $H\psi = E\psi$, and for the momentum eigenvalue equation (11.14).

The eigenvalue equation (12.26) can be rewritten as

$$\sum_j (H_{ij} - q\delta_{ij}) v_j = 0, \tag{12.27}$$

where δ_{ij} is the Kronecker delta function (eq. [10.5]). This is a set of M homogeneous linear equations (there are M possible choices of the index i in eq. [12.27]) for the M unknowns v_j. It has a non-trivial solution (that is, not all v_j equal to zero) if the determinant of the matrix of coefficients $H_{ij} - q\delta_{ij}$ vanishes. (If the determinant does not vanish the matrix of the coefficients has an inverse, and we can multiply the equations by the inverse to get $v_i = 0$ for all i.) Thus we require

$$\det |H_{ij} - q\delta_{ij}| = 0. \tag{12.28}$$

This is a polynomial of order M, which has M solutions, q^a. For each q^a we get an eigenvector v_i^a. The eigenvalues q^a are real; the reason will be seen in the next section. To simplify the discussion, it will be assumed also that the q^a all are different. (The notation is confusing until one has some practice at it. The index a means that q^a is the ath solution to eqs. [12.26] and [12.28], with v_i^a the column vector belonging to this solution.)

Returning now to the normal mode problem, we will choose the coefficients in equation (12.24) to satisfy

$$\sum_{j=1,3N} V_{ij} L_j^a = q^a L_i^a. \tag{12.29}$$

This is an eigenvalue equation for the $3N$-by-$3N$ symmetric matrix V_{ij}. There are $3N$ solutions; the ath eigenvalue is q^a, the components of the ath eigenvector being L_i^a.

An orthogonality relation follows by multiplying equation (12.29) by L_i^b and summing over i:

$$\begin{aligned} \sum_{ij} V_{ij} L_i^b L_j^a &= q^a \sum_i L_i^a L_i^b \\ &= q^b \sum_i L_i^a L_i^b. \end{aligned} \tag{12.30}$$

The two equations are obtained by applying equation (12.29) either to the eigenvector a or b, and recalling that V_{ij} is symmetric. Since the eigenvalues belonging to different eigenvectors are supposed to be different, the sum has to vanish when $a \neq b$. Thus we can write

$$\sum_i L_i^a L_i^b = \delta_{ab}. \tag{12.31}$$

Since the eigenvalue equation is linear, we can normalize the eigenvectors L_i^a so the sum is unity when $a = b$, as indicated by the Kronecker delta function. An integral version of this orthogonality relation is shown in equation (10.4).

A technical point might be noted: equation (12.31) is the matrix product of the matrix L and the transpose of L. Since the determinant of a matrix product is the product of the determinants, and the determinant of the transpose of L is the same as the determinant of L, and the determinant of δ_{ab} is unity, the determinant of L is ± 1. That means the matrix inverse of L_i^a exists, which is a Good Thing, because it means we can invert equation (12.24) to write the new coordinates y_a as functions of the old coordinates x_i.

On multiplying equation (12.31) by L_j^a and summing over a, we get

$$\sum_i (\sum_a L_j^a L_i^a) L_i^b = L_j^b. \tag{12.32}$$

A solution to this equation is

$$\sum_a L_j^a L_i^a = \delta_{ji}. \tag{12.33}$$

Because the matrix L has an inverse, we know this solution is unique. This is another orthogonality relation, similar to equation (12.31).

Now we are ready to change variables in the Hamiltonian (12.23). The potential energy term is

$$\begin{aligned}\sum V_{ij}x_i x_j &= \sum V_{ij}L_i^a L_j^b y_a y_b \\ &= \sum q^a L_j^a L_j^b y_a y_b \\ &= \sum q^a (y_a)^2.\end{aligned} \tag{12.34}$$

The first equation is the change of variables in equation (12.24), the second follows from the eigenvalue equation (12.29), and the last from the orthogonality equation (12.31).

The change of variables in the partial derivatives follows as in equation (12.14). With equation (12.24) we have

$$\begin{aligned}\frac{\partial}{\partial y_a} &= \sum_i \frac{\partial x_i}{\partial y_a}\frac{\partial}{\partial x_i} \\ &= \sum_i L_i^a \frac{\partial}{\partial x_i}.\end{aligned} \tag{12.35}$$

Thus we can define a momentum operator belonging to the variable y_a:

$$P_a = -i\hbar\frac{\partial}{y_a} = \sum L_i^a p_i, \tag{12.36}$$

with $p_i = -i\hbar\partial/\partial x_i$ the usual momentum operator associated with the ith cartesian position variable (eq. [12.8]). Using the second of the orthogonality relations (eq. [12.33]), we see that the sum of the squares of the new momenta is

$$\begin{aligned}\sum_a (P_a)^2 &= \sum_{aij} L_i^a L_j^a p_i p_j \\ &= \sum_i (p_i)^2.\end{aligned} \tag{12.37}$$

With equations (12.34) and (12.37), the Hamiltonian (12.23) expressed in the new variables y_a is

$$H = \sum \left[\frac{(P_a)^2}{2m} + \frac{q^a}{2} (y_a)^2 \right]. \tag{12.38}$$

A very similar calculation in classical physics leads to a similar result: the classical Hamiltonian can be written as a sum over one-dimensional simple harmonic oscillator Hamiltonians, each of the form $P_a^2/2m + q^a y_a^2/2$, one for each coordinate $y_a(t)$. That means $y_a(t)$ oscillates with time with frequency $\omega_a = (q^a/m)^{1/2}$. (If any of the q^a were negative, the equilibrium point $y_a = 0$ would be unstable, which we can assume does not happen.) Each $y_a(t)$ is a normal mode of oscillation of the solid. In a normal mode of oscillation the position coordinates $x_i(t)$ of the particles in the solid are $x_i(t) = L_i^a y_a(t)$. The general solution is a sum over the normal modes, as in equation (12.24).

In quantum mechanics, the Hamiltonian in equation (12.38) is a sum of operators H_a, each of which involves a different single coordinate, y_a. As in equation (12.19), this means we can find eigenfunctions of H that are products of functions of one variable, of the form

$$\psi = \phi_1(y_1)\phi_2(y_2)\ldots\phi_N(y_N), \tag{12.39}$$

where the $\phi_a(y)$ satisfy

$$H_a \phi_a = E_a \phi_a. \tag{12.40}$$

This is a solution to $H\psi = E\psi$, with $H = \sum H_a$, and with energy

$$E = \sum E_a. \tag{12.41}$$

Because $H_a = P_a^2/2m + q^a y_a^2/2$ is a one-dimensional simple harmonic oscillator Hamiltonian, we know the eigenvalues E_a in equation (12.40) are $(n_a + 1/2)\hbar\omega_a$, where ω_a is the classical frequency of the normal mode. Apart from the constant $\hbar\omega_a/2$, this is what was assumed in the Debye theory in section 1. The quanta $\hbar\omega_a$ of excitation of the normal modes of vibration of the solid are called phonons, in analogy to photons as the quanta of excitation of modes of oscillation of the electromagnetic field.

Problems

Some of the problems in this chapter (and on occasion through the rest of the book) are meant to help establish a feeling for orders of magnitude. To the uninitiated these order of magnitude estimates may seem silly in a supposedly exact science, but they are in fact an important art by which one ferrets out of a complex situation the effects that determine what is happening, and by which one simplifies a problem to the point that it is feasible to compute in detail. Where a numerical result is called for you should carry the analysis through to the final quantities, of course with a sensible number of significant figures. Some useful physical constants are:

$$\text{Planck's constant: } \hbar = 1.054 \times 10^{-27}\,\text{erg s}$$
$$\text{Velocity of light: } c = 3.00 \times 10^{10}\,\text{cm s}^{-1}$$
$$\text{Boltzmann's constant: } k = 1.38 \times 10^{-16}\,\text{ergs K}^{-1}$$
$$\text{proton mass: } m_p = 1.67 \times 10^{-24}\,\text{g}$$
$$\text{electron mass: } m_e = 9.11 \times 10^{-28}\,\text{g}$$
$$\text{electron volt: } 1\,\text{eV} = 1.602 \times 10^{-12}\,\text{erg}$$
$$\text{Ångstrom: } 1\,\text{Å} = 1 \times 10^{-8} cm$$
$$\text{Bohr radius: } a_o = 5.3 \times 10^{-9}\,\text{cm} = 0.53\,\text{Å}$$

I.1) Use the density of solid hydrogen, $\rho = 0.07\,\text{g cm}^{-3}$, and the mass of a hydrogen atom, to estimate the radius of a hydrogen atom, and compare the result to the Bohr radius.

I.2) As discussed in section 1, the heat capacity of gaseous molecular hydrogen is roughly constant at $3k/2$ per molecule at $T < 200\,\text{K}$, and roughly constant at $5k/2$ at $200 < T \ll 1000\,\text{K}$. Use this to estimate the energy in eV of the first rotationally excited state of a hydrogen molecule relative to the ground state energy.

I.3) As a model for the rotationally excited states of a hydrogen molecule that is a little more accurate than the one in the text, imagine a dumbbell with two proton masses m_p at separation equal to the Bohr radius, a_o. The dumbbell is rotating about the center of mass along an axis perpendicular to the line joining the two masses. Following the Bohr-Ehrenfest rule (eq. [4.5]), assume the allowed values for the angular momentum are integral multiples of $\hbar$. Using classical mechanics from here on, estimate the energy of the first rotationally excited state of a hydrogen molecule in this model, and compare the result to what is found in problem (I.2).

I.4) In the derivation of the Debye specific heat in section 1, the sum over normal modes is approximated as an integral. Find an order of magnitude estimate of the temperature at which this approximation fails for a cube of material 1 cm on a side, in which the velocity of sound is $c_s = 5 \times 10^5\,\mathrm{cm\,sec^{-1}}$.

I.5) Planck's expression for the mean energy per mode of oscillation of the electromagnetic field at temperature T is given by equation (1.27). Following Einstein, suppose this energy consists of photons, each with energy $h\nu = \hbar\omega$. The number of photons in each mode is then

$$N_\omega = \frac{1}{e^{\hbar\omega/kT} - 1}. \tag{I.1}$$

By repeating the calculation that led to the blackbody energy density in equation (2.18), find an expression for the mean number of photons per unit volume in blackbody radiation at temperature T. Reduce the integral you obtain to a dimensionless constant.

Space is observed to be filled with blackbody radiation at a temperature $T \sim 3\,\mathrm{K}$. (This is radiation left over from the hot Big Bang at early epochs of the expanding universe.) Using the fact that the dimensionless integral in the expression for the photon number density is on the order of unity, estimate the number density of photons in this radiation.

I.6) In interstellar space, hydrogen atoms can get hung up in states with large values of the principal quantum number n in equation (4.10), and such that the selection rules to be discussed in section 36 and problem V.21 require that the atoms decay toward the ground state by the chain $n \to n-1 \to n-2 \ldots$. Use the Bohr model to show that in this case, and in the limit of large n, the frequency of the radiation emitted by the atom agrees with the frequency with which the electron rotates around the proton.

Use the wavelength of the light emitted in the transition $n = 2 \to n = 1$, $\lambda = 1200\,\text{Ångstroms} = 1.2 \times 10^{-5}\,\mathrm{cm}$, to find the wavelength emitted in a case observed in the interstellar medium, where $n \to n-1$ with $n = 100$.

I.7) Estimate the de Broglie wavelengths of

a) a thermal neutron: the kinetic energy of the neutron is $3kT/2$, where $T \sim 300\,\text{K}$ is room temperature, and the neutron mass can be taken to be the same as that of a proton;

b) an electron with energy $E = 1\,\text{Gev} = 10^9\,\text{eV}$, where the momentum in this case is related to the energy by the expression

$$E^2 = p^2c^2 + m^2c^4. \tag{I.2}$$

I.8) The ground state wave function for a one-dimensional simple harmonic oscillator is fixed by the differential equation (eq. [6.48])

$$a\psi_0 = 0. \tag{I.3}$$

a) Using the expression for the lowering operator a as a differential operator (eq. [6.34]), find a solution for this differential equation for $\psi_0(x)$. Use the integral

$$\int_{-\infty}^{\infty} dy\; e^{-y^2} = \pi^{1/2} \tag{I.4}$$

to normalize ψ_0 to the condition

$$\int_{-\infty}^{\infty} |\psi_0|^2 dx = 1. \tag{I.5}$$

b) Use equation (6.34) to express x as a linear combination of the lowering and raising operators a and $a^\dagger$, and then use operator methods to show that $x\psi_0$ is the (not normalized) wave function for the first excited state.

c) Use the properties of the raising operator to argue that the wave function for the nth energy level, with $E_n = (n + 1/2)\hbar\omega$, for the oscillator is a polynomial multiplied by a Gaussian (e^{-cx^2}), and find the order of the polynomial. Show that the result agrees with the node count argument in section 7.

I.9) The Hamiltonian for a three-dimensional simple harmonic oscillator is

$$H = -\frac{\hbar^2}{2m}\nabla^2 + \frac{1}{2}Kr^2. \tag{I.6}$$

a) By expressing ∇^2 and r^2 in terms of the cartesian position components x, y, and z and their derivatives, show that H can be written as the sum of three operators,

$$H = H_x + H_y + H_z, \tag{I.7}$$

where H_x involves only x and derivatives with respect to x, and so on.

b) Use equation (I.7) to show that wave functions of the form

$$\psi(\mathbf{r}) = \phi_i(x)\phi_j(y)\phi_k(z) \tag{I.8}$$

where the functions $\phi_i(x)$ are the energy eigenfunctions for a one-dimensional simple harmonic oscillator, satisfy $H\psi = E\psi$, and find the allowed values of E for the three-dimensional oscillator.

c) Using the result from problem (I.8a), show the ground state wave function (I.8) is spherically symmetric.

I.10) A particle of mass m moves in one dimension in the potential well $V(x)$. The potential energy vanishes at distances $|x| > a$ from the origin, and is constant and negative at smaller $|x|$:

$$\begin{aligned} V(x) &= 0 \quad \text{at} \quad |x| > a, \\ &= -V_o \quad \text{at} \quad -a \leq x \leq a. \end{aligned} \tag{I.9}$$

The ground state energy is negative, $E_0 = -B$, where B is of course positive.

a) Following the arguments in section 7, sketch a graph of the shape of the ground state wave function, $\psi_0(x)$. High art is not expected in this exercise; the object is to work out the appropriate curvature of the wave function in the classically allowed and forbidden regions, to indicate how the function behaves at the boundaries between classically allowed and forbidden regions, and to indicate the behavior at $|x| \to \infty$.

b) Write down Schrödinger's equation for this system, and find the solutions for the ground state wave function (up to undetermined constants of proportionality) in terms of B in the classically allowed and forbidden regions, and state the joining conditions on the solutions.

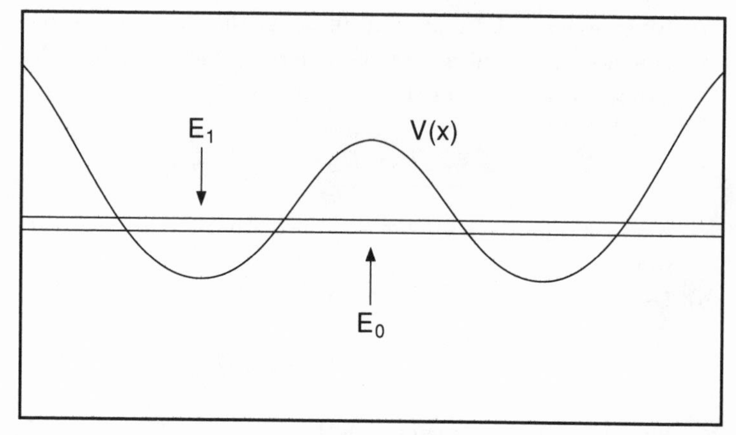

Fig. I.1 Model for the low energy states of an ammonia molecule. The potential well has minima for the position of the nitrogen atom on either side of the triangle defined by the three hydrogen atoms.

I.11) In a streamlined model for the low energy states of an ammonia molecule (NH_3), imagine that a nitrogen atom moves in one dimension in the potential $V(x)$ sketched in figure I.1. The potential has two minima, one on each side of the triangle defined by the three hydrogen atoms, and a relatively high peak between the minima, at the plane of the three hydrogen atoms. Thus in classical physics the nitrogen atom in the ground state would sit in one of the minima. Work is required to pull the nitrogen atom away from the molecule, and in classical physics work is required to push the nitrogen atom into the plane of the three hydrogen atoms at $x = 0$. Following the arguments in section 7, sketch the shapes of the wave functions $\psi_0(x)$ and $\psi_1(x)$ for the ground and first excited states of motion of the nitrogen atom.

The energies E_0 and E_1 of the ground and first excited states in this system are very nearly equal. Explain how this is to be understood.

I.12) A particle that moves in three dimensions is trapped in a deep spherically symmetric potential $V(r)$:

$$\begin{aligned} V(r) &= 0 \quad \text{at} \quad r < r_o, \\ &\to \infty \quad \text{at} \quad r \geq r_o, \end{aligned} \tag{I.10}$$

where r_o is a positive constant. The ground state wave function is spherically symmetric, so the radial wave function $u(r)$ satisfies the one-dimensional Schrödinger energy eigenvalue equation (6.17), with the boundary condition $u(0) = 0$ (eq. [6.18]).

a) Explain why, in the potential well in equation (I.10), the wave function is forced to vanish at $r = r_0$.

b) Using the known boundary conditions on the radial wave function $u(r)$ at $r = 0$ and $r = r_o$, find the ground state energy of the particle in this potential well.

I.13) A particle of mass m moves in three dimensions in the potential well

$$\begin{aligned} V(r) &= -V_o \quad \text{at} \quad r < r_o, \\ &= 0 \quad \text{at} \quad r \geq r_o, \end{aligned} \tag{I.11}$$

where V_o and r_o are positive constants. If there exists a state in which the particle is bound to the potential well, the wave function for the bound state with the lowest energy is spherically symmetric and the radial wave function satisfies equations (6.17) and (6.18).

Find the minimum value of the depth V_o for which there exists a bound state. (Recall that the radial function satisfies the condition $u(0) = 0$, because $\psi(r) = u(r)/r$ has to be regular at the origin.)

I.14) A deuteron is a bound state of a neutron and proton. In the center of mass coordinates discussed in section 12, one considers a single particle with the reduced mass moving in a fixed potential well $V(r)$, where r is the separation of the neutron and proton. To a first approximation, the nuclear interaction can be taken to be a square potential well,

$$\begin{aligned} V(r) &= -V_n \qquad \text{at} \qquad r < r_n = 1 \times 10^{-13}\,\text{cm}, \\ &= 0 \qquad \text{at} \qquad r \geq r_n. \end{aligned} \tag{I.12}$$

There is only one bound state. In the approximation of equation (I.12), the wave function ψ in this bound state is spherically symmetric.

The depth, V_n, of the potential well is such that the deuteron is only just bound: if the radius r_n of the nuclear interaction were only slightly smaller, there would not be a deuteron bound state. Use this to estimate V_n in MeV.

I.15) The normalized wave function of a particle that moves in three dimensions is

$$\psi(\mathbf{r}) = A(\mathbf{r})e^{i\mathbf{k}\cdot\mathbf{r}}, \tag{I.13}$$

where $\mathbf{k}$ is a real constant vector, and A is a real function of position $\mathbf{r}$.

a) In terms of the given constants and functions, what is the probability distribution in the position of the particle?

b) In terms of the given constants and functions, what is the probability flux density, $\mathbf{j}$ (eq. [8.13])?

I.16) Suppose that, at a given instant of time, the wave function of a particle that moves in one dimension is

$$\psi(x) \propto e^{-x^2/x_o^2}, \tag{I.14}$$

where x_o is a real constant. This is the form of the ground state wave function for a simple harmonic oscillator, but of course the function need not represent a simple harmonic oscillator: equation (I.14) is an initial value for Schrödinger's equation.

a) Write down the normalization for ψ, as in problem I.8.

b) Find the expectation value $\langle x \rangle$ of the result of measuring the position of the particle at the given instant of time, when the wave function is given by equation (I.14)

c) Find the expectation value of the square of the position, $\langle x^2 \rangle$.

d) The standard deviation σ_a of the random variable a is defined by the equation

$$\begin{aligned} \sigma_a^2 &\equiv \langle (a - \langle a \rangle)^2 \rangle \\ &= \langle a^2 \rangle - \langle a \rangle^2. \end{aligned} \tag{I.15}$$

That is, σ_a is the root mean square difference between the measured value of a and the expectation value of a. Show how the second line in equation (I.15) follows from the first, and use the result to compute the standard deviation σ_x in the position of the particle with the wave function in equation (I.14).

e) Find the probability distribution dP/dp (eqs. [10.39], [10.41]) for the result of measuring the momentum of the particle with the wave function in equation (I.14). The integral can be evaluated by completing squares, as in equation (10.28).

f) Use the result from part (e) to find the standard deviation σ_p in the result of measuring the momentum. If all goes well, you should find

$$\sigma_x \sigma_p = \hbar/2. \tag{I.16}$$

As will be seen in problem (III.3), this is the minimum possible value for this product allowed by the Heisenberg uncertainty principle.

g) Find the probability distribution dP/dp in the momentum of the particle with the wave function of the first excited state of a one-dimensional simple harmonic oscillator,

$$\phi(x) \propto x\, e^{-x^2/x_o^2}. \tag{I.17}$$

The easy trick here is to notice that the derivative with respect to k of the Fourier transform of the function in equation (I.14) is proportional to the Fourier transform of the function in equation (I.17).

h) Find the probability distribution dP/dp in the momentum of the particle with the wave function

$$\chi(x) \propto e^{ik_o x - x^2/x_o^2}, \tag{I.18}$$

where k_o and x_o are real constants. You should find that this is an easy generalization of the calculation in part (e).

I.17) The cold fusion calculation in section 9 assumed a spherically symmetric wave function, $\psi(r)$, so the radial wave function $u(r) = r\psi(r)$ satisfies a one-dimensional Schrödinger equation with the same potential $V(r)$ as in the three-dimensional problem. Then $u(r)$ was used to calculate the probability distribution and probability flux in the radial direction, as if this were a true one-dimensional situation. The purpose of this problem is to work through the logic of this approach.

a) The wave function ψ is normalized to

$$\int dV\, |\psi(r)|^2 = 1. \tag{I.19}$$

By writing the volume element dV in polar coordinates, find the normalization condition on $u(r) = r\psi(r)$ as an integral over the radius r. (Polar coordinates are defined in fig. 17.1.)

b) The probability that the particle is found at position $\mathbf{r}$ in the volume dV is

$$dP = dV\,|\psi|^2. \tag{I.20}$$

By using polar coordinates again, and integrating over the direction of $\mathbf{r}$, find the probability that the particle is found to be at radius r in the interval dr. Express the result in terms of the radial wave function $u(r)$.

c) Find the probability flux density $\mathbf{j}$ for the wave function $\psi(r)$. Use equation (6.11) to express the result in terms of $u(r)$ and du/dr.

d) Use the result from part (c) to find the net flux, f, of probability through a shell of fixed radius r, again expressed in terms of the radial wave function $u(r)$ and its derivative.

CHAPTER 2

WAVE MECHANICS

The last chapter showed how one can understand the energy quantization of an isolated system of particles, by associating a wave function with the system. We introduced linear combinations of wave functions, to get new functions (as in eq. [5.9]), we introduced a normalizing condition, that can be written as $(\psi, \psi) = 1$ (eq. [11.7]), and we saw that it is useful to extend this notation to a sort of scalar product of functions, (ψ, ϕ) (eqs. [11.8] and [11.13]). The purpose of the first two sections in this chapter is to formalize these manipulations, and to write down the set of general assumptions of wave mechanics. The way the formalism is used will be illustrated by application to the symmetry arguments by which one derives the laws of conservation of parity and linear and angular momentum and the electromagnetic interaction.

13 Linear Space of Wave Functions

Wave Functions and Operators

We used linear combinations of wave functions, as in equation (5.9), to describe the possible states of a system of particles. This is formalized by introducing a linear space of functions of m variables (corresponding to the m coordinates for the system). Linearity means that if ψ and ϕ are elements of the space of functions so is the function

$$\chi = c\psi + d\phi, \tag{13.1}$$

where c and d are constants. One sometimes adds the condition that the functions be square integrable,

$$0 < \int |\psi|^2 d^m r < \infty, \tag{13.2}$$

but we will ignore that when convenient. The word space is used here to express an analogy between the set of wave functions and the set of vectors in ordinary three-dimensional space. The vectors $\mathbf{r}$ and $\mathbf{s}$ can be multiplied by constants and added to get a new vector, $c\mathbf{r}+d\mathbf{s}$. Equation (13.1) expresses the same operation for functions.

We also have an analog of the dot product $\mathbf{r}\cdot\mathbf{s}$ of two vectors in three-dimensional space. The corresponding measure in the space of functions is the inner product

$$(\psi, \phi) = \int d^m r \, \psi^* \phi. \tag{13.3}$$

As usual, the star means the complex conjugate. The integral is over the m position coordinates of the system. This form was used in writing out the expectation values of position and momentum (eqs. [11.8] and [11.13]).

The inner product satisfies the relations

$$\begin{aligned} (\psi, \phi)^* &= (\phi, \psi), \\ (\psi, c\phi + d\chi) &= c(\psi, \phi) + d(\psi, \chi), \end{aligned} \tag{13.4}$$

where c and d are constants. It follows that

$$(c\phi + d\chi, \psi) = c^*(\phi, \psi) + d^*(\chi, \psi). \tag{13.5}$$

A linear operator Q maps any element of the space of functions into another element,

$$Q\psi = \phi. \tag{13.6}$$

Linearity means that, for functions ψ and ϕ and constants c and d,

$$Q(c\psi + d\phi) = cQ\psi + dQ\phi. \tag{13.7}$$

Examples of the linear operators of quantum mechanics include the components of the cartesian position vector, r_α, the components of the momentum operator, p_α (eq. [11.10]), the Hamiltonian, H (eq. [11.15]), and the ladder operators a and $a^\dagger$ that were used in section 6 to find the energy levels of a simple harmonic oscillator (eq. [6.34]). For example, the component x of the position vector $\mathbf{r}$, multiplied by the function $\psi(\mathbf{r})$, gives a new function of position, $x\psi(\mathbf{r})$.

If P and Q are linear operators then the sum $P+Q$ and the product PQ are linear operators:

$$(P+Q)\psi = P\psi + Q\psi, \qquad PQ\psi = P(Q\psi). \tag{13.8}$$

Any three operators satisfy the law $(PQ)R = P(QR)$. It will be noted that in general operators need not commute: PQ need not be the same as QP. An example was seen in section 6, in the commutator of x and the derivative ∂_x with respect to x (eq. [6.33]).

The Hamiltonian H is an operator function of operators. Any function $f(Q)$ of the operator Q that can be expressed as a power series can be defined as an operator, for the term Q^n in the expansion just means the operator Q applied n times. Thus, for example, if

$$U = e^{iQ}, \tag{13.9}$$

with Q an operator, then

$$U\psi = \sum \frac{(iQ)^n \psi}{n!}. \tag{13.10}$$

Adjoint of an Operator

The linear operator $Q^\dagger$ adjoint to the linear operator Q is defined by the equation

$$(Q^\dagger \psi, \phi) = (\psi, Q\phi), \tag{13.11}$$

for all ψ and ϕ. By taking the complex conjugate of this equation, and using equation (13.4), we see that this is equivalent to

$$(\psi, Q\phi)^* = (\phi, Q^\dagger \psi). \tag{13.12}$$

A self-adjoint operator satisfies

$$Q^\dagger = Q. \tag{13.13}$$

It follows from equation (13.12) that $(\psi, Q\psi)$ is real if Q is self-adjoint. It is left as an exercise to show the converse, that if $(\phi, Q\phi)$ is real for all ϕ, then $(\psi, Q\phi) = (Q\psi, \phi)$ for all ψ and ϕ, so Q is self-adjoint.

Because expectation values of observable quantities are of the form $(\psi, Q\psi)$, as in the examples in equations (11.8) and (11.13), and measurements yield real numbers, the operators Q representing observables are self-adjoint.

Let us consider some examples. If c is a constant, we evidently have

$$c^\dagger = c^*, \tag{13.14}$$

that is, the adjoint of a constant is its complex conjugate. If P and Q are operators, with adjoints $P^\dagger$ and $Q^\dagger$, the product PQ satisfies

$$(\psi, PQ\phi) = (P^\dagger\psi, Q\phi) = (Q^\dagger P^\dagger\psi, \phi), \tag{13.15}$$

Thus the adjoint of the product PQ is the product of the adjoints in the reverse order:

$$(PQ)^\dagger = Q^\dagger P^\dagger. \tag{13.16}$$

It is left as an exercise to show that

$$\begin{aligned} (P+Q)^\dagger &= P^\dagger + Q^\dagger, \\ Q^{\dagger\dagger} \equiv (Q^\dagger)^\dagger &= Q. \end{aligned} \tag{13.17}$$

To find the adjoint of the derivative operator

$$\partial_x = \frac{\partial}{\partial x}, \tag{13.18}$$

where x is one of the position coordinates, write

$$\psi^* \frac{\partial}{\partial x}\phi = \frac{\partial}{\partial x}(\psi^*\phi) - \frac{\partial\psi^*}{\partial x}\phi, \tag{13.19}$$

and integrate over all coordinates. The integral over the total derivative yields the integrand at the upper and lower limits; this vanishes because the functions have to approach zero at infinity (or with periodic boundary conditions because the upper and lower limits are the same point). This gives

$$(\psi, \partial_x\phi) = -(\partial_x\psi, \phi), \tag{13.20}$$

so the adjoint defined in equation (13.11) is

$$\partial_x^\dagger = -\partial_x. \tag{13.21}$$

The momentum operator is (eq. [11.10])

$$p_\alpha = -i\hbar\partial_\alpha. \tag{13.22}$$

This is self-adjoint, $p_\alpha^\dagger = p_\alpha$, because the adjoint of the factor i is $-i$ (eq. [13.14]). Therefore, the expectation values of the momentum, and of any function of the momentum, are real. In particular, the energy operator,

$$H = \frac{p^2}{2m} + V, \tag{13.23}$$

is self-adjoint if the potential energy is real, because $(p_x p_x)^\dagger = p_x p_x$.

The adjoint of the operator $U = e^{iQ}$ in equation (13.10) is

$$U^\dagger = e^{-iQ} = U^{-1}, \tag{13.24}$$

as we see by taking the adjoint of the series expansion of U. That is, the adjoint of this operator is equal to its inverse:

$$U^\dagger U = 1. \tag{13.25}$$

This is said to be a unitary operator, and the mapping

$$\begin{aligned} \psi &\to U\psi, \\ Q &\to UQU^{-1}, \end{aligned} \tag{13.26}$$

for all wave functions ψ and operators Q is a unitary transformation. This is of particular interest because all predictions of quantum mechanics are in expressions of the form $(\psi, Q\phi)$. It is left as an exercise to check that $(\psi, Q\phi)$ is unaffected by a unitary transformation of all the wave functions and operators.

Eigenfunctions and Eigenvalues

Consider a linear self-adjoint operator Q. As we have discussed on several occasions, the eigenvalue equation for Q is

$$Q\psi = q\psi, \tag{13.27}$$

where ψ is a nonzero function ($(\psi, \psi) \neq 0$ and $(\psi, \psi) \neq \infty$). The number q is the eigenvalue belonging to the eigenfunction ψ. The eigenvalues can be discrete, as are the energy levels of a particle bound in a potential well, or continuous, as the energy of a free particle that can spread over an unbounded space. Only discrete eigenvalues are considered here; generalization to the continuous case is discussed in section 20 below.

As an example, the eigenfunctions and eigenvalues of the momentum operator (eq. [13.22]) with periodic boundary conditions in a region of volume $V = L^3$ are (eqs. [1.54], [1.55] and [11.14])

$$\psi_{\mathbf{p}}(\mathbf{r}) = e^{i\mathbf{p}\cdot\mathbf{r}/\hbar}/V^{1/2}, \qquad p_\alpha = 2\pi\hbar n_\alpha/L. \tag{13.28}$$

The n_α are integers. This wave function has been normalized to $(\psi_{\mathbf{p}}, \psi_{\mathbf{p}}) = 1$.

Here are some properties of the eigenvalues and eigenfunctions of self-adjoint operators.

1) The eigenvalues are real. For a self-adjoint operator, $Q^\dagger = Q$, we have (eq. [13.11])

$$(\psi, Q\psi) = (Q\psi, \psi), \tag{13.29}$$

and the eigenvalue equation (13.27) gives

$$\begin{aligned} (\psi, q\psi) &= (q\psi, \psi), \\ q(\psi, \psi) &= q^*(\psi, \psi), \end{aligned} \tag{13.30}$$

where the second step follows from equation (13.4). Since ψ is nonzero, that is, $0 < (\psi, \psi) < \infty$, we see from this equation that $q^* = q$.

2) The eigenfunctions with different eigenvalues are orthogonal. Consider the two eigenfunctions

$$\begin{aligned} Q\psi_1 &= q_1\psi_1, \\ Q\psi_2 &= q_2\psi_2. \end{aligned} \tag{13.31}$$

Because we are taking Q to be self-adjoint, we have (from eq. [13.11] with $Q^\dagger = Q$)

$$(\psi_1, Q\psi_2) = (Q\psi_1, \psi_2), \tag{13.32}$$

and, since we have seen that the q_i are real, we conclude from the eigenvalue equations (13.31) that

$$q_2(\psi_1, \psi_2) = q_1(\psi_1, \psi_2). \tag{13.33}$$

This means either $q_1 = q_2$ or

$$(\psi_1, \psi_2) = 0 \quad \text{if} \quad q_1 \neq q_2. \tag{13.34}$$

The two functions are said to be orthogonal, following the analogy of dot products of vectors.

3) Different eigenfunctions with the same eigenvalue can be chosen to be orthogonal. Eigenfunctions with the same eigenvalue are said to be degenerate. If ψ_1 and ψ_2 are degenerate then another degenerate eigenfunction is the linear combination $c_1\psi_1 + c_2\psi_2$. We avoid these trivial degenerate eigenfunctions by considering only a linearly independent set of functions, where linear independence means that the only solution to the equation

$$\sum_{n=1}^{m} c_n \psi_n = 0, \tag{13.35}$$

where the c_n are a set of constants, is $c_n = 0$ for all n.

Suppose there are m linearly independent eigenfunctions, all with the same eigenvalue,

$$Q\psi_n = q\psi_n, \quad \text{for} \quad n = 1, \ldots m. \tag{13.36}$$

We can generate an orthogonal set of functions ϕ_n as follows. The first of the new functions is just

$$\phi_1 = \psi_1. \tag{13.37}$$

The second is

$$\phi_2 = \psi_2 + c\phi_1, \tag{13.38}$$

where the constant c is chosen so ϕ_2 is orthogonal to ϕ_1:

$$(\phi_1, \phi_2) = (\psi_1, \psi_2) + c(\psi_1, \psi_1) = 0. \tag{13.39}$$

This can be solved for c because $(\psi_1, \psi_1) \neq 0$. We know ϕ_2 is nonzero because ψ_1 and ψ_2 are linearly independent. Thus ϕ_1 and ϕ_2 are orthogonal eigenfunctions of Q. To get the next orthogonal eigenfunction, write

$$\phi_3 = \psi_3 + d\phi_2 + e\phi_1, \tag{13.40}$$

and seek values for the constants d and e so ϕ_3 is orthogonal to ϕ_1 and ϕ_2. Since ϕ_1 is orthogonal to ϕ_2,

$$\begin{aligned}(\phi_1, \phi_3) &= (\phi_1, \psi_3) + e(\phi_1, \phi_1) = 0,\\ (\phi_2, \phi_3) &= (\phi_2, \psi_3) + d(\phi_2, \phi_2) = 0.\end{aligned} \tag{13.41}$$

These equations yield d and e. Continuing in this way, we can arrange that all m eigenfunctions satisfy

$$(\phi_i, \phi_j) = \delta_{ij}, \tag{13.42}$$

where δ_{ij} is the Kronecker delta function (eq. [10.5]). The new orthogonal eigenfunctions ϕ_i have been normalized by dividing the original ϕ_i from the above procedure by $(\phi_i, \phi_i)^{1/2}$.

4) If the operators P and Q commute, that is,

$$[P, Q] \equiv PQ - QP = 0, \tag{13.43}$$

then it can be arranged that all the eigenfunctions of P are eigenfunctions of Q and all the eigenfunctions of Q are eigenfunctions of P.

Suppose first there are no degenerate eigenfunctions (so each different eigenfunction of P has a different eigenvalue p). Then on applying the operator Q to the eigenvalue equation for an eigenfunction of P,

$$P\psi = p\psi, \tag{13.44}$$

and using the fact that P and Q commute, we get

$$P(Q\psi) = p(Q\psi). \tag{13.45}$$

This says that $Q\psi$ is an eigenfunction of P. Since ψ is not degenerate, $Q\psi$ must be proportional to ψ,

$$Q\psi = q\psi, \tag{13.46}$$

where q is the constant of proportionality. But this equation says ψ is an eigenfunction of Q, with eigenvalue q, as well as an eigenfunction of P with eigenvalue p. That is, ψ is a simultaneous eigenfunction of P and Q.

Suppose next P has m ($m < \infty$) linearly independent eigenfunctions all with the same eigenvalue p. As discussed above, these functions can be chosen to be orthogonal and normalized,

$$P\psi_i = p\psi_i, \qquad (\psi_i, \psi_j) = \delta_{ij}. \tag{13.47}$$

Since the list of eigenfunctions ψ_i is supposed to be complete, any eigenfunction of P with eigenvalue p must be a linear combination of these ψ_i. Now if $PQ = QP$ we get from the first of equations (13.47)

$$P(Q\psi_i) = p(Q\psi_i). \tag{13.48}$$

If the function $Q\psi_i$ vanishes then it is an eigenfunction of Q with eigenvalue $q = 0$. If $Q\psi_i$ is nonzero, equation (13.48) says it is an eigenfunction of P, with eigenvalue p, so it has to be a linear combination of the complete set of eigenfunctions ψ_i with eigenvalue p:

$$Q\psi_i = \sum C_{ij}\psi_j. \tag{13.49}$$

By the orthogonality relation (13.47), the inner product of (13.49) with ψ_k is

$$C_{ik} = (\psi_k, Q\psi_i). \tag{13.50}$$

Now let us seek eigenfunctions of Q that are linear combinations of the ψ_i:

$$Q \sum D_i\psi_i = q \sum D_k\psi_k. \tag{13.51}$$

The D_i are constants that will be chosen to satisfy the eigenvalue equation (13.51) for Q. We will see that there are m different solutions. This is what allows us to classify the eigenfunctions of P as eigenfunctions of Q.

On using equation (13.49) for $Q\psi_i$, we see that equation (13.51) is

$$\sum_{ij} D_i C_{ij}\psi_j = q \sum_k D_k\psi_k. \tag{13.52}$$

As in equation (13.50), the orthogonality condition (13.47) gives

$$\sum_i D_i C_{ij} = q D_j. \tag{13.53}$$

This is a matrix eigenvalue equation for the row vector D_j, where the matrix elements, C_{ij}, are given by equation (13.50). As discussed in section 12, there are m solutions to this eigenvalue equation, each yielding a set of coefficients D_i and an eigenvalue q. This set of D_i yields a function, $\sum D_i\psi_i$, that by equation (13.51) is an eigenfunction of Q with eigenvalue q. Since $\sum D_i\psi_i$ is a linear combination of the eigenfunctions of P all with the same eigenvalue p, we see that $\sum D_i\psi_i$ is a simultaneous eigenfunction of the commuting operators P and Q. Since there are m different solutions to equation (13.53), there are m different linear combinations $\sum D_i\psi_i$ that are eigenfunctions of Q as well as P. The eigenfunctions belonging to different values of q are orthogonal; degenerate eigenfunctions of Q can be made orthogonal in the way described above.

The result of this operation is that the set of degenerate eigenfunctions ψ_i of P has been replaced with a set of linear combinations of the ψ_i, with the combinations chosen so each is an orthogonal eigenfunction of the operators Q and P. For example, the three momentum operators p_x, p_y and p_z commute (because partial derivatives may be computed in any order: $\partial^2\phi/\partial x\partial y = \partial^2\phi/\partial y\partial x$). As in equation (13.28), the eigenfunctions of the operator p_x with given eigenvalue $2\pi n_x\hbar/L$ labeled by the integer n_x, with periodic boundary conditions, are of the form

$$\psi = f(y, z)\, e^{2\pi i n_x x/L}, \tag{13.54}$$

where f is any (reasonably well behaved) periodic function of y and z. That is, there are many linearly independent eigenfunctions of p_x all with the same eigenvalue. Because p_x, p_y and p_z commute, one can seek simultaneous eigenfunctions of these operators. The simultaneous eigenfunctions given in equation (13.28) are uniquely specified (up to the usual multiplicative factor) by the three eigenvalues $2\pi\mathbf{n}\hbar/L$ (that are fixed by the three integers n_x, n_y and n_z).

Completeness

To every measurable attribute of a physical system, quantum mechanics associates a self-adjoint operator, Q, with eigenfunctions ψ_i. As we have seen, we can choose a complete set of linearly independent eigenfunctions of Q to be orthogonal and normalized,

$$(\psi_i, \psi_j) = \delta_{ij}. \tag{13.55}$$

The completeness assumption is that any wave function ϕ that describes an allowed state of the system can be expanded as a series in (linear combination of) these eigenfunctions,

$$\phi = \sum c_i \psi_i. \tag{13.56}$$

The c_i are constant expansion coefficients. On taking the inner product of this sum with ψ_j, and using the orthogonality relation (13.55), we see that the jth expansion coefficient is

$$c_j = (\psi_j, \phi). \tag{13.57}$$

Thus a compact expression for the expansion is

$$\phi = \sum \psi_i (\psi_i, \phi). \tag{13.58}$$

For an example of the completeness expression, consider the set of simultaneous eigenfunctions of the components of the single particle momentum operator (eq. [13.28]). The completeness expression here is

$$\psi(\mathbf{r}) = \sum c_{\mathbf{p}} \psi_{\mathbf{p}}(\mathbf{r}), \qquad c_{\mathbf{p}} = (\psi_{\mathbf{p}}, \psi). \tag{13.59}$$

The sum is over the integers $\mathbf{n}$ in equation (13.28). This sum is the Fourier series expansion of $\psi(\mathbf{r})$, as discussed in section 10.

The expansion in equation (13.58) can be compared to the representation of a vector in a three-dimensional space by its components in a cartesian coordinate system. Let $\mathbf{i}$, $\mathbf{j}$, and $\mathbf{k}$ be the unit vectors along the orthogonal x, y, z axes in a three-dimensional space. The orthogonality relations analogous to equation (13.55) are $\mathbf{i} \cdot \mathbf{i} = 1$, $\mathbf{i} \cdot \mathbf{j} = 0$, and so on. The component of the vector $\mathbf{r}$ along the x axis is $x = \mathbf{i} \cdot \mathbf{r}$, which can be compared to equation (13.57). The vector may be written as a linear combination of the unit vectors with coefficients equal to the components x, y, z:

$$\mathbf{r} = \mathbf{i}x + \mathbf{j}y + \mathbf{k}z. \tag{13.60}$$

Equation (13.58) generalizes equation (13.60) to a linear space with an infinite number of dimensions. Following the usual convention in linear spaces, the ψ_i are called a basis for the space of functions.

*Proof of Completeness**

The general completeness theorem will not be proved here, but it is fairly easy to work through a special case. Suppose the self-adjoint operator Q has eigenvalues q_i that are not degenerate, and arrange the eigenvalues so

$$q_{i+1} > q_i. \tag{13.61}$$

Suppose also that there is a minimum value, q_0, but no maximum, so $q_i \to \infty$ as $i \to \infty$. Without further loss of generality we can take it that the q_i are not negative (by subtracting a suitable constant from Q). As usual, the eigenfunction ψ_i belonging to q_i will be normalized to $(\psi_i, \psi_j) = \delta_{ij}$.

Consider the expression

$$q = \frac{(\psi, Q\psi)}{(\psi, \psi)}. \tag{13.62}$$

The first step is to show that the function that minimizes q is the wave function ψ_0 belonging to the smallest eigenvalue q_0.

For any ψ, consider the change of q under a small change in the function,

$$\psi \to \psi + \delta\psi. \tag{13.63}$$

The result of applying this to equation (13.62) is

$$\delta q = \frac{(\delta\psi, Q\psi)}{(\psi, \psi)} - \frac{(\delta\psi, \psi)(\psi, Q\psi)}{(\psi, \psi)^2} + \text{cc}. \tag{13.64}$$

Here cc means the complex conjugate of the preceding terms (in which $\delta\psi$ appears in the right-hand side of the inner product). Using the definition of q, we can simplify this to

$$\delta q = \frac{(\delta\psi, (Q - q)\psi)}{(\psi, \psi)} + \text{cc}. \tag{13.65}$$

Now suppose ψ minimizes q. This means δq must vanish to first order in $\delta\psi$ for any $\delta\psi$. So choose a particular form for $\delta\psi$,

$$\delta\psi \propto (Q - q)\psi \equiv f. \tag{13.66}$$

This brings equation (13.65) to

$$\delta q \propto (f, f)/(\psi, \psi). \qquad (13.67)$$

If f is nonzero then $\delta q > 0$, which contradicts the assumption that ψ minimizes q. The contradiction is avoided if $f = 0$. But then equation (13.66) says

$$Q\psi = q\psi. \qquad (13.68)$$

That is, the ψ that minimizes q has to be an eigenfunction of Q. The smallest eigenvalue of Q is q_0, so ψ_0 minimizes q.

Next, for any ψ let

$$\phi = \psi - \psi_0(\psi_0, \psi). \qquad (13.69)$$

This projects the part ψ_0 out of ψ, so ϕ is orthogonal to ψ_0:

$$(\psi_0, \phi) = 0. \qquad (13.70)$$

Consider for this new function the expression in equation (13.62):

$$q = \frac{(\phi, Q\phi)}{(\phi, \phi)}. \qquad (13.71)$$

By the above argument, if ϕ minimizes q then

$$Q\phi = q\phi. \qquad (13.72)$$

If q were equal to q_0 then ϕ would be proportional to ψ_0, which would contradict equation (13.70). Thus the minimum value of q in equation (13.71) is obtained when ϕ is the first eigenvalue above the minimum one, with $q = q_1$. One similarly finds that if

$$\phi_m = \psi - \sum_{i=0,m-1} \psi_i(\psi_i, \psi) \qquad (13.73)$$

then

$$\frac{(\phi_m, Q\phi_m)}{(\phi_m, \phi_m)} \geq q_m. \qquad (13.74)$$

We need one last inequality. We have from equation (13.73) after a little rearrangement and application of the orthogonality of the ψ_i,

$$(\phi_m, Q\phi_m) = (\psi, Q\psi) - \sum_{0,m} q_i|(\psi_i, \psi)|^2. \tag{13.75}$$

The last term is negative, because the q_i are positive, so

$$(\phi_m, Q\phi_m) < (\psi, Q\psi). \tag{13.76}$$

The inequalities (13.74) and (13.76) say

$$(\phi_m, \phi_m) \leq \frac{(\phi_m, Q\phi_m)}{q_m} < \frac{(\psi, Q\psi)}{q_m}. \tag{13.77}$$

In the limit $m \to \infty$, the eigenvalue $q_m \to \infty$, so $(\phi_m, \phi_m) \to 0$. And as ϕ_m approaches zero we see from equation (13.73) that ψ becomes arbitrarily well approximated by the series

$$\psi = \sum \psi_i(\psi_i, \psi). \tag{13.78}$$

This is the completeness expression (eq. [13.58]).

14 Principles of Quantum Mechanics

Now we are in a position to write down the general assumptions of quantum wave mechanics. It is hoped that these are seen to be reasonable generalizations from the considerations in chapter 1, and that these considerations are a reasonable response to the puzzle of energy quantization. But of course the ultimate justification is that the resulting theory is experimentally very successful. In section 21 the assumptions are repeated in the notation of an abstract linear space (rather than the space of wave functions used here).

Observables and Probability Distributions

In wave mechanics the possible states of a physical system are represented by the functions in a linear space with the properties discussed in the last section. If the system is described by m coordinates, then the wave functions are functions of m variables plus time. It should be noted that the wave function is not to be identified as the physical system. The wave function is what is used to predict what the system will be observed to do.

To every measurable attribute of the system, wave mechanics associates a self-adjoint linear operator in the space of wave functions. The operator is called an observable. Examples are the momentum and energy operators (13.22) and (12.6); others to be discussed in sections below include parity and angular momentum.

The possible results of a measurement of the observable Q are its eigenvalues, q_n, in the equation

$$Q\psi_n = q_n\psi_n. \tag{14.1}$$

The operators representing observables are required to be self-adjoint so the eigenvalues are real (as shown in section 13), as is appropriate for an observable. An example is Schrödinger's equation $H\psi = E\psi$, which was used in section 6 to find energy levels E for a simple harmonic oscillator and a spherically symmetric hydrogen atom.

Within quantum mechanics, the state wave function ψ determines everything that can be predicted about the result of a measurement of the system. If the wave function for the system is the eigenfunction ψ_n of Q in equation (14.1), then the result of measuring Q certainly is the eigenvalue q_n. If ψ is not an eigenfunction of Q the theory specifies the probability distribution in the result of measuring Q, as follows.

Suppose the eigenvalues of Q are not degenerate (all q_n are different). As discussed in the last section, the ψ_n can be taken to be orthogonal and normalized to

$$(\psi_m, \psi_n) = \delta_{mn}. \tag{14.2}$$

If the quantity represented by the observable Q is measured in the system represented by the wave function ψ, the possible results of the measurement are the eigenvalues q_n, and the probability that the result is the particular value q_n is

$$P_n = |(\psi_n, \psi)|^2, \tag{14.3}$$

where ψ_n is the eigenfunction of Q with eigenvalue q_n.

The probabilities for obtaining any possible result must sum to unity:

$$\sum P_n = 1. \tag{14.4}$$

To see what this implies, consider the expansion of the wave function ψ in the eigenfunctions ψ_i of Q (the completeness relation in eq. [13.58]):

$$\psi = \sum \psi_n(\psi_n, \psi). \tag{14.5}$$

The inner product of this sum with itself is, by the orthogonality relation (14.2),

$$(\psi, \psi) = \sum |(\psi, \psi_n)|^2. \tag{14.6}$$

By equations (14.3) and (14.4), this means

$$(\psi, \psi) = 1. \tag{14.7}$$

Thus we see that if ψ satisfies the normalization condition discussed in section 8 (eq. [11.7]) then the probabilities P_i for any observable properly sum to unity.

As an example of the application of the probability assumption, consider a momentum measurement in a system consisting of a single particle that moves in one dimension in a space with periodic boundary conditions with period L. The momentum eigenfunctions (eq. [11.14]) are

$$\phi_p(x) = \frac{e^{ipx/\hbar}}{L^{1/2}}, \qquad p = 2\pi\hbar n/L, \tag{14.8}$$

with $n = 0, \pm 1, \pm 2, \ldots$. This is the one-dimensional version of equation (13.28). The eigenfunctions ϕ_p are orthogonal, as in equation (10.4), and normalized. If the wave function for the system is $\psi(x)$, the wave function expansion in equation (14.5) is

$$\psi(x) = \sum_n c_p \frac{e^{ipx/\hbar}}{L^{1/2}}. \tag{14.9}$$

The orthonormality of the eigenfunctions ϕ_p allows us to write the expansion coefficients as

$$c_p = \int_0^L dx\, \psi(x) \frac{e^{-ipx/\hbar}}{L^{1/2}}. \tag{14.10}$$

The probability that the momentum measurement yields the value p is (eq. [14.3])

$$P_p = |c_p|^2. \tag{14.11}$$

Since $p = 2\pi\hbar n/L$ (eq. [14.8]), the number of integers in the range of values of momenta between p and $p + \delta p$ is

$$\delta n = \frac{L}{2\pi\hbar}\delta p. \tag{14.12}$$

The probability that the momentum is in the range p to $p + \delta p$ is the product of the probability P_p per momentum eigenvalue (eq. [14.11]) with the number δn of eigenvalues in the wanted range of momentum (eq. [14.12]). This gives

$$\delta P = P_p \delta n = \delta p \left| \int_0^L dx\, \psi(x) \frac{e^{-ipx/\hbar}}{(2\pi\hbar)^{1/2}} \right|^2 . \tag{14.13}$$

This is equations (10.39) and (10.41). That is, the probability expression (14.3) is consistent with the argument in section 10 for the measurement of the momentum of a free particle by time of flight.

Expectation values or ensemble average values were discussed in section 11. The expectation value of the result of measuring the attribute associated with the observable Q in the state of the system represented by ψ is

$$\langle Q \rangle = (\psi, Q\psi), \tag{14.14}$$

as may be checked by applying equations (14.1) to (14.5):

$$\begin{aligned} \langle Q \rangle &= (\psi, Q\{\textstyle\sum \psi_n(\psi_n, \psi)\}) \\ &= \textstyle\sum (\psi, Q\psi_n)(\psi_n, \psi) \\ &= \textstyle\sum (\psi, \psi_n) q_n (\psi_n, \psi) \\ &= \textstyle\sum q_n P_n. \end{aligned} \tag{14.15}$$

The first line uses the expansion (14.5) of the wave function, ψ. The second line uses the linearity of the operator Q and of the inner product (eq. [13.4]). The last line uses the probability expression (14.3). The result is the definition of an ensemble average value, as in equation (11.2).

It will be recalled that the last line of equation (14.15) means that if we had a statistical ensemble of systems all with the same wave function ψ, then $\langle Q \rangle$ would be the arithmetic mean value of the results of the measurements of Q averaged over the members of the ensemble. Examples of equation (14.14) for the expectation values of position and momentum were seen in section 11. The probability expression (14.3) generalizes this relation to any observable.

If the eigenfunctions of Q are degenerate, it is assumed that there is another observable P that commutes with Q. As discussed in section 13, we can find a complete set of simultaneous eigenfunctions of P and Q. These functions are labeled by two indices, $\psi_{m,n}$, to represent the eigenvalues p_m and q_n of P and Q. This may remove the degeneracy, in the sense that m and n uniquely fix the eigenfunction (up to the usual multiplicative factor). If not, we can seek simultaneous eigenfunctions of P, Q, and another observable R that commutes with P and with Q, and so on until the degeneracy is removed. This gives a complete set of simultaneous eigenfunctions $\psi_{m,n,\ldots}$ of the commuting observables $P, Q, \ldots$, the functions being uniquely labeled by the quantum numbers $m, n, \ldots$.

If the wave function of the system is the simultaneous eigenfunction $\psi_{m,n,\ldots}$ of the commuting observables $P, Q, \ldots$, the result of measuring P certainly is p_m, the result of measuring Q is q_n, and so on. If the wave function ψ that represents the state of the system is not one of the simultaneous eigenfunctions $\psi_{m,n,\ldots}$, the completeness condition (14.5) is that ψ may be expanded in the basis $\psi_{m,n,\ldots}$ as

$$\psi = \sum_{m,n,\ldots} \psi_{m,n,\ldots}(\psi_{m,n,\ldots}, \psi). \tag{14.16}$$

As in equation (14.6), we have, from the orthonormality of the basis functions,

$$\sum_{m,n,\ldots} |(\psi, \psi_{m,n,\ldots})|^2 = (\psi, \psi) = 1, \tag{14.17}$$

assuming the wave function is normalized to $(\psi, \psi) = 1$. If the system is in the state represented by ψ, the results of simultaneous measurements of the observables $P, Q, R, \ldots$ are the eigenvalues $p_m, q_n, \ldots$, with probability

$$P_{m,n,\ldots} = |(\psi_{m,n,\ldots}, \psi)|^2. \tag{14.18}$$

We see from equation (14.17) that this probability distribution is correctly normalized.

If only the observable P is measured, then the probability for the result to be the eigenvalue p_m is found by summing the probabilities

$P_{m,n,\ldots}$ in equation (14.18) over all possible values of the unmeasured attributes:

$$P_m = \sum_{n,\ldots} |(\psi_{m,n,\ldots}, \psi)|^2. \tag{14.19}$$

The expectation value of the observable Q is

$$\langle Q \rangle = \sum_{m,n,\ldots} q_n P_{m,n,\ldots} = (\psi, Q\psi). \tag{14.20}$$

It is left as an exercise to use the expansion (14.17) to show how the last expression follows from the probability law (14.8). The expectation value of the nth power of the observable Q is similarly found to be $\langle Q^n \rangle = (\psi, Q^n \psi)$, and the expectation value of the operator function $f(Q)$ is $\langle f(Q) \rangle = (\psi, f(Q)\psi)$.

Finally, we have to specify what happens to the wave function when an observable of the system is measured. This follows the discussion in section 8. After a measurement, we have to use a new wave function that takes account of the information from the measurement. In particular, if the complete set of commuting observables $P, Q, \ldots$ are simultaneously measured, and the results are $p_m, q_n, \ldots$, then the the new wave function has to be the eigenfunction $\psi_{m,n,\ldots}$ with the observed eigenvalues $p_m, q_n, \ldots$, for by equation (14.18) this new wave function says that if $P, Q, \ldots$ were immediately remeasured the results would be what was just found. This is the observed behavior of physical systems. For example, if a particle position is measured in a spark chamber, experience shows that a measurement repeated closely after will localize the particle in almost the same place. The significance of this assumption is discussed further in chapter 4 below.

This measurement prescription shows how the initial condition for Schrödinger's time-dependent equation (14.21) to be discussed next can be set: if the commuting observables $P, Q, \ldots$ are measured at time t_i, with the result $p_m, q_n, \ldots$, then the initial condition for Schrödinger's equation at $t = t_i$ is $\psi(t_i) = \psi_{m,n,\ldots}$.

The commuting observables $P, Q, \ldots$, are said to be compatible in the sense that a simultaneous measurement of these observables leaves the system in a state $\psi_{m,n,\ldots}$ in which the values of all of the observables are known. The x components of position and momentum are not compatible because these observables do not commute (as we see in eq. [6.33]). If the x component of position of a particle is measured to high precision, it leaves the wave function for the particle with a small spread in the x direction, and, as discussed in section 10, in a state in which there is a large uncertainty in the momentum of the particle. That is, we cannot imagine a measurement that leaves these incompatible observables in a state in which both position and momentum are sharply defined.

Equations of Motion

Let us consider now how a system evolves with time. In the Schrödinger representation, the operators belonging to observables are independent of time (unless some externally applied field changes them). As discussed in section 6, the wave function representing the system evolves according to Schrödinger's equation

$$i\hbar \frac{\partial}{\partial t}\psi(t) = H\psi(t), \tag{14.21}$$

where the Hamiltonian (energy operator) is H.

If H does not contain the time (as the result of some externally applied time variable field), then we can write a formal solution to the time evolution implied by Schrödinger's equation (14.21) in terms of the eigenfunctions ψ_i of H, as follows. Since H is not supposed to depend on time, we can take the eigenfunctions ψ_i to be time-independent. Using the completeness relation (eq. [14.5]), we can write the wave function for the system as

$$\psi(t) = \sum c_i(t)\psi_i. \tag{14.22}$$

The expansion coefficients $c_i(t)$ vary with time, reflecting the time-evolution of the wave function. The result of substituting this expression into Schrödinger's equation (14.21) is

$$i\hbar \sum \frac{dc_i}{dt}\psi_i = H \sum c_i\psi_i = \sum c_i E_i \psi_i. \tag{14.23}$$

The last step follows because the ψ_i are eigenfunctions of H with eigenvalues E_i. In the usual way, the orthogonality of the ψ_i means the coefficients of ψ_i on each side of the equation have to agree:

$$i\hbar \frac{dc_i}{dt} = E_i c_i. \tag{14.24}$$

The solution to this equation is

$$c_i(t) = d_i e^{-iE_i t/\hbar}, \tag{14.25}$$

where the d_i are constants. On substituting this result into the expansion (14.22), we get the general expression for the time-evolution of the wave function,

$$\psi(t) = \sum d_i \psi_i e^{-iE_i t/\hbar}. \tag{14.26}$$

The constants d_i are fixed by the initial value of the wave function ψ.

Another formal expression for the time evolution of the wave function is

$$\psi(t) = U(t)\psi(0), \qquad U(t) = e^{-iHt/\hbar}, \tag{14.27}$$

where $\psi(0)$ in the right-hand side of the first equation is the initial value of the function at time $t = 0$. The meaning of the exponential operator function U of the operator H was discussed in the last section, in the power series expansion in equation (13.10). It is left as an exercise to use this series expansion to show that the operator U satisfies the differential equation

$$i\hbar \frac{\partial U}{\partial t} = HU, \tag{14.28}$$

and to use this result to check that equation (14.27) for $\psi(t)$ agrees with the Schrödinger equation (14.21).

The meaning of equation (14.27) is that $U(t) = e^{-iHt/\hbar}$ is a time translation operator: it maps the wave function at time t_i into the wave function at time $t + t_i$.

The operator $U(t)$ is unitary, that is, its adjoint is equal to its inverse (eq. [13.25]). This gives an easy way to check that the normalization of $\psi(t)$ does not change with time. The normalization is

$$\begin{aligned}(\psi(t), \psi(t)) &= (U(t)\psi(0), U(t)\psi(0)) \\ &= (\psi(0), U(t)^\dagger U(t)\psi(0)) \\ &= (\psi(0), \psi(0)).\end{aligned} \tag{14.29}$$

The second line uses the definition (13.11) of the adjoint $U^\dagger$ of U. We see the normalization is independent of time because $U^\dagger U = 1$.

It is also worth noting that the mapping

$$\begin{aligned} \psi(t) &\rightarrow \psi \equiv U(t)^\dagger \psi(t), \\ Q &\rightarrow Q(t) \equiv U(t)^\dagger Q U(t), \end{aligned} \tag{14.30}$$

applied to all wave functions and all operators, is a unitary transformation, of the kind discussed in the last section (eq. [13.26]). This is a transformation from the time-dependent wave functions and constant operators of the Schrödinger representation to the constant wave functions and time-dependent operators of what is called the Heisenberg representation. As you were invited to check, a unitary transformation has no effect on the quantities $(\psi, Q\phi)$ that are used to express the physical predictions of the theory. Since this unitary transformation changes none of the physics, the Schrödinger and Heisenberg representations are equivalent.

The time derivative of the observable $Q(t)$ in the Heisenberg representation is, from equations (14.28) and (14.30),

$$i\hbar \frac{dQ(t)}{dt} = [Q(t), H]. \tag{14.31}$$

This applied to all observables is the equation of motion in the Heisenberg representation.

Ehrenfest's Theorem

A useful relation is obtained by writing down the time derivative of the expectation value of the observable Q (eq. [14.20]). In the Schrödinger representation, we have

$$\begin{aligned} i\hbar \frac{\partial \langle Q \rangle}{\partial t} &= i\hbar \frac{\partial}{\partial t} (\psi(t), Q\psi(t)) \\ &= i\hbar \left((\psi(t), Q\frac{\partial \psi(t)}{\partial t}) + (\frac{\partial \psi(t)}{\partial t}, Q\psi(t) \right) \\ &= (\psi(t), QH\psi(t)) - (H\psi(t), Q\psi(t)). \end{aligned} \tag{14.32}$$

The change of sign in the second term in the last line comes from putting the factor i into the left side of the inner product. Then Schrödinger's equation (14.21) allows us to replace the time derivative with the operator H. Using the fact that H is self-adjoint, we can bring H to the right-hand side of the inner product, to get Ehrenfest's theorem,

$$i\hbar\frac{\partial\langle Q\rangle}{\partial t} = (\psi(t), (QH - HQ)\psi(t)) = (\psi(t), [Q, H]\psi(t)). \qquad (14.33)$$

Another way to write this expression is

$$\frac{\partial\langle Q\rangle}{\partial t} = \langle -i[Q, H]/\hbar\rangle. \qquad (14.34)$$

As is readily checked (using eq. [13.16]), the operator $-i[Q, H]/\hbar$ is self-adjoint. Ehrenfest's theorem says the expectation value of the observable $-i[Q, H]/\hbar$ is the time rate of change of the expectation value of the observable Q.

In the Heisenberg representation, wave functions are constant and observables vary with time. In this representation, Ehrenfest's theorem follows immediately from equation (14.31).

Let us apply Ehrenfest's theorem to the expectation values of the position and momentum of a particle moving in three dimensions in a potential well, $V(\mathbf{r})$. We have

$$\frac{\partial}{\partial x}\{V(\mathbf{r})\psi(\mathbf{r}, t)\} = \frac{\partial V}{\partial x}\psi + V\frac{\partial\psi}{\partial x}. \qquad (14.35)$$

This yields the commutation relation

$$[\frac{\partial}{\partial x}, V] = \frac{\partial V}{\partial x}. \qquad (14.36)$$

The momentum operator $p_\alpha = -i\hbar\partial_\alpha$ therefore satisfies

$$[p_\alpha, V] = -i\hbar\frac{\partial V}{\partial r_\alpha}. \qquad (14.37)$$

Since the Hamiltonian is $H = p^2/2m + V$, and $\mathbf{p}$ commutes with p^2, the commutator of the momentum with the Hamiltonian is

$$[p_\alpha, H] = -i\hbar\frac{\partial V}{\partial r_\alpha}. \qquad (14.38)$$

Thus Ehrenfest's theorem (14.33) says

$$i\hbar\frac{\partial}{\partial t}\langle p_\alpha\rangle = (\psi, [p_\alpha, H]\psi) = -i\hbar(\psi, \frac{\partial V}{\partial r_\alpha}\psi), \tag{14.39}$$

which simplifies to

$$\partial\langle p_\alpha\rangle/\partial t = -\langle\partial V/\partial r_\alpha\rangle. \tag{14.40}$$

In classical mechanics, the time rate of change of the momentum is the force, which is the gradient of the potential: $dp_\alpha/dt = -\partial V/dr_\alpha$. Quantum mechanics replaces this relation with the expectation values in equation (14.40). It is left as an exercise to check that Ehrenfest's theorem says the time rate of change of the expectation value of the position of the particle is

$$\frac{\partial\langle r_\alpha\rangle}{\partial t} = \frac{\langle p_\alpha\rangle}{m}. \tag{14.41}$$

This also replaces the classical relation with expectation values.

The Details

To apply these general assumptions to a given physical system we must give a specific prescription for the observables and their algebra, and we must adopt a definite form for the Hamiltonian as a function of the observables. The equivalent in Newtonian mechanics is the specification of the force law in a given physical situation, such as the inverse square law for gravity. Just as the inverse square law was guessed at from fragmentary evidence and then justified by the success of many tests, the prescription for the Hamiltonian and the algebra of the observables for a quantum system is to be considered a guess, to be justified if it produces experimentally successful predictions. A formal guide to the algebra, canonical quantization, is discussed in section 19 below.

Summary

The general prescriptions of wave mechanics may be summarized as follows. The state of a physical system is represented by a wave function, ψ, and each measurable attribute of the system is represented by a linear self-adjoint operator Q in the space of functions. The ensemble average value of the result of measuring Q in the state ψ is $\langle Q\rangle = (\psi, Q\psi)$

(eq. [14.20]). In particular, if ψ is the eigenfunction ψ_n of Q, with eigenvalue q_n, then the result of the measurement of Q certainly is q_n. The observables Q are self-adjoint (eqs. [13.12] and [13.13]); that guarantees that the q_n are real (eq. [13.30]).

The eigenfunctions ψ_n of the observable Q can be arranged to be normalized and orthogonal (eq. [13.42]). If the observables P and Q commute, there is a complete set of simultaneous eigenfunctions of P and Q (that are normalized and orthogonal). For a sufficiently large set of observables, $P, Q, \ldots$, all of which commute with each other, the eigenvalues $p_m, q_n, \ldots$, uniquely fix the eigenfunction, $\psi_{m,n,\ldots}$ (up to a multiplicative phase factor). These eigenfunctions are a basis in terms of which any state wave function ψ can be expressed as a linear combination

$$\psi = \sum_{m,n,\ldots} c_{m,n,\ldots}\psi_{m,n,\ldots}. \tag{14.42}$$

The expansion coefficients $c_{m,n,\ldots}$ determine the wave function in terms of the basis functions. The squares of the coefficients, $|c_{m,n,\ldots}|^2$, are the probabilities that the values of the observables are measured to be the eigenvalues $p_m, q_n, \ldots$ (eq. [14.18]). A measurement that yields the results $p_m, q_n, \ldots$, leaves the system in the state represented by the eigenfunction $\psi_{m,n,\ldots}$.

The time evolution of the wave function between measurements is given by Schrödinger's equation (14.21) (in the Schrödinger representation), or by the time translation operator $e^{-iHt/\hbar}$ (eq. [14.27]), or by the expansion of the wave function in the basis of energy eigenstates (eq. [14.26]).

We will consider next an application of these principles in the simple but important case of the parity observable, Π.

15 Parity

Eigenfunctions and Eigenvalues

Parity is an observable attribute of a particle or of the orbital motion of a system of particles; the latter is considered here.

The parity operator Π changes the sign of each component of each particle position variable in the wave function. Thus for a single-particle wave function in three dimensions the operator is

$$\Pi\psi(\mathbf{r}) = \psi(-\mathbf{r}). \tag{15.1}$$

This is equivalent to a mirror image plus a rotation: the view in a mirror in the $x - y$ plane gives

$$\begin{aligned} x &\to x \\ y &\to y \\ z &\to -z, \end{aligned} \tag{15.2}$$

and a 180° rotation around the z axis gives

$$\begin{aligned} x &\to -x \\ y &\to -y \\ -z &\to -z, \end{aligned} \tag{15.3}$$

which is the effect of the parity operator Π.

To see that Π is self-adjoint, consider the one-dimensional case, where

$$(\psi, \Pi\phi) = \int_{-\infty}^{\infty} dx\ \psi(x)^* \Pi\phi(x) = \int_{-\infty}^{\infty} dx\ \psi(x)^* \phi(-x). \tag{15.4}$$

The change of variables $x' = -x$ gives

$$(\psi, \Pi\phi) = -\int_{\infty}^{-\infty} dx'\ \psi(-x')^* \phi(x'). \tag{15.5}$$

Exchanging the limits of integration gets rid of the minus sign in front of the integral. On writing

$$\psi(-x')^* = \Pi\psi(x')^*, \tag{15.6}$$

we get

$$(\psi, \Pi\phi) = (\Pi\psi, \phi). \tag{15.7}$$

This means Π is self-adjoint, $\Pi^\dagger = \Pi$ (eq. [13.11]). The argument is readily generalized to m dimensions.

Now consider the eigenfunctions and eigenvalues of Π. Since $\Pi^2 = \Pi \times \Pi$ sends $\mathbf{r} \to -\mathbf{r} \to \mathbf{r}$, it is the identity operator:

$$\Pi^2 = 1. \tag{15.8}$$

Thus if ψ is an eigenfunction of Π with eigenvalue π, so $\Pi\psi = \pi\psi$, then

$$\psi = \Pi^2\psi = \Pi\pi\psi = \pi^2\psi. \tag{15.9}$$

This says $\pi^2 = 1$, so the eigenvalues π have to be

$$\pi = \pm 1. \tag{15.10}$$

If $\pi = +1$ the eigenfunction is even under inversion:

$$\Pi\psi(\mathbf{r}) = \psi(-\mathbf{r}) = \psi(\mathbf{r}). \tag{15.11}$$

If $\pi = -1$ the eigenfunction is odd:

$$\Pi\psi(\mathbf{r}) = \psi(-\mathbf{r}) = -\psi(\mathbf{r}). \tag{15.12}$$

The completeness expression is easy to write down. For any function,

$$\psi(\mathbf{r}) = [\psi(\mathbf{r}) + \psi(-\mathbf{r})]/2 + [\psi(\mathbf{r}) - \psi(-\mathbf{r})]/2. \tag{15.13}$$

This is the sum of an even and odd eigenfunction of Π.

Commutation Relations and Conservation of Parity

For any function $\psi(\mathbf{r})$, the product of ψ with the x component of the position vector $\mathbf{r}$ satisfies

$$\Pi x\psi(\mathbf{r}) = -x\psi(-\mathbf{r}) = -x\Pi\psi(\mathbf{r}). \tag{15.14}$$

Since this is true for any ψ, we have

$$\Pi\mathbf{r} = -\mathbf{r}\Pi. \tag{15.15}$$

These operators are said to anticommute.

The momentum vector operator also anticommutes with Π. To see this, let

$$\frac{\partial\psi(\mathbf{r})}{\partial x} = \phi(\mathbf{r}), \tag{15.16}$$

for any function ψ. Also, write

$$\partial_x = \frac{\partial}{\partial x}, \tag{15.17}$$

and let $\mathbf{r}' = -\mathbf{r}$. Then we have

$$\Pi\partial_x\psi(\mathbf{r}) = \Pi\phi(\mathbf{r}) = \phi(-\mathbf{r}) = \phi(\mathbf{r}'). \tag{15.18}$$

The result of applying the operators Π and ∂_x in the opposite order is

$$\begin{aligned}\partial_x\Pi\psi(\mathbf{r}) &= \partial_x\psi(-\mathbf{r}) \\ &= \frac{\partial\psi(\mathbf{r}')}{\partial x} \\ &= \frac{\partial\psi(\mathbf{r}')}{\partial x'}\frac{\partial x'}{\partial x} \\ &= -\phi(\mathbf{r}').\end{aligned} \tag{15.19}$$

Equations (15.18) and (15.19) say $\Pi\partial_x\psi = -\partial_x\Pi\psi$ for any ψ, so

$$\Pi\partial_x = -\partial_x\Pi. \tag{15.20}$$

Since the momentum operator is $p_x = -i\hbar\partial_x$ (eq. [11.10]), we see that parity anticommutes with each component of momentum:

$$\Pi p_\alpha = -p_\alpha\Pi. \tag{15.21}$$

For the square of the momentum, we have

$$\Pi\mathbf{p}^2 = -\mathbf{p}\cdot\Pi\mathbf{p} = \mathbf{p}^2\Pi, \tag{15.22}$$

so $\mathbf{p}^2$ and Π commute:

$$[\Pi, \mathbf{p}^2] = 0. \tag{15.23}$$

A particle in a central potential moves in a potential $V(r)$ that is a function only of distance r from the origin, at $\mathbf{r} = 0$. Since r is the length of the position vector $\mathbf{r}$, it does not change under the parity operation. Therefore Π commutes with $V(r)$. Since equation (15.23) says Π commutes with the kinetic energy $p^2/2m$, we conclude that Π commutes with the Hamiltonian operator in equation (11.15) for a particle with potential energy $V(r)$:

$$[\Pi, H] = [\Pi, \mathbf{p}^2/2m + V(r)] = 0. \tag{15.24}$$

The result $[\Pi, H] = 0$ applies to high accuracy in atomic physics. It has some important consequences.

1) If ψ is a solution to Schrödinger's time-dependent equation,

$$i\hbar\frac{\partial\psi}{\partial t} = H\psi, \tag{15.25}$$

then the result of operating on both sides of this equation by Π and using $[\Pi, H] = 0$ is

$$i\hbar\frac{\partial\Pi\psi}{\partial t} = H\Pi\psi. \tag{15.26}$$

That is, if ψ represents a possible state of evolution of the system, then the mirror image (plus rotation) of the wave function, $\Pi\psi$, is another possible state of evolution of the system. This means the physics is symmetric under a mirror reflection. Until the 1950s this symmetry was taken to be more or less self-evident: why would Nature care about a reflection? Nature does care: the mirror image of a neutrino produced in a nuclear β decay is not observed. But the parity symmetry violation of the weak interactions has a very small effect on atomic structure, so the equation $[H, \Pi] = 0$ is an excellent approximation here. To this approximation, the mirror image of an atomic system is another allowed state of the system.

2) As discussed in section 13, if $[H, \Pi] = 0$ we can find a complete set of simultaneous eigenfunctions of H and Π. This means that each energy eigenfunction can be assigned a parity quantum number, $\pi = \pm 1$. For example, the one-dimensional simple harmonic oscillator discussed in section 6 has the potential $Kx^2/2$ that is symmetric under reflection through $x = 0$, so the Hamiltonian commutes with the one-dimensional parity operator Π. Therefore energy eigenfunctions can be classified by their parity. The ground state wave function is $\psi_0 \propto e^{-\beta x^2}$, where β is a constant; this state has even parity because it is symmetric under $x \to -x$. The wave function for the first excited state is $\psi_1 \propto xe^{-\beta x^2}$; it has odd parity because it changes sign under $x \to -x$.

3) Parity is conserved if $[H, \Pi] = 0$. If at time $t = 0$ the system is in the state $\psi_{\pi,n}$ with definite energy E_n and definite parity π, then the solution to Schrödinger's time-dependent equation is

$$\psi(t) = \psi_{\pi,n}e^{-iE_nt/\hbar}. \tag{15.27}$$

Therefore at a later time t the state still has energy E_n and still has the same parity, π. It is left as an exercise to show that if the state $\psi(t)$

does not have definite parity the expectation value of the parity, $\langle \Pi \rangle$, is independent of time when Π commutes with the Hamiltonian.

The theme of this discussion is worth summarizing. One considers an observable, here parity, Π. The observable maps any state wave function, ψ, into a new wave function, $\Pi\psi$. If Π commutes with the Hamiltonian, then $\Pi\psi$ also is an allowed physical state function for the system. In this case physics is symmetric under the mapping. This symmetry implies the existence of a conserved quantum number, here the parity π.

16 Linear Momentum

Momentum and Space Translations

We saw in section 12 that a two-body system can be assigned a fixed total momentum, leaving a one-body problem for the relative motion. The total momentum of a many-body system will be considered here.

Suppose a system of N particles has a wave function $\psi(t, \mathbf{r}_1, \ldots \mathbf{r}_N)$. We can produce another wave function by shifting this one in space by the distance $\delta\mathbf{r}$, to get

$$\psi'(t, \mathbf{r}_1, \mathbf{r}_2, \ldots, \mathbf{r}_N) = \psi(t, \mathbf{r}_1 - \delta\mathbf{r}, \mathbf{r}_2 - \delta\mathbf{r}, \ldots, \mathbf{r}_N - \delta\mathbf{r}). \tag{16.1}$$

The function ψ' represents a system shifted relative to the original one by the distance $\delta\mathbf{r}$, because the arguments of ψ' have to be larger than the arguments of ψ by the amount $\delta\mathbf{r}$ to get the same value of the function. Equation (16.1) also is mapping of every function ψ into a new function ψ'. As we will now discuss, this mapping is generated by the total momentum operator. If the Hamiltonian is unaffected by the shift in position of the system, then H commutes with the total momentum operator, so momentum is conserved.

For an infinitesimal shift $\delta\mathbf{r}$, we can rewrite equation (16.1) as the first terms in a Taylor series expansion in $\delta\mathbf{r}$:

$$\psi'(t, \mathbf{r}_1, \ldots) = \psi(t, \mathbf{r}_1, \ldots) - \sum_i \delta\mathbf{r} \cdot \nabla_i \psi. \tag{16.2}$$

Here the gradient operator ∇_i means the derivatives with respect to the position components for particle i. The momentum operator for the ith particle is $\mathbf{p}_i = -i\hbar\nabla_i$, and the total momentum operator is defined to be

$$\mathbf{P} = \sum \mathbf{p}_i = -i\hbar \sum_i \nabla_i. \tag{16.3}$$

Using the operator $\mathbf{P}$, we can rewrite equation (16.2) as

$$\psi'(t, \mathbf{r}_1, \ldots) = (1 - i\delta\mathbf{r} \cdot \mathbf{P}/\hbar)\psi(t, \mathbf{r}_1, \ldots). \tag{16.4}$$

This shows how the momentum operator shifts the system by the infinitesimal distance $\delta\mathbf{r}$.

A shift of the system by the finite distance $\Delta\mathbf{r}$ can be generated by applying equation (16.4) many times. Consider N sequentially applied shifts of amount $\Delta\mathbf{r}/N$. As $N \to \infty$ this is

$$\psi'(\mathbf{r}_i) = \left(1 - \frac{i\Delta\mathbf{r} \cdot \mathbf{P}}{N\hbar}\right)^N \psi(\mathbf{r}_i). \tag{16.5}$$

The usual expression, $(1 - a/N)^N \to e^{-a}$ at $N \to \infty$, applies here because the components of $\mathbf{P}$ commute among themselves, so the usual algebra applies. Thus a finite shift of the system by the vector displacement $\Delta\mathbf{r}$ is

$$\psi'(\mathbf{r}_i) = e^{-i\Delta\mathbf{r}\cdot\mathbf{P}/\hbar}\psi(\mathbf{r}_i). \tag{16.6}$$

As discussed in section 13, an exponential function of an operator is defined by the power series expansion (eq. [13.10]). It is left as an exercise to check that the power series expansion of the exponential in equation (16.6) gives the usual Taylor expansion of the function $\psi(\mathbf{r} - \Delta\mathbf{r})$ as a power series in $\Delta\mathbf{r}$.

Equation (16.6) shows that the unitary operator $U = e^{-i\Delta\mathbf{r}\cdot\mathbf{P}/\hbar}$ moves the system by the displacement $\Delta\mathbf{r}$. In the argument of the exponential, the components of the displacement $\Delta\mathbf{r}$ are multiplied by the components of the total momentum operator $\mathbf{P}$. The operator $\mathbf{P}$ is said to generate the translation. In the next section, it will be seen that there is a similar result for rotations: the components of the total angular momentum operator, $\mathbf{L}$, generate rotations in the same way $\mathbf{P}$ generates translations.

Momentum Conservation

Suppose the Hamiltonian for the N-particle system is

$$H = \sum_i \frac{\mathbf{p}_i^2}{2m_i} + \sum_{ij} V_{ij}(|\mathbf{r}_i - \mathbf{r}_j|). \tag{16.7}$$

This has the same form as in equation (12.22), but we have added the condition the potential is a function of the relative positions of the particles. This means the expression is unaffected by a shift of position of the system of particles. Let us see how this symmetry leads to momentum conservation.

The operator $\mathbf{P}$ defined in equation (16.3) applied to the function $V_{ij}(\mathbf{r}_i - \mathbf{r}_j)$ gives zero, because the derivatives of $V_{ij}(\mathbf{r}_i - \mathbf{r}_j)$ with respect to $\mathbf{r}_i$ and $\mathbf{r}_j$ are equal in magnitude and opposite in sign, so the derivatives add to zero. This means $\mathbf{P}$ commutes with each of the potential energy terms in the Hamiltonian (16.7). Since $\mathbf{P}$ commutes with each of the $\mathbf{p}_i$ (partial derivatives can be taken in either order), we see that $\mathbf{P}$ commutes with the Hamiltonian in equation (16.7):

$$[H, \mathbf{P}] = 0. \tag{16.8}$$

As for parity, this has several very useful consequences:

1) By repeating the calculation in equations (15.25) and (15.26), one immediately sees that if $\psi(t, \mathbf{r}_j)$ is a solution to Schrödinger's equation, so is the shifted function $e^{-i\Delta\mathbf{r}\cdot\mathbf{P}/\hbar}\psi$. This means the system can be translated in space to get another physically realizable system, because the Hamiltonian (16.7) is unaffected by a space translation.

2) We can find a complete set of simultaneous eigenfunctions of H and $\mathbf{P}$. The simultaneous eigenfunctions can be written as

$$\psi = \Psi \exp i\mathbf{P}_c \cdot \sum m_i \mathbf{r}_i/(M\hbar), \tag{16.9}$$

with $M = \sum m_i$. Here Ψ is a function of relative positions $\mathbf{r}_i - \mathbf{r}_j$ only, so the total momentum operator $\mathbf{P}$ operating on Ψ gives zero (as was discussed for the function $V(\mathbf{r}_i - \mathbf{r}_j)$). Thus the operator $\mathbf{P}$ applied to equation (16.9) passes through Ψ and yields the constant (not operator) vector $\mathbf{P}_c$ when operating on the exponential:

$$\mathbf{P}\psi = \mathbf{P}_c\psi. \tag{16.10}$$

The state (16.9) thus is an eigenfunction of $\mathbf{P}$ with eigenvalue $\mathbf{P}_c$; it represents a system with total momentum $\mathbf{P}_c$. It is left as an exercise to show that, if Ψ is an eigenfunction of H in equation (16.7) with energy E_i, then the wave function ψ in equation (16.9) is an eigenfunction of H with eigenvalue $E_i + P_c^2/2M$. This is the familiar expression for the energy as the sum of the internal energy, E_i, and the kinetic energy of translation, $P_c^2/2M$.

3) Momentum is conserved, by the same arguments used for parity. This conservation law follows from a symmetry assumption, that space is homogeneous. That means a physical system is unaffected by a spatial translation, so the three generators $\mathbf{P}$ of translations in three dimensions commute with the Hamiltonian. It follows that the quantities associated with $\mathbf{P}$, the components of the total momentum, are conserved.

17 Orbital Angular Momentum

Operator Algebra

We can follow classical mechanics in defining the orbital angular momentum operator for a single particle moving in three dimensions as the cross product

$$\mathbf{L} = \mathbf{r} \times \mathbf{p}, \qquad \mathbf{p} = -i\hbar\nabla. \tag{17.1}$$

The position vector is $\mathbf{r}$, and $\mathbf{p}$ is the linear momentum operator introduced in section 11 (eq. [11.10]) and discussed in the last section. The components of this equation are

$$\begin{aligned} \mathbf{L}_x &= yp_z - zp_y, \\ \mathbf{L}_y &= zp_x - xp_z, \\ \mathbf{L}_z &= xp_y - yp_x. \end{aligned} \tag{17.2}$$

It will be noted that we get the same operators from the classical expression $\mathbf{L} = -\mathbf{p} \times \mathbf{r}$, because y commutes with p_z and so on.

The adjoint of the x component of $\mathbf{L}$ is

$$\begin{aligned} L_x^\dagger &= (yp_z - zp_y)^\dagger = p_z^\dagger y^\dagger - p_y^\dagger z^\dagger \\ &= p_z y - p_y z = yp_z - zp_y \\ &= L_x. \end{aligned} \tag{17.3}$$

This uses the properties of the adjoint of an observable (eqs. [13.16] and [13.17]), and that position and momentum are self-adjoint observables such that p_z commutes with y and p_y commutes with z. The result is that the operator L_x is self-adjoint. Since there is nothing special about the x component, we conclude that $\mathbf{L}$ is self-adjoint,

$$L_\alpha^\dagger = L_\alpha. \tag{17.4}$$

Next let us consider the commutation relations for the angular momentum observables L_α. We have $[x, \partial_x] = -1$ (eq. [6.33]), and $p_x = -i\hbar\partial_x$ (eq. [17.1]). The commutation relations for position and linear momentum thus are

$$[r_\alpha, p_\beta] = i\hbar\delta_{\alpha\beta}. \tag{17.5}$$

The position components r_α commute among themselves, because they are just numbers, and the momentum components p_β commute among themselves, because partial derivatives can be taken in either order. We have then

$$\begin{aligned} [L_x, L_y] &= [(yp_z - zp_y), (zp_x - xp_z)] \\ &= [yp_z, zp_x] + [zp_y, xp_z] \\ &= (yp_x - xp_y)[p_z, z] \\ &= i\hbar(xp_y - yp_x) \\ &= i\hbar L_z. \end{aligned} \tag{17.6}$$

Thus we arrive at the equations

$$\begin{aligned} [L_x, L_y] &= i\hbar L_z, \\ [L_y, L_z] &= i\hbar L_x, \\ [L_z, L_x] &= i\hbar L_y. \end{aligned} \tag{17.7}$$

These are the fundamental angular momentum commutation relations. They will be applied in section 23 below to intrinsic angular momentum (spin), and in the present section to orbital angular momentum. One way to remember these commutation relations, and the relations for orbital angular momentum in equation (17.2), is to note the cyclic progression $x \to y \to z \to x \ldots$.

It is often convenient to express the angular momentum commutation relations (17.7) in a more compact way by means of the permutation symbol, $\epsilon_{\alpha\beta\gamma}$. The Greek indices range from 1 to 3 to represent the three cartesian position coordinates, x, y, and z. The permutation symbol $\epsilon_{\alpha\beta\gamma}$ changes sign under a permutation of the indices (exchange of any neighboring pair) and $\epsilon_{123} = 1$. Thus $\epsilon_{213} = -1$, and $\epsilon_{231} = 1$. Using this symbol, we can write the components of the cross product of two vectors $\mathbf{r}$ and $\mathbf{p}$, as in equation (17.2), as

$$\begin{aligned}(\mathbf{r} \times \mathbf{p})_\alpha = L_\alpha &= \sum_{\beta\gamma} \epsilon_{\alpha\beta\gamma} r_\beta p_\gamma \\ &= \epsilon_{\alpha\beta\gamma} r_\beta p_\gamma.\end{aligned} \tag{17.8}$$

As indicated in the second line, we can make the notation a little more compact by using the Einstein summation convention, that the repeated indices β and γ are summed (here, from 1 to 3).

In terms of the permutation symbol, the angular momentum commutation relations (17.7) are

$$[L_\alpha, L_\beta] = i\hbar\epsilon_{\alpha\beta\gamma} L_\gamma. \tag{17.9}$$

This says for example $[L_1, L_2] = i\hbar\epsilon_{123} L_3 = i\hbar L_3$, which is the first of equations (17.7).

The square of the angular momentum operator is defined to be

$$L^2 = L_x^2 + L_y^2 + L_z^2. \tag{17.10}$$

To see that the operator L^2 commutes with each component of $\mathbf{L}$, consider first

$$\begin{aligned}[L_x, L_y^2] &= [L_x, L_y]L_y + L_y[L_x, L_y] \\ &= i\hbar(L_z L_y + L_y L_z).\end{aligned} \tag{17.11}$$

The first line can be checked by writing out the commutators. The second line follows from the fundamental commutation relations (17.7). One similarly finds

$$[L_x, L_z^2] = -i\hbar(L_y L_z + L_z L_y). \tag{17.12}$$

Since L_x obviously commutes with L_x^2, we see that $[L^2, L_x] = 0$, and more generally that

$$[L^2, L_\alpha] = 0. \tag{17.13}$$

In a system of n particles, the ith particle is assigned position vector $\mathbf{r}_i$, momentum $\mathbf{p}_i$, and angular momentum $\mathbf{L}_i = \mathbf{r}_i \times \mathbf{p}_i$. The total angular momentum is

$$\mathbf{L} = \sum \mathbf{L}_i. \tag{17.14}$$

Because the observables belonging to different particles commute, the above calculations immediately lead us to the commutation relations in equations (17.7) and (17.13) for the components and square of the total angular momentum. That means the following algebra applies to the total angular momentum of a system of particles as well as to the angular momentum of a single particle.

Eigenvalues

Equation (17.7) tells us we cannot find a complete set of simultaneous eigenfunctions of the three components of $\mathbf{L}$, because the three components do not commute. Equation (17.13) says we can seek a complete set of simultaneous eigenfunctions of L^2 and one component of $\mathbf{L}$:

$$\begin{aligned} L^2\psi_{ab} &= a\psi_{ab}, \\ L_z\psi_{ab} &= b\psi_{ab}. \end{aligned} \tag{17.15}$$

The allowed values of the eigenvalues a and b are computed here.

Some inequalities will be needed. Since the components of $\mathbf{L}$ are self-adjoint, any wave function ψ for the system satisfies

$$(\psi, L_x^2\psi) = (L_x\psi, L_x\psi) \geq 0. \tag{17.16}$$

For the eigenfunction ψ_{ab}, the definition (17.10) of L^2 gives

$$(\psi_{ab}, L^2\psi_{ab}) = (\psi_{ab}, L_x^2\psi_{ab}) + (\psi_{ab}, L_y^2\psi_{ab}) + (\psi_{ab}, L_z^2\psi_{ab}). \tag{17.17}$$

Using equation (17.15), we get

$$a(\psi_{ab}, \psi_{ab}) = (\psi_{ab}, L_x^2\psi_{ab}) + (\psi_{ab}, L_y^2\psi_{ab}) + b^2(\psi_{ab}, \psi_{ab}). \tag{17.18}$$

The first two terms on the right side are not negative, and $(\psi_{ab}, \psi_{ab}) \neq 0$, so

$$a \geq b^2. \tag{17.19}$$

Consider next the operators

$$L_+ = L_x + iL_y,$$
$$L_- = L_x - iL_y. \tag{17.20}$$

Since the components of $\mathbf{L}$ are self-adjoint, $L_+^\dagger = L_-$. Since the components of $\mathbf{L}$ commute with L^2 (eq. [17.13]), L_+ and L_- commute with L^2:

$$[L^2, L_\pm] = 0. \tag{17.21}$$

The angular momentum commutation equations (17.7) give

$$\begin{aligned}[L_z, L_+] &= [L_z, L_x] + i[L_z, L_y] \\ &= i\hbar(L_y - iL_x) \\ &= \hbar L_+.\end{aligned} \tag{17.22}$$

The adjoint of this equation gives the commutator of L_z with L_-. The results are

$$[L_z, L_\pm] = \pm\hbar L_\pm. \tag{17.23}$$

Now we can play a game much like that for the simple harmonic oscillator ladder operators a and $a^\dagger$ in section 6 (eq. [6.46]). Because L^2 commutes with $L_\pm$ (eq. [17.21]), the result of applying L_+ to the first of the eigenvalue equations (17.15) is

$$L^2(L_+\psi_{ab}) = a(L_+\psi_{ab}). \tag{17.24}$$

This means that if $L_+\psi_{ab}$ does not vanish it is an eigenfunction of L^2 with eigenvalue a. The result of applying L_+ to the second of the eigenvalue equations (17.15) is

$$L_+L_z\psi_{ab} = ([L_+, L_z] + L_zL_+)\psi_{ab} = b(L_+\psi_{ab}). \tag{17.25}$$

The middle step introduces a commutator that can be replaced using equation (17.23). We can rewrite the result as

$$L_z(L_+\psi_{ab}) = (b + \hbar)(L_+\psi_{ab}). \tag{17.26}$$

This says that if $L_+\psi_{ab}$ does not vanish, it is an eigenfunction of L_z with eigenvalue $b+\hbar$:

$$L_+\psi_{ab} \propto \psi_{a,b+\hbar}. \tag{17.27}$$

One similarly finds that L_- lowers the eigenvalue b of L_z by $\hbar$:

$$L_-\psi_{ab} \propto \psi_{a,b-\hbar}. \tag{17.28}$$

That is, L_+ and L_- are ladder operators for the z component of angular momentum, just as the operators a and $a^\dagger$ in section 6 are ladder operators for the simple harmonic oscillator Hamiltonian.

Repeated application of $L_\pm$ yields a sequence of eigenvalues of L_z,

$$b, \quad b\pm\hbar, \quad b\pm 2\hbar, \ldots, \tag{17.29}$$

all belonging to the eigenvalue a for L^2. This set of eigenvalues b_m for L_z must terminate above and below, because $b^2 \leq a$ (eq. [17.19]). Thus there must be maximum and minimum values of b such that

$$\begin{aligned} L_+\psi_{\max} &= 0, \qquad L_z\psi_{\max} = b_{\max}\psi_{\max}, \\ L_-\psi_{\min} &= 0, \qquad L_z\psi_{\min} = b_{\min}\psi_{\min}. \end{aligned} \tag{17.30}$$

Next, we need the identity

$$\begin{aligned} L_-L_+ &= (L_x - iL_y)(L_x + iL_y) \\ &= L_x^2 + L_y^2 + i[L_x, L_y] \\ &= L_x^2 + L_y^2 - \hbar L_z. \end{aligned} \tag{17.31}$$

The commutator appears on multiplying out the product, and is simplified using the angular momentum commutation relations (17.7). We can rewrite this expression as

$$L^2 = L_-L_+ + L_z^2 + \hbar L_z. \tag{17.32}$$

A similar calculation gives

$$L^2 = L_+L_- + L_z^2 - \hbar L_z. \tag{17.33}$$

The result of applying equation (17.32) to $\psi_{\rm max}$ in equation (17.30), and recalling that $\psi_{\rm max}$ is an eigenfunction of L^2 and L_z with eigenvalues a and $b_{\rm max}$, and that $L_+\psi_{\rm max} = 0$, is

$$a\psi_{\rm max} = (b_{\rm max}^2 + \hbar b_{\rm max})\psi_{\rm max}. \tag{17.34}$$

The application of equation (17.33) to $\psi_{\rm min}$ similarly gives

$$a\psi_{\rm min} = (b_{\rm min}^2 - \hbar b_{\rm min})\psi_{\rm min}. \tag{17.35}$$

These two equations give

$$a = b_{\rm max}^2 + \hbar b_{\rm max} = b_{\rm min}^2 - \hbar b_{\rm min}. \tag{17.36}$$

The difference of these two equations is

$$0 = (b_{\rm max} - b_{\rm min} + \hbar)(b_{\rm max} + b_{\rm min}). \tag{17.37}$$

Since $b_{\rm max} \geq b_{\rm min}$ the first factor cannot vanish, so the second factor has to vanish:

$$b_{\rm min} = -b_{\rm max}. \tag{17.38}$$

The final step is to note that

$$b_{\rm max} - b_{\rm min} = n\hbar, \tag{17.39}$$

where $n \geq 0$ is an integer, because we can get from $b_{\rm min}$ to $b_{\rm max}$ by applying the raising operator L_+ an integral number of times, n.

Equations (17.38) and (17.39) say

$$b_{\rm max} = n\hbar/2 \equiv l\hbar. \tag{17.40}$$

The quantum number l is $l = n/2$. Finally, equation (17.36) yields $a = b_{\rm max}^2 + \hbar b_{\rm max} = \hbar^2 l(l+1)$.

This calculation shows that the allowed eigenvalues of simultaneous eigenfunctions of the angular momentum operators L^2 and L_z are

$$\begin{aligned} L^2: &\quad \hbar^2 l(l+1), \quad l = 0, \frac{1}{2}, 1, \frac{3}{2}, \ldots, \\ L_z: &\quad m\hbar, \quad m = -l, -l+1, \ldots, l-1, l. \end{aligned} \tag{17.41}$$

These relations follow from the algebra (17.7) for the components of the angular momentum operator. Since the total angular momentum operator (17.14) satisfies the same angular momentum commutation relations, these forms for the eigenvalues are the same for a single particle and for a system of particles. Since n in equation (17.40) is an integer, the allowed values of the quantum number l are integers and half odd integers. As discussed next, orbital angular momentum has integer values of l; half-integer values will be seen in section 23 for the spin of an electron.

Single-Particle Angular Momentum Eigenfunctions

To compute the eigenfunctions of L^2 and L_z for a single particle moving in three dimensions, it is convenient to go from the cartesian coordinates x, y, z used in equation (17.1) to polar coordinates r, θ ,ϕ.

As indicated in figure 17.1, the distance from the origin is r, the angular distance from the z axis is the polar angle θ, and the position projected onto the xy plane is at azimuthal angle ϕ measured from the x axis and swinging toward the y axis. The relations to the cartesian coordinates are

$$\begin{aligned} x &= r \sin\theta \cos\phi, \\ y &= r \sin\theta \sin\phi, \\ z &= r \cos\theta. \end{aligned} \tag{17.42}$$

The derivative of a wave function $\psi(x, y, z)$ with respect to ϕ at fixed θ and r is

$$\begin{aligned} \frac{\partial\psi}{\partial\phi} &= \frac{\partial\psi}{\partial x}\frac{\partial x}{\partial\phi} + \frac{\partial\psi}{\partial y}\frac{\partial y}{\partial\phi} + \frac{\partial\psi}{\partial z}\frac{\partial z}{\partial\phi} \\ &= -r\sin\theta\sin\phi\frac{\partial\psi}{\partial x} + r\sin\theta\cos\phi\frac{\partial\psi}{\partial y} \\ &= x\frac{\partial\psi}{\partial y} - y\frac{\partial\psi}{\partial x}. \end{aligned} \tag{17.43}$$

On multiplying this expression by $-i\hbar$, and using the definition (17.1) of the linear momentum operator, we get

$$-i\hbar\frac{\partial\psi}{\partial\phi} = (xp_y - yp_x)\psi. \tag{17.44}$$

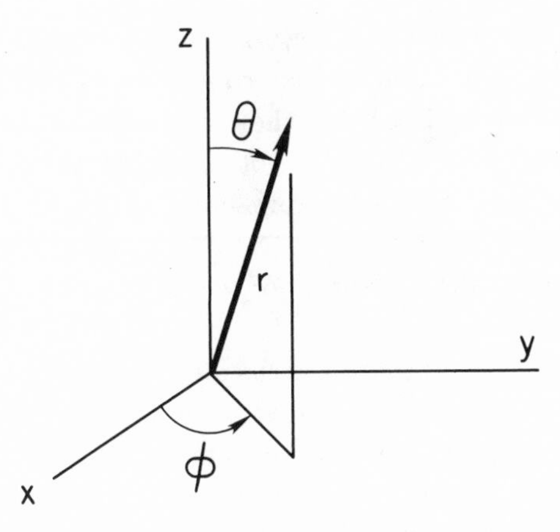

Fig. 17.1 Polar coordinates: the distance from the origin is r, the polar angle is θ, and the azimuthal angle is ϕ.

The operator on the right-hand side is the z component of the single-particle orbital angular momentum (eq. [17.2]). Since this result applies for any ψ, we have

$$L_z = -i\hbar\frac{\partial}{\partial\phi}. \tag{17.45}$$

The L_z eigenvalue equation for a simultaneous eigenfunction ψ_l^m of L^2 and L_z is

$$L_z\psi_l^m = -i\hbar\frac{\partial\psi_l^m}{\partial\phi} = m\hbar\psi_l^m. \tag{17.46}$$

The eigenvalue has been written as $m\hbar$, as in equation (17.41). The solution to this equation is of the form

$$\psi_l^m \propto e^{im\phi}. \tag{17.47}$$

The constant of proportionality is some function of θ and r.

Since the wave function ψ_l^m in equation (17.47) has to be a single-valued function of position, m must be an integer. Since m ranges from $-l$ to l, the quantum number l in equation (17.41) must be an integer in the case of the orbital angular momentum of a particle. As discussed in section 17 below, that means the total orbital angular momentum of a system of particles has to have integer l.

A similar but more lengthy calculation of the change of variables from x, y, z to r, θ, ϕ gives

$$L_\pm = \hbar e^{\pm i\phi}\left(\pm\frac{\partial}{\partial\theta} + i\cot\theta\frac{\partial}{\partial\phi}\right), \tag{17.48}$$

where $L_\pm$ are the raising and lowering operators for L_z in equation (17.20). Since L_x and L_y are linear combinations of $L_\pm$, we see from equations (17.45) and (17.48) that the components of $\mathbf{L}$ involve only the angular polar coordinates θ and ϕ, not the radius r.

It might be noted that the expressions for $L_\pm$ in equation (17.48) do not obviously satisfy $L_+^\dagger = L_-$. (Note that $\partial/\partial\phi$ does not commute with $\exp i\phi$.) An explicit derivation of the relation $L_+^\dagger = L_-$ follows the lines of problem II.25.

Equation (17.47) shows how the eigenfunctions ψ_l^m of L^2 and L_z vary with ϕ. The θ dependence is computed as follows. For given total angular momentum quantum number l, the maximum value of the z component quantum number is $m = l$ (eq. [17.41]). The raising operator L_+ applied to the eigenfunction ψ_l^l with $m = l$ thus has to vanish. This condition with equation (17.47) for the ϕ-dependence of the eigenfunction gives a differential equation for the θ-dependence:

$$L_+\psi_l^l = 0 = \frac{\partial\psi_l^l}{\partial\theta} - l\cot\theta\,\psi_l^l, \tag{17.49}$$

by equation (17.48). The solution is

$$\psi_l^l \propto \sin^l\theta\; e^{il\phi}. \tag{17.50}$$

The lowering operator L_- in equation (17.48) applied to this expression for ψ_l^l gives the eigenfunctions ψ_l^m for $m = l-1, l-2, \ldots, -l$.

The radius r does not enter these expressions for the functions ψ_l^m. That is, the simultaneous eigenfunctions of L^2 and L_z for a single particle are definite functions of the angles θ and ϕ multiplied by any function of r:

$$\psi_l^m = f(r)\,Y_l^m(\theta,\phi). \tag{17.51}$$

The angular functions Y_l^m are the spherical harmonics. The result of applying the lowering ladder operator L_- to the function in equation (17.50) is to generate spherical harmonics of the form

$$Y_l^m(\theta,\phi) = P_l^m(\theta)e^{im\phi}, \tag{17.52}$$

where the P_l^m can be taken to be real functions of the polar angle θ (because the factor i in eq. [17.48] is eliminated by the factor im from the derivative of $e^{im\phi}$ with respect to ϕ). The standard normalization of the spherical harmonics is

$$\int d\Omega \, |Y_l^m|^2 = 1, \tag{17.53}$$

where $d\Omega = \sin\theta d\theta d\phi$ is the element of solid angle, and the integral is over all directions.

It is left as an exercise to show that the spherical harmonics satisfy the orthogonality relation

$$\int d\Omega \, (Y_l^m)^* Y_{l'}^{m'} = \int d\Omega \, Y_l^{-m} Y_{l'}^{m'} = \delta_{ll'}\delta_{mm'}. \tag{17.54}$$

The parity operator Π discussed in section 15 anticommutes with the components of the position and momentum observables (eqs. [15.15] and [15.21]), so it commutes with the products of the form xp_y that appear in the components of the single-particle angular momentum observable. Thus Π commutes with orbital angular momentum:

$$[\Pi, \mathbf{L}] = 0. \tag{17.55}$$

This means the simultaneous eigenfunctions Y_l^m of L^2 and L_z must have definite parity, π. We see from equations (17.42) that $\mathbf{r} \to -\mathbf{r}$ is equivalent to $\theta \to \pi - \theta$ and $\phi \to \phi + \pi$. Since Y_0^0 is independent of θ and ϕ (eqs. [17.47] and [17.50]), it is even under parity, $\pi = 1$. It is left as an exercise to show that Y_l^m has parity $\pi = (-1)^l$.

Generating Rotations

Just as the momentum operators P_α generate translations (eq. [16.6]), the angular momentum operators L_α generate rotations. This will be discussed for a one-particle system; the generalization to a system of N particles is just a matter of adding indices.

A physical system represented by the function ψ is rotated around the z axis by the infinitesimal amount $\delta\phi$ in the direction of increasing ϕ by the mapping

$$\psi(\theta, \phi) \to \psi'(\theta, \phi) = \psi(\theta, \phi - \delta\phi) = \psi(\theta, \phi) - \delta\phi \frac{\partial \psi}{\partial \phi}. \tag{17.56}$$

The argument is the same as in equations (16.1) and (16.2) for an infinitesimal spatial translation. We can use equation (17.45) for L_z to write equation (17.56) as

$$\psi'(\theta, \phi) = (1 - i\delta\phi L_z/\hbar)\psi(\theta, \phi). \tag{17.57}$$

This can be compared to equation (16.4) for an infinitesimal translation.

The result of iterating the operation in equation (17.57), as was done in section 16 for a translation, is the finite rotation

$$\psi' = e^{-i\Delta\phi L_z/\hbar}\psi. \tag{17.58}$$

This can be compared to equation (16.6) for a translation.

Since there is nothing special about the z axis, a rotation by angle χ around the x axis is

$$\psi'' = e^{-i\chi L_x/\hbar}\psi, \tag{17.59}$$

and a rotation by angle θ around the axis with unit normal $\mathbf{n}$ is

$$\psi''' = U\psi, \qquad U = e^{-i\theta\mathbf{n}\cdot\mathbf{L}}. \tag{17.60}$$

The sense of rotation is that the mapping in equation (17.58) with positive $\Delta\phi$ swings the system around the z axis in the direction of increasing ϕ, from the positive x axis toward the positive y axis in figure 17.1. In the same way, the operation in equation (17.59) with positive χ swings the system around the x axis from the positive y axis toward the positive z axis.

It will be noted that a rotation about the z axis followed by a rotation about the x axis is not the same as the same two rotations applied in the opposite order, so we know that in general

$$e^{-i\chi L_x/\hbar}e^{-i\Delta\phi L_z/\hbar} \neq e^{-i\Delta\phi L_z/\hbar}e^{-i\chi L_x/\hbar}. \tag{17.61}$$

The algebra of the angular momentum commutation relations (eq. [17.7]) describes this difference between rotations applied in different order.

Conservation of Angular Momentum

The symmetry of the Hamiltonian under rotations implies conservation of angular momentum, by the same argument used for parity and linear momentum. Consider the single-particle Hamiltonian

$$H = \frac{p^2}{2m} + V(r). \tag{17.62}$$

The potential is a function only of distance r from the origin, so it is unaffected by a rotation. Let us check that this means H commutes with the components of $\mathbf{L}$.

We have from equations (17.2)

$$\begin{aligned}[L_z, p_x] &= [xp_y - yp_x, p_x] = i\hbar p_y, \\ [L_z, p_y] &= -i\hbar p_x, \\ [L_z, p_z] &= 0.\end{aligned} \tag{17.63}$$

Then manipulations that are by now familiar yield

$$[L_z, p^2] = 0. \tag{17.64}$$

Since $\mathbf{L}$ generates rotations about $\mathbf{r} = 0$, and $V(r)$ is rotationally symmetric, V must commute with $\mathbf{L}$:

$$[\mathbf{L}, V(r)] = 0. \tag{17.65}$$

Therefore, the components of $\mathbf{L}$ commute with the Hamiltonian (17.62):

$$[H, \mathbf{L}] = 0. \tag{17.66}$$

The arguments used for parity and linear momentum show that this rotational symmetry implies that energy eigenfunctions can be assigned angular momentum quantum numbers, l, m, that are conserved.

*Systems of Particles**

As discussed in section 12, a system of N particles moving in three dimensions is described by a function of time plus $3N$ variables $\mathbf{r}_i$ for the $3N$ position coordinates. Associated with each particle are the three components of the position operator $\mathbf{p}_i$ (eq. [17.1]). The angular momentum operator for the ith particle is

$$\mathbf{L}_i = \mathbf{r}_i \times \mathbf{p}_i, \tag{17.67}$$

and the total angular momentum of the system is

$$\mathbf{L} = \sum \mathbf{L}_i. \tag{17.68}$$

As in equation (17.10), the square of the total angular momentum is

$$L^2 = L_x^2 + L_y^2 + L_z^2. \tag{17.69}$$

The components of a single-particle angular momentum operator $\mathbf{L}_i$ satisfy the standard angular momentum commutation relations in equation (17.7), and the angular momentum operators belonging to different particles commute with each other, so the components of the total angular momentum $\mathbf{L}$ satisfy the angular momentum commutation relations. Therefore the algebra of equations (17.15) to (17.41) shows that there is a complete set of simultaneous eigenfunctions of L^2 and L_z, with eigenvalues $\hbar^2 l(l+1)$ and $\hbar m$, respectively. We will see that l has to be a nonnegative integer. As usual, m can assume the values $l, l-1, l-2, \ldots -l$.

By repeating the discussion in equation (17.56) for the generation of rotations, one sees that the components of the total angular momentum operator $\mathbf{L}$ generate rotations of the many-particle system. The N-body Hamiltonian H in equation (16.7) is rotationally invariant, so it should be no surprise that this H commutes with the components of $\mathbf{L}$. Thus the eigenfunctions of H can be assigned conserved total angular momentum quantum numbers l and m.

As an example of how single-particle angular momenta add, let us consider a two-body system. The total angular momentum is

$$\mathbf{L} = \mathbf{L}(1) + \mathbf{L}(2). \tag{17.70}$$

Since $\mathbf{L}(1)$ commutes with $\mathbf{L}(2)$, the square of this sum is

$$L^2 = L(1)^2 + 2\mathbf{L}(1) \cdot \mathbf{L}(2) + L(2)^2. \tag{17.71}$$

Since $L(1)^2$ commutes with the components of $\mathbf{L}(1)$ (eq. [17.13]), it commutes with $L_z = L_z(1) + L_z(2)$, and we see from equation (17.71) that it commutes with L^2. Thus the four operators

$$L^2, \quad L_z, \quad L(1)^2, \quad L(2)^2 \tag{17.72}$$

all commute with each other. (Note that $L_z(1)$ does not commute with L^2, because eq. [17.71] contains the y and z components of $\mathbf{L}(1)$.)

Because the four operators in equation (17.72) commute with each other, they have a complete set of simultaneous eigenfunctions. These eigenfunctions are labeled by the quantum numbers l, m, l_1, and l_2, with $|m| \leq l$ as usual.

Another complete set of commuting observables is

$$L(1)^2, \quad L_z(1), \quad L(2)^2, \quad L_z(2). \tag{17.73}$$

The simultaneous eigenfunctions of these operators have eigenvalues l_1, m_1, l_2, and m_2.

Now the question arises, if the system has quantum numbers l_1 and l_2 for the angular momenta of the individual particles, what are the possible values of the total angular momentum quantum number, l? The answer is that l can be in the range

$$|l_1 - l_2| \leq l \leq l_1 + l_2. \tag{17.74}$$

This is sometimes called the triangle rule. It resembles the classical picture, in which the length of the sum of two vectors, $\mathbf{A} + \mathbf{B}$, must be between the sum and difference of the lengths of the two vectors, $A + B$ and $|A - B|$. (Of course this is not literally correct; the square of $\mathbf{L}$ assumes the values $\hbar^2 l(l+1)$, not $\hbar^2 l^2$.) The triangle rule is derived by the following construction.

Simultaneous eigenfunctions of the operators (17.72) may be labeled as $\psi(l, m, l_1, l_2)$. Since we are going to hold l_1 and l_2 fixed, let us drop these quantum numbers and write the eigenfunctions as $\psi(l, m)$. The eigenfunctions of the operators (17.73) similarly will be written $\phi(m_1, m_2)$.

Since the $\phi(m_1, m_2)$ are a complete set of eigenfunctions, we can expand the $\psi(l, m)$ as a linear combination of the $\phi(m_1, m_2)$. Since $L_z = L_z(1) + L_z(2)$, L_z operating on $\phi(m_1, m_2)$ gives $\hbar(m_1 + m_2)$. As usual, L_z operating on $\psi(l, m)$ gives $\hbar m$. Therefore, the expansion of $\psi(l, m)$ can only contain terms with $m = m_1 + m_2$:

$$\psi(l, m) = \sum_{m_1} c(m_1)\phi(m_1, m - m_1). \tag{17.75}$$

Of course, we could equally well write $\phi(m_1, m_2)$ as a linear combination of the $\psi(l, m)$ with $m = m_1 + m_2$:

$$\phi(m_1, m_2) = \sum_{l} d(l)\psi(l, m_1 + m_2). \tag{17.76}$$

The expansion coefficients $c(m_1)$ or $d(l)$ can be found by using the ladder operators for total angular momentum. These ladder operators are defined in equation (17.20). Since $\mathbf{L} = \mathbf{L}(1) + \mathbf{L}(2)$ (eq. [17.70]), these ladder operators are sums of the single-particle ladder operators:

$$L_\pm = L_\pm(1) + L_\pm(2). \tag{17.77}$$

Consider the function $\phi(l_1, l_2)$, where m_1 and m_2 have their largest possible values. We have from $L_z = L_z(1) + L_z(2)$ and equation (17.77)

$$L_z \phi(l_1, l_2) = \hbar(l_1 + l_2)\phi(l_1, l_2), \qquad L_+\phi(l_1, l_2) = 0. \tag{17.78}$$

The first equation says $m = l_1 + l_2$, so $\phi(l_1, l_2)$ expressed as a sum over the $\psi(l, m)$ could only contain terms with $l \geq l_1 + l_2$ (because in general $l \geq m$). The second equation follows because m_1 and m_2 are as large as they can be. But since the z quantum number m cannot be raised, l must be just equal to $l_1 + l_2$:

$$\phi(l_1, l_2) = \psi(l_{\text{max}}, l_{\text{max}}), \quad l_{\text{max}} = l_1 + l_2. \tag{17.79}$$

As indicated, the maximum value of l is $l_{\text{max}} = l_1 + l_2$, because that is the maximum value of m that can be produced out of the $\phi(m_1, m_2)$. This is the right-hand side of the triangle rule (17.74).

The result of operating on equation (17.79) with L_- is

$$\psi(l_{\text{max}}, l_{\text{max}} - 1) = a\phi(l_1 - 1, l_2) + b\phi(l_1, l_2 - 1). \tag{17.80}$$

Here $m = l_{\text{max}} - 1$, because L_- lowers the value of m in ψ. By equation (17.77), the right-hand side is a linear combination of terms with m_1 or m_2 lowered by unity. The constants a and b are determined by the way the $L_\pm(1, 2)$ operate. (Specific examples are to be found in problems [II.18], [II.19], and [V.14], and in section 25.) We can choose another pair of constants, c and d, to get a second linear combination that is orthogonal to equation (17.80):

$$\psi(l_{\text{max}} - 1, l_{\text{max}} - 1) = c\phi(l_1 - 1, l_2) + d\phi(l_1, l_2 - 1). \tag{17.81}$$

This has $m = l_{\text{max}} - 1$, so that is the minimum value of l in this linear combination. That is also the maximum value, because (17.81) is orthogonal to the function (17.80) with $l = l_{\text{max}}$. Therefore (17.81) is the eigenfunction with $l = m = l_{\text{max}} - 1$, as indicated in the left side of the equation.

Next, application of L_- to the wave functions in equations (17.80) and (17.81) yields $\psi(l_{\text{max}}, l_{\text{max}} - 2)$ and $\psi(l_{\text{max}} - 1, l_{\text{max}} - 2)$. Each is a linear combination of the three terms $\phi(l_1 - 2, l_2)$, $\phi(l_1 - 1, l_2 - 1)$, and $\phi(l_1, l_2 - 2)$ that $L_- = L_-(1) + L_-(2)$ generates when it acts on the right-hand sides of equations (17.80) and (17.81). Since there are three coefficients in these linear combinations, we can find a third linear combination orthogonal both to $\psi(l_{\text{max}}, l_{\text{max}} - 2)$ and to $\psi(l_{\text{max}} - 1, l_{\text{max}} - 2)$. As before, we see that this new combination has $m = l_1 + l_2 - 2$, and it cannot contain any terms with $l = l_1 + l_2$ or $l = l_1 + l_2 - 1$, because it is orthogonal to these functions, so this new linear combination must be the eigenfunction $\psi(l_{\text{max}} - 2, l_{\text{max}} - 2)$.

This continues until we get to $l = l_{\text{min}} = l_1 - l_2$ if $l_1 \geq l_2$. The eigenfunction $\psi(l_{\text{min}}, l_{\text{min}})$ is a linear combination of $\phi(l_1 - 2l_2, l_2)$, $\phi(l_1 - 2l_2 + 1, l_2 - 1)$, and so on to $\phi(l_1, -l_2)$. When L_- is applied to this linear combination, it does not increase the number of terms in the sum, because in the last term $m_2 = -l_2$ is as low as it can be. Thus, we cannot find a new linear combination orthogonal to the ones already obtained by applications of L_-, so we cannot find a new l less than l_{min}. This is the second part of the triangle rule (17.74).

18 Single Particle in a Central Potential

It was shown in section 12 that Schrödinger's equation for a two-body system such as a hydrogen atom, where the potential is a function only of the distance between the particles, can be separated into an equation describing the free motion of the center of mass and a one-body Schrödinger equation with the Hamiltonian

$$H = \frac{p^2}{2m} + V(\mathbf{r}), \qquad \mathbf{p} = -i\hbar\nabla. \tag{18.1}$$

The reduced mass is m, and the derivative is with respect to the relative position $\mathbf{r}$ of the two particles. As discussed in the last section, if the potential V is spherically symmetric H commutes with the angular momentum, $\mathbf{L} = \mathbf{r} \times \mathbf{p}$. That means we can seek a complete set of simultaneous eigenfunctions of H, L^2, and L_z for the wave function for the relative position of the two particles. It will be shown here that this leaves us with a one-dimensional Schrödinger equation in the variable $r = |\mathbf{r}|$, with potential energy that is the sum of $V(r)$ and a centrifugal term, in close analogy to what is done in classical mechanics.

Hamiltonian in Polar Coordinates

The first step is to write the kinetic energy part of the Hamiltonian as the sum of radial and angular parts. It will be seen that the latter is proportional to the square of the angular momentum, just as in classical physics, though here L^2 is an operator. We need to express the Laplacian ∇^2 in the polar coordinates of equation (17.42). This can be worked out using the differential calculus, but we will use an operator method, as follows.

To shorten the equations, let

$$\Lambda = \mathbf{r} \times \nabla, \tag{18.2}$$

so $\mathbf{L} = -i\hbar\Lambda$. As discussed in section 2 (eqs. [2.7] and [2.8]), the vector identities

$$\begin{aligned} \mathbf{A} \cdot (\mathbf{B} \times \mathbf{C}) &= (\mathbf{A} \times \mathbf{B}) \cdot \mathbf{C}, \\ \mathbf{A} \times (\mathbf{B} \times \mathbf{C}) &= \mathbf{B}(\mathbf{A} \cdot \mathbf{C}) - \mathbf{C}(\mathbf{A} \cdot \mathbf{B}), \end{aligned} \tag{18.3}$$

apply to vector functions of position and the gradient operator ∇ if we are careful not to change the order of factors. Also,

$$\mathbf{r} \times \nabla = -\nabla \times \mathbf{r}, \tag{18.4}$$

because the products of the components of position, $\mathbf{r}$, and of ∇ in this cross product always belong to different coordinate axes. With these identities, we can write the square of Λ as

$$\begin{aligned} \Lambda^2 &= -\nabla \times \mathbf{r} \cdot \mathbf{r} \times \nabla \\ &= -\nabla \cdot \mathbf{r} \times \ (\mathbf{r} \times \nabla) \\ &= -\nabla \cdot [\mathbf{r}(\mathbf{r} \cdot \nabla) - r^2 \nabla]. \end{aligned} \tag{18.5}$$

Now differentiate out the products of the derivative operators, remembering that, for any function $f(\mathbf{r})$,

$$\frac{\partial}{\partial x} x f(\mathbf{r}) = x \frac{\partial f(\mathbf{r})}{\partial x} + f(\mathbf{r}), \tag{18.6}$$

and that

$$\nabla r^2 = 2\mathbf{r}, \qquad \nabla \cdot \mathbf{r} = 3, \tag{18.7}$$

to get

$$\Lambda^2 = -3\mathbf{r}\cdot\nabla - (\mathbf{r}\cdot\nabla)(\mathbf{r}\cdot\nabla) + 2\mathbf{r}\cdot\nabla + r^2\nabla^2. \tag{18.8}$$

In polar coordinates, the components of the gradient of a function in the directions of the angles θ and ϕ are perpendicular to the radius vector $\mathbf{r}$, so

$$\mathbf{r}\cdot\nabla f(r,\theta,\phi) = r\frac{\partial f}{\partial r}, \tag{18.9}$$

for any function f. This means

$$\mathbf{r}\cdot\nabla = r\frac{\partial}{\partial r}. \tag{18.10}$$

With this identity, equation (18.8) is

$$\Lambda^2 = r^2\nabla^2 - r\frac{\partial}{\partial r}r\frac{\partial}{\partial r} - r\frac{\partial}{\partial r}. \tag{18.11}$$

On rearranging this and noting that

$$r\frac{\partial}{\partial r}r\frac{\partial}{\partial r} = r^2\frac{\partial^2}{\partial r^2} + r\frac{\partial}{\partial r}, \tag{18.12}$$

we get finally

$$\begin{aligned}\nabla^2 &= \frac{\Lambda^2}{r^2} + \frac{\partial^2}{\partial r^2} + \frac{2}{r}\frac{\partial}{\partial r} \\ &= \frac{\Lambda^2}{r^2} + \frac{1}{r}\frac{\partial^2}{\partial r^2}r.\end{aligned} \tag{18.13}$$

The second line follows if you recall that

$$\frac{\partial^2 AB}{\partial r^2} = \frac{\partial^2 A}{\partial r^2}B + 2\frac{\partial A}{\partial r}\frac{\partial B}{\partial r} + A\frac{\partial^2 B}{\partial r^2}. \tag{18.14}$$

With $\mathbf{p} = -i\hbar\nabla$ and $\mathbf{L} = -i\hbar\Lambda$ (eq. [18.2]), equation (18.13) is

$$\mathbf{p}^2 = \frac{\mathbf{L}^2}{r^2} - \frac{\hbar^2}{r}\frac{\partial^2}{\partial r^2}r. \tag{18.15}$$

The Hamiltonian (18.1) is then

$$H = -\frac{\hbar^2}{2mr}\frac{\partial^2}{\partial r^2}r + \frac{\mathbf{L}^2}{2mr^2} + V(r). \tag{18.16}$$

This is the Hamiltonian operator expressed in polar coordinates. The factor $\mathbf{L}^2$ contains only the angle variables θ and ϕ and derivatives with respect to the angle variables, as one sees in equations (17.45) and (17.48). The radial derivatives all are in the first term in the right-hand side of equation (18.16).

Radial Wave Equation

Since the components of the angular momentum operator, $\mathbf{L}$, contain only angular coordinates and derivatives, it is evident that $\mathbf{L}$ commutes with the first and last terms in the right-hand side of equation (18.16), because these terms contain only the radial coordinate. Since L^2 commutes with the components of $\mathbf{L}$ (eq. [17.13]), H commutes with L^2 and L_z. Therefore, we can seek simultaneous eigenfunctions ψ_{nlm} of H, L^2, and L_z, with eigenvalues E_n, $\hbar^2 l(l+1)$, and $m\hbar$. For these eigenfunctions Schrödinger's equation $H\psi = E\psi$ with H given by equation (18.16) is

$$-\frac{\hbar^2}{2mr}\frac{\partial^2}{\partial r^2}r\psi_{nlm} + \frac{\hbar^2}{2m}\frac{l(l+1)}{r^2}\psi_{nlm} + V(r)\psi_{nlm} = E_n\psi_{nlm}. \tag{18.17}$$

This differential equation involves only the radial coordinate r, so the solutions are functions of r multiplied by arbitrary functions of the angles θ and ϕ. Because ψ_{nlm} is an eigenfunction of L^2 and L_z, the angular function is the spherical harmonic $Y_l^m(\theta, \phi)$ discussed in section 17 (eqs. [17.51] and [17.52]).

It is standard and convenient to write $\psi_{nlm} = Y_l^m(\theta, \phi)u(r)/r$, where $u(r)$ is the radial wave function used in section 6. On multiplying equation (18.17) by r, and making this change of functions, we get

$$-\frac{\hbar^2}{2m}\frac{d^2u}{dr^2} + \frac{\hbar^2}{2m}\frac{l(l+1)}{r^2}u + V(r)u = E_n u. \tag{18.18}$$

This is a one-dimensional Schrödinger equation with the effective potential

$$V_{\text{eff}}(r) = V(r) + \frac{\hbar^2}{2m}\frac{l(l+1)}{r^2}. \tag{18.19}$$

The boundary conditions for a bound state are that the radial wave function $u(r)$ must go to zero at infinity, as usual, and that

$$u(0) = 0, \tag{18.20}$$

because the wave function $\psi \propto u(r)/r$ has to be nonsingular at $r = 0$.

The potential V_{eff} in equation (18.19) is the same as the form one finds in classical mechanics for the radial equation of motion of a particle, with the square of the angular momentum of the particle equal to $\hbar^2 l(l+1)$. If l is not zero, the effective potential diverges at $r \to 0$, and this suppresses $u(r)$ near the origin. This is the analog of centrifugal repulsion in classical mechanics.

Hydrogen Atom

The potential energy here is $-e^2/r$, so the radial Schrödinger equation (18.18) is

$$-\frac{\hbar^2}{2m}\frac{d^2u}{dr^2} + \frac{\hbar^2}{2m}\frac{l(l+1)}{r^2}u - \frac{e^2u}{r} = Eu = -Uu(r). \tag{18.21}$$

Since bound state energies are negative in this potential well, the last step sets $U = -E$. This generalizes equation (6.20) to states that are not spherically symmetric.

We can rewrite the Schrödinger equation (18.21) as

$$\frac{d^2u}{dr^2} - \frac{l(l+1)u}{r^2} + \frac{2me^2u}{\hbar^2 r} = \frac{2mUu}{\hbar^2}. \tag{18.22}$$

The eigenvalues U are found by the following trick. Write the radial wave function as

$$u(r) = v(r)\exp -r(2mU)^{1/2}/\hbar, \tag{18.23}$$

where v is a new function of r (that will be seen to be a polynomial). This brings equation (18.22) to

$$\frac{d^2v}{dr^2} - \frac{2(2mU)^{1/2}}{\hbar}\frac{dv}{dr} - \frac{l(l+1)v}{r^2} + \frac{2me^2}{\hbar^2}\frac{v}{r} = 0. \tag{18.24}$$

Next, seek a series solution for the function $v(r)$:

$$v(r) = \sum_{p>0} A_p r^p. \tag{18.25}$$

The series has to start at $p = 1$ because of the boundary condition (18.20). Substitute this series into equation (18.24), collect common powers of r, and note that the coefficient of each power of r has to vanish. The coefficient of r^{p-1} is

$$[p(p+1) - l(l+1)]A_{p+1} = \left[\frac{2p(2mU)^{1/2}}{\hbar} - \frac{2me^2}{\hbar^2}\right] A_p. \tag{18.26}$$

This is an iteration equation for the coefficients A_p.

We see from equation (18.26) that $A_l = 0$, because the left side vanishes for $p = l$. By iteration of this equation it follows that $A_p = 0$ for all $p \leq l$.[1]

Now let us choose A_{l+1} to be some nonzero number. Then equation (18.26) fixes A_{l+2}, that fixes A_{l+3}, and so on. If the right-hand side of equation (18.26) does not vanish for any $p > l$, then the iteration equation at large p approaches

$$A_p = \frac{2(2mU)^{1/2}}{\hbar} \frac{A_{p-1}}{p}. \tag{18.27}$$

This says

$$A_p \propto \left(\frac{2(2mU)^{1/2}}{\hbar}\right)^p \frac{1}{p!}, \tag{18.28}$$

at large p, so the power series (18.25) is

$$v \sim \sum \left(\frac{2r(2mU)^{1/2}}{\hbar}\right)^p \frac{1}{p!} = \exp 2r(2mU)^{1/2}/\hbar. \tag{18.29}$$

The positive argument in this exponential is twice the magnitude of the negative argument in the exponential in equation (18.23), so the radial wave function $u(r)$ in equation (18.23) diverges at $r \to \infty$, which is not allowed for a bound state.

The iteration of equation (18.26) must therefore terminate because at some integer $p = n > l$ the right-hand side of equation (18.26) vanishes:

$$\frac{2n(2mU)^{1/2}}{\hbar} = \frac{2me^2}{\hbar^2}. \tag{18.30}$$

[1] The other possibility is that the expression in brackets in the right-hand side of equation (18.26) vanishes at some value of $p < l$, but we want to save that case to terminate the iteration series at some value of p greater than $l + 1$.

This equation says the energy levels are

$$U = \frac{e^4 m}{2\hbar^2}\frac{1}{n^2} \qquad n = 1, 2, 3, \ldots, \tag{18.31}$$

in agreement with the Bohr model in section 4, and, apart from small relativistic corrections to be discussed in chapters 5 and 8, with the measured energies.

With equation (18.31) for U, equation (18.23) for the radial wave function becomes

$$u(r) = v(r)e^{-r/(na_o)}, \tag{18.32}$$

where the Bohr radius is

$$a_o = \hbar^2/(me^2). \tag{18.33}$$

The iteration of equation (18.26) terminates at $A_p = 0$ for $p > n$, so $v(r)$ in equation (18.32) is a polynomial, with the powers of r ranging from r^{l+1} to r^n. Thus the wave function $\psi \propto u(r)/r$ is the product of an exponential and a polynomial with powers of r ranging from r^l to r^{n-1}.

The hydrogen wave functions for $n = 1$ and $n = 2$ are

$$\begin{aligned}
&\psi_{1,0,0} \propto e^{-r/a_o}, \\
&\psi_{2,0,0} \propto [1 - r/(2a_o)]e^{-r/(2a_o)}, \\
&\psi_{2,1,1} \propto r\sin\theta e^{i\phi}e^{-r/(2a_o)} = (x + iy)e^{-r/(2a_o)}, \\
&\psi_{2,1,0} \propto r\cos\theta e^{-r/(2a_o)} = ze^{-r/(2a_o)}, \\
&\psi_{2,1,-1} \propto r\sin\theta e^{-i\phi}e^{-r/(2a_o)} = (x - iy)e^{-r/(2a_o)}.
\end{aligned} \tag{18.34}$$

The energy eigenstates are labeled by three quantum numbers, n, l, and m. The integer n is called the principal quantum number; it determines the energy (eq. [18.31]), apart from relativistic corrections. The second and third indices are the angular momentum quantum numbers, l and m, for L^2 and L_z. The wave function $\psi_{1,0,0}$ was obtained in section 6. The scale length in the exponential in the $n = 2$ functions is two Bohr radii (eq. [18.32]). The polynomials in r multiplying the exponentials are found by working out the iteration of equation (18.26). For $n = 2$ the polynomial is a simple linear expression. The angular functions are the spherical harmonics obtained from equations (17.48) and (17.50). It is sometimes useful to know that the functions have simple forms in terms of the cartesian position components x, y, and z.

It is traditional to call the states with $l = 0$ s-wave states, the s standing for "sharp," for reasons that are lost in the mists of time. States with $l = 1$ are p-waves, for "principal," and those with $l = 2$ are d-waves, for "diffuse." States with $l = 3$ and 4 are labeled f and g; these do not seem to stand for anything. The ground state of a hydrogen is called the $1S$ state. There are four degenerate states with principal quantum number $n = 2$, that is, four states with the same energy in the nonrelativistic Hamiltonian in equation (18.1). One is the spherically symmetric $2S$ state with $l = 0$, and the other three are the $2P$ states with $l = 1$ and $m = -1$, 0, and 1. The degeneracy of these four states is partially removed by relativistic corrections to the Hamiltonian, and can be completely removed by the effects of external perturbations, examples of which are discussed beginning in section 30 below. At principal quantum number $n = 3$ there is a $3S$ state, three $3P$ states, and five $3D$ states.

19 Particle in an Electromagnetic Field*

The discussion of the Hamiltonian for a charged particle in a given electromagnetic field requires rather a lengthy excursion into classical mechanics and electromagnetism; it is presented here for several reasons. First, this Hamiltonian is needed to understand the behavior of atoms and molecules, whose structures are determined almost solely by the electromagnetic interaction. Second, the Hamiltonian is a good example of canonical quantization. Third, it is a beautiful example of a symmetry that has no classical analog, that reflects the fact that all the wave functions for a system can be multiplied by a phase factor $e^{i\chi(\mathbf{r})}$ without affecting the inner products, (ψ, ϕ), out of which one computes the predictions of the theory.

One arrives at the operators and operator algebra for a particle in an electromagnetic field by the so-called canonical quantization procedure, in which the canonical momenta of classical Hamiltonian mechanics are replaced with derivative operators. We must begin therefore with a review of the classical theory of the motion of a charged particle in an electromagnetic field.

Equations of Motion in an Electromagnetic Field

The classical Lagrangian of a nonrelativistic particle moving in a scalar potential field $V(\mathbf{r})$ is the difference between the kinetic and potential energies,

$$L = m\dot{\mathbf{r}}^2/2 - V(\mathbf{r}). \tag{19.1}$$

More generally, the Lagrangian is some function of position, $\mathbf{r}$, velocity, $\dot{\mathbf{r}} = d\mathbf{r}/dt$, and time, t. The action of a particle moving from given initial to final positions is the time integral of the Lagrangian along the orbit $\mathbf{r}(t)$,

$$I = \int_i^f dt\ L(\mathbf{r}, \dot{\mathbf{r}}, t). \tag{19.2}$$

The action principle of classical mechanics says the path $\mathbf{r}(t)$ the particle actually follows is the one that makes the action stationary, unchanged to first order by small changes of the orbit.

Suppose $\mathbf{r}(t)$ is the true orbit, and consider the neighboring path, $\mathbf{r}(t) + \delta\mathbf{r}(t)$, with $\delta\mathbf{r} = 0$ at the initial and final times (because the initial and final positions are supposed to be fixed). The action principle says the difference between the values of I for the true and neighboring paths vanishes to first order in $\delta\mathbf{r}$, because the action is at a stationary point. The difference in I is, to order $\delta\mathbf{r}$,

$$\begin{aligned} \delta I &= \int dt\ [L(\mathbf{r} + \delta\mathbf{r}, \dot{\mathbf{r}} + \delta\dot{\mathbf{r}}, t) - L(\mathbf{r}, \dot{\mathbf{r}}, t)] \\ &= \int dt\ \left[\frac{\partial L}{\partial r_\alpha}\delta r_\alpha + \frac{\partial L}{\partial \dot{r}_\alpha}\delta\dot{r}_\alpha\right]. \end{aligned} \tag{19.3}$$

As usual, the repeated index α is summed from 1 to 3, for the three components of position and velocity. Because derivatives can be taken in either order we have

$$\delta\dot{r}_\alpha = \frac{d\delta r_\alpha}{dt}. \tag{19.4}$$

With this equation, we can write the second term in the right-hand side of equation (19.3) as

$$\frac{\partial L}{\partial \dot{r}_\alpha}\delta\dot{r}_\alpha = \frac{d}{dt}\left[\frac{\partial L}{\partial \dot{r}_\alpha}\delta r_\alpha\right] - \delta r_\alpha \frac{d}{dt}\frac{\partial L}{\partial \dot{r}_\alpha}. \tag{19.5}$$

The first term on the right side of this equation is a total time derivative, so its time integral is the difference of values at the end points. This vanishes because $\delta\mathbf{r}$ vanishes at the end points. We are left with

$$0 = \delta I = \int dt \left(\frac{\partial L}{\partial r_\alpha} - \frac{d}{dt}\frac{\partial L}{\partial \dot{r}_\alpha} \right) \delta r_\alpha. \tag{19.6}$$

Since δI has to vanish for arbitrary infinitesimal $\delta\mathbf{r}(t)$, the quantity multiplying $\delta\mathbf{r}(t)$ has to vanish. This gives the Euler-Lagrange equation of motion,

$$\frac{\partial L}{\partial r_\alpha} = \frac{d}{dt}\frac{\partial L}{\partial \dot{r}_\alpha}. \tag{19.7}$$

The canonical momentum belonging to the coordinate r_α is defined as

$$p_\alpha = \frac{\partial L}{\partial \dot{r}_\alpha}, \tag{19.8}$$

and the Euler-Lagrange equation (19.7) says

$$\frac{dp_\alpha}{dt} = \frac{\partial L}{\partial r_\alpha}. \tag{19.9}$$

The Hamiltonian is defined to be

$$H(\mathbf{r}, \mathbf{p}, t) = p_\alpha \dot{r}_\alpha - L. \tag{19.10}$$

As indicated, one is supposed to solve equation (19.8) for the velocities as functions of the positions and momenta, so the Hamiltonian can be written as a function of positions and momenta. To get Hamilton's equations of motion, one writes down the effect on equation (19.10) of infinitesimal changes $\delta r_\alpha(t)$ and $\delta p_\alpha(t)$ in the positions and momenta:

$$\delta H = \dot{r}_\alpha \delta p_\alpha + p_\alpha \delta \dot{r}_\alpha - \frac{\partial L}{\partial r_\alpha}\delta r_\alpha - \frac{\partial L}{\partial \dot{r}_\alpha}\delta \dot{r}_\alpha. \tag{19.11}$$

On using equation (19.8) to eliminate the second and last terms on the right-hand side of this equation, and equation (19.7) to rewrite the third, we get

$$\delta H = \dot{r}_\alpha \delta p_\alpha - \dot{p}_\alpha \delta r_\alpha. \tag{19.12}$$

This gives Hamilton's equations of motion,

$$\dot{r}_\alpha = \frac{\partial H}{\partial p_\alpha}, \qquad \dot{p}_\alpha = -\frac{\partial H}{\partial r_\alpha}. \tag{19.13}$$

The classical Lagrangian for a charged particle in an electromagnetic field is written in terms of the scalar and vector potentials of electromagnetism, so we should review next how these are defined.

Maxwell's equation $\nabla \cdot \mathbf{B} = 0$ (eq. [2.3]) is satisfied by writing the magnetic field as the curl of the vector potential field $\mathbf{A}$:

$$\mathbf{B} = \nabla \times \mathbf{A}, \tag{19.14}$$

because the divergence of a curl vanishes identically. The induction equation,

$$\nabla \times \mathbf{E} + \frac{1}{c}\frac{\partial \mathbf{B}}{\partial t} = 0, \tag{19.15}$$

with equation (19.14), is

$$\nabla \times \left(\mathbf{E} + \frac{1}{c}\frac{\partial \mathbf{A}}{\partial t} \right) = 0. \tag{19.16}$$

This equation is satisfied by writing the vector field in the parenthesis as the gradient of a scalar potential field, ϕ, because the curl of a gradient identically vanishes. The result is

$$\mathbf{E} = -\nabla \phi - \frac{1}{c}\frac{\partial \mathbf{A}}{\partial t}. \tag{19.17}$$

Equations (19.14) and (19.17) give the electric and magnetic fields in terms of the scalar and vector potentials, ϕ and $\mathbf{A}$. The potentials are not unique: the gauge transformation

$$\phi' = \phi + \frac{\partial \chi}{\partial t}, \qquad \mathbf{A}' = \mathbf{A} - c\nabla \chi, \tag{19.18}$$

where χ is any function of position and time, give the same $\mathbf{E}$ and $\mathbf{B}$ as ϕ and $\mathbf{A}$.

Now let us write down the Lagrangian for a particle with mass m and charge q in an electromagnetic field, and then check that it gives the familiar equation of motion. The form for the action (that gives orbits that agree with experiment) is

$$I = \int \left(-mc^2 ds + \frac{q}{c} A_i dx^i \right), \tag{19.19}$$

where A_i is the four-vector potential of electromagnetism. As discussed in chapter 8, in special relativity the invariant time interval is (eq. [52.3])

$$ds = (dt^2 - dr^2/c^2)^{1/2} = dt\,(1 - \dot{r}^2/c^2)^{1/2}, \tag{19.20}$$

with $\dot{\mathbf{r}} = d\mathbf{r}/dt$. This is the proper time interval that would be measured by an observer at rest on the particle. The expression $A_i dx^i$, which is summed over $i = 0$ (for time) to 3, also is an invariant, so the action (19.19) is invariant under Lorentz transformations.

In the nonrelativistic limit, we can expand ds as a series in $\dot{r}^2/c^2$, and keep only the first nontrivial term. Dropping the constant term mc^2, which only adds a constant to the action, we can write the action (19.19) in this limit as

$$I = \int dt \left(\frac{1}{2} m\dot{r}^2 + \frac{q}{c}(\mathbf{A} \cdot \dot{\mathbf{r}} + A_0) \right), \tag{19.21}$$

where $\mathbf{A}$ is the space part of the four-vector A_i, and A_0 is the time part. On setting $A_0 \equiv -\phi c$, we see that the integrand is

$$L = \frac{m\dot{\mathbf{r}}^2}{2} - q\phi + \frac{q}{c}\mathbf{A} \cdot \dot{\mathbf{r}}. \tag{19.22}$$

This is the Lagrangian for a nonrelativistic particle of mass m and charge q in the scalar and vector potentials ϕ and $\mathbf{A}$ for the electromagnetic field.

The canonical momentum (eq. [19.8]) is

$$p_\alpha = \frac{\partial L}{\partial \dot{r}_\alpha} = m\dot{r}_\alpha + qA_\alpha/c. \tag{19.23}$$

The Euler-Lagrange equations (19.9) are

$$\frac{dp_\alpha}{dt} = \frac{d}{dt}(m\dot{r}_\alpha + qA_\alpha/c) = \frac{\partial L}{\partial r_\alpha} = -q\phi_\alpha + qA_{\beta,\alpha}\dot{r}_\beta/c. \tag{19.24}$$

To shorten the notation a little, partial derivatives with respect to position have been written as

$$\frac{\partial \phi}{\partial r_\alpha} = \phi_{,\alpha}. \tag{19.25}$$

Equation (19.24) contains the total derivative with respect to time of A_α. Since A_α is a function of time and of position $\mathbf{r}(t)$, we have

$$\frac{d}{dt}A_\alpha(t,\mathbf{r}(t)) = \frac{\partial A_\alpha}{\partial t} + \frac{\partial A_\alpha}{\partial r_\beta}\frac{dr_\beta}{dt}. \tag{19.26}$$

This brings equation (19.24) to the wanted form

$$m\frac{d^2 r_\alpha}{dt^2} = -q(\phi_{,\alpha} + \dot{A}_\alpha/c) + q(A_{\beta,\alpha} - A_{\alpha,\beta})v_\beta/c, \tag{19.27}$$

where $v_\beta = \dot{r}_\beta$ is the particle velocity.

The first term on the right-hand side of this equation is the electric field (eq. [19.17]). We can use the permutation symbol (eq. [17.8]) to write the second term as

$$A_{\beta,\alpha} - A_{\alpha,\beta} = \epsilon_{\alpha\beta\gamma}B_\gamma. \tag{19.28}$$

Here $\mathbf{B}$ is the magnetic field (eq. [19.14]), as can be checked by looking at cases. For example, $\alpha = 1$ and $\beta = 2$ gives

$$B_z = \frac{\partial A_y}{\partial x} - \frac{\partial A_x}{\partial y}, \tag{19.29}$$

which is the z component of equation (19.14). Thus the last term in the equation of motion is

$$q\epsilon_{\alpha\beta\gamma}v_\beta B_\gamma/c. \tag{19.30}$$

This is the α component of the cross product $q\mathbf{v}\times\mathbf{B}/c$ (as in eq. [17.8]). The equation of motion (19.27) is therefore

$$m\frac{d^2\mathbf{r}}{dt^2} = q\,(\mathbf{E} + \mathbf{v}\times\mathbf{B}/c). \tag{19.31}$$

This is the Lorentz force law (eq. [2.5]). That is, the Lagrangian (19.22) gives the wanted equation of motion.

Finally, we need the Hamiltonian. The result of substituting equation (19.22) for the Lagrangian into equation (19.10) for the Hamiltonian, and using equation (19.23) to replace the velocities with the canonical momenta, is

$$H = \frac{1}{2m}(\mathbf{p} - q\mathbf{A}/c)^2 + q\phi. \tag{19.32}$$

This is the Hamiltonian for a nonrelativistic particle with mass m and charge q in an electromagnetic field with vector potential $\mathbf{A}$ and scalar potential ϕ.

Canonical Quantization

A formal guide to the operator algebra in a quantum system is canonical quantization, in which the canonical momenta in the classical theory are replaced with operators that satisfy the canonical commutation relations,

$$[x_i, p_j] = i\hbar\delta_{ij}. \tag{19.33}$$

Here the x_i are the coordinates, and the operators p_i replace the canonical momenta (eq. [19.8]) of the classical theory. In the general formalism presented in chapter 3 below, the coordinates and momenta both are treated as operators. In the wave mechanics of this chapter, the x_i are the particle coordinates x, y, and z, and the canonical commutation relations are obtained by writing the components of the momentum operator in terms of the derivative operators, as

$$\mathbf{p} = -i\hbar\nabla. \tag{19.34}$$

As we have seen (eq. [17.5]), this gives the canonical commutation relations (19.33). Finally, the quantum Hamiltonian is obtained by replacing the momenta in the classical Hamiltonian with the momentum operators, with the factors so arranged that the quantum Hamiltonian is self-adjoint. The motivation behind this canonical prescription is that it automatically makes the forms for time derivatives of expectation values agree with the classical expressions we know are experimentally successful in the classical limit. An example of how this works will be presented shortly.

The quantum Hamiltonian for a particle in an electromagnetic field with vector and scalar potentials $\mathbf{A}$ and ϕ is obtained by replacing the momenta $\mathbf{p}$ in equation (19.32) with the momentum operators in equation (19.34). If there is no magnetic field, we can choose a gauge in which $\mathbf{A} = 0$, and the Hamiltonian is the form already discussed for a particle with potential energy $V = q\phi$. If $\mathbf{A}$ is not zero, the kinetic energy part of the Hamiltonian could be multiplied out in two ways, as

$$\mathbf{p}^2 - \frac{q}{c}(\mathbf{p} \cdot \mathbf{A} + \mathbf{A} \cdot \mathbf{p}) + \frac{q^2}{c^2}\mathbf{A}^2, \tag{19.35}$$

or as

$$\mathbf{p}^2 - 2\frac{q}{c}\mathbf{p}\cdot\mathbf{A} + \frac{q^2}{c^2}\mathbf{A}^2. \tag{19.36}$$

In classical physics these two forms are the same; in quantum mechanics they can be different because the derivative operator need not commute with $\mathbf{A}$. The form that works is the first, which is equivalent to writing the kinetic energy term as the square of the operator $(\mathbf{p}-q\mathbf{A}/c)$ divided by $2m$. The second form (19.36) is unacceptable because it is not self-adjoint in general.

A few of the many applications of the Hamiltonian (19.32) are presented later. As a first example, let us compare the time rates of change of the expectation values of position and momentum with the corresponding classical expressions. Ehrenfest's theorem (eq. [14.33]) says the time rate of change of the expectation value of the observable Q is

$$i\hbar\frac{d\langle Q\rangle}{dt} = \langle[Q,H]\rangle. \tag{19.37}$$

To evaluate the commutators of the position and momentum operators $\mathbf{r}$ and $\mathbf{p}$ with H we need the identities

$$\begin{aligned} [A,B^2] &= [A,B]B + B[A,B], \\ [\mathbf{p},F(\mathbf{r})] &= -i\hbar\nabla F(\mathbf{r}), \end{aligned} \tag{19.38}$$

where F is a function of position $\mathbf{r}$, and $\mathbf{p}$ is the derivative operator in equation (19.34). The first commutation relation can be checked by writing out the commutators. The second follows by the calculation in equation (14.36). With the first of these identities and equation (19.32) for H one finds, after a little rearranging,

$$[r_\alpha, H] = \frac{i\hbar}{m}(p_\alpha - qA_\alpha/c), \tag{19.39}$$

so Ehrenfest's theorem says

$$\frac{d\langle r_\alpha\rangle}{dt} = \langle p_\alpha - qA_\alpha/c\rangle/m. \tag{19.40}$$

If the wave function for the particle is significantly different from zero in a patch of space smaller than the length scale over which the potential $\mathbf{A}(\mathbf{r})$ varies, it is a good approximation to write

$$\langle\mathbf{A}\rangle = \mathbf{A}(\langle\mathbf{r}\rangle, t). \tag{19.41}$$

Then with $\langle \mathbf{r} \rangle \equiv \mathbf{r}(t)$, one sees that equation (19.40) can be rewritten as

$$\langle \mathbf{p} \rangle = m\frac{d\mathbf{r}}{dt} + \frac{q}{c}\mathbf{A}(\mathbf{r}(t), t), \tag{19.42}$$

which can be compared to the classical equation (19.23).

It is left as an exercise to show that Ehrenfest's equation for $d\langle \mathbf{p} \rangle / dt$ agrees with equation (19.24) when the values and gradients of the scalar and vector potentials are close to constant over the region of space occupied by the wave function of the particle, so expectation values may be simplified as in equation (19.41). Once this is done we have checked that the quantum Hamiltonian (19.32) gives sensible equations of motion in the case that the wave function can be treated as a compact wave packet.

Gauge Invariance

Scalar and vector potentials related by a gauge transformation (eq. [19.18]) describe the same electromagnetic field. That is, potentials related by a gauge transformation describe the identical classical situation. This gauge symmetry applies also in quantum mechanics, though it is clear that something new is needed because the Hamiltonian operator (19.32) certainly looks different if the potentials are changed by a gauge transformation. We will see that the new element is that a gauge transformation multiplies the wave function by a phase factor.

Suppose a particle with charge q is described by the wave function ψ, and introduce a new wave function, ψ', by the equation

$$\psi = e^{iq\xi/\hbar}\psi', \tag{19.43}$$

where ξ is a real function of position and time, and the factors q and $\hbar$ have been introduced to simplify the following equations. The momentum operator (19.34) applied to this product gives

$$\mathbf{p}\psi = -i\hbar\nabla(e^{iq\xi/\hbar}\psi') = e^{iq\xi/\hbar}(q\nabla\xi + \mathbf{p})\psi'. \tag{19.44}$$

We have therefore

$$(\mathbf{p} - q\mathbf{A}/c)\psi = e^{iq\xi/\hbar}(\mathbf{p} - q(\mathbf{A} - c\nabla\xi)/c)\psi'. \tag{19.45}$$

The effect of applying this equation twice is

$$(\mathbf{p} - q\mathbf{A}/c)^2\psi = e^{iq\xi/\hbar}(\mathbf{p} - q(\mathbf{A} - c\nabla\xi)/c)^2\psi'. \tag{19.46}$$

The time derivative of ψ is

$$i\hbar\frac{\partial\psi}{\partial t} = e^{iq\xi/\hbar}\left(i\hbar\frac{\partial\psi'}{\partial t} - q\frac{\partial\xi}{\partial t}\psi'\right). \tag{19.47}$$

With the Hamiltonian (19.32), Schrödinger's equation for the original wave function is

$$i\hbar\frac{\partial\psi}{\partial t} = \frac{1}{2m}\left[\mathbf{p} - \frac{q}{c}\mathbf{A}\right]^2\psi + q\phi\psi. \tag{19.48}$$

On using equations (19.46) and (19.47) in this equation, we see that Schrödinger's equation for the new function is

$$i\hbar\frac{\partial\psi'}{\partial t} = \frac{1}{2m}\left[\mathbf{p} - \frac{q}{c}(\mathbf{A} - c\nabla\xi)\right]^2\psi' + q\left(\phi + \frac{\partial\xi}{\partial t}\right)\psi'. \tag{19.49}$$

That is, if ψ satisfies Schrödinger's equation with potentials ϕ and $\mathbf{A}$, then ψ' satisfies Schrödinger's equation with potentials

$$\phi' = \phi + \frac{\partial\xi}{\partial t}, \qquad \mathbf{A}' = \mathbf{A} - c\nabla\xi. \tag{19.50}$$

This is a gauge transformation of the potentials (eq. [19.18]).

The conclusion is that a gauge transformation, $\xi(\mathbf{r}, t)$, of the potentials transforms all the wave functions by a phase factor, $\psi' = e^{-iq\xi/\hbar}\psi$. This phase factor has no effect on inner products (χ, ψ). It has no effect on observables constructed out of the velocity, for we have from equations (19.23), (19.45), and (19.50)

$$(\chi, m\dot{\mathbf{r}}\psi) = (\chi, (\mathbf{p} - q\mathbf{A}/c)\psi) = (\chi', (\mathbf{p} - q\mathbf{A}'/c)\psi'). \tag{19.51}$$

The last expression is $m\dot{\mathbf{r}}$ evaluated in the new gauge for the vector potential and using the new phase factor for the wave functions. That is, the experimental predictions of the theory are not changed by a gauge transformation, as in classical mechanics.

In classical mechanics one can go further: physics can be expressed in terms of the electric and magnetic fields $\mathbf{E}$ and $\mathbf{B}$ alone, eliminating the potentials altogether. This is not so in quantum mechanics, as is illustrated by the Bohm-Aharonov effect, that goes as follows.

Suppose a static magnetic field is confined to the long straight cylinder shown in figure 19.1. Outside the cylinder the electric and magnetic

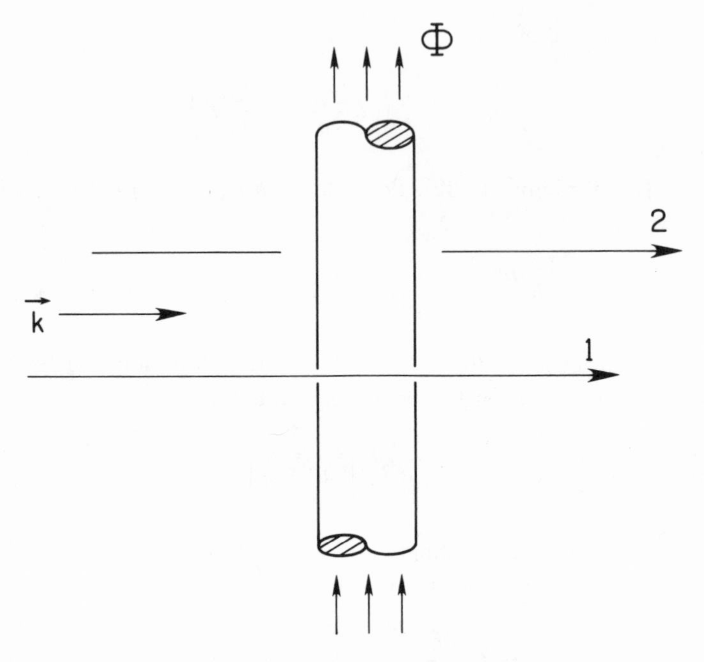

Fig. 19.1 Bohm-Aharonov effect. The magnetic flux Φ is confined to the cylinder. Part of a wave packet with momentum $\sim \hbar\mathbf{k}$ is scattered by the cylinder; we can neglect this scattered wave. Parts of the wave packet move past the cylinder to the left and to the right, on the paths indicated as 1 and 2.

fields are negligibly small. The path integral of the vector potential around a closed path that is outside the cylinder is

$$\oint d\mathbf{r} \cdot \mathbf{A} = \int d\mathbf{a} \cdot \nabla \times \mathbf{A} = \int d\mathbf{a} \cdot \mathbf{B} = \Phi. \qquad (19.52)$$

The second expression uses Stokes's theorem to change the path integral to an integral over a surface bounded by the path. The magnitude of $d\mathbf{a}$ is the element of surface area, and the direction is the normal to the surface. The curl of the vector potential is the magnetic field. As indicated in the last expression, if the path is completely outside the cylinder the path integral is the total magnetic flux Φ within the cylinder, independent of the path. The curl of $\mathbf{A}$ vanishes outside the cylinder, because the magnetic field vanishes, but the field $\mathbf{A}$ itself cannot vanish: under axial symmetry, the magnitude of $\mathbf{A}$ varies inversely with the distance from the cylinder, because the path length is proportional to the distance.

Now consider a wave function representing a particle with charge q moving normal to the cylinder. The particle energy is E, so Schrödinger's equation is

$$E\psi = \frac{1}{2m}(\mathbf{p} - q\mathbf{A}/c)^2\psi. \tag{19.53}$$

We want to compare the phases of the wave computed along the parallel paths 1 and 2 that pass perpendicular to the cylinder and pass on opposite sides of it.

It will be assumed that the de Broglie wavelength is much smaller than the distance between the paths, and that $\mathbf{A}$ acts as a small perturbation to the wave function. In this case, the spatial part of the wave function along each path is well approximated as

$$\psi = e^{i(\mathbf{k}\cdot\mathbf{r}+\chi)}, \tag{19.54}$$

where $\mathbf{k}$ is a constant, that points normal to the cylinder, and χ is the perturbation to the phase due to $\mathbf{A}$. This resembles the WKB approximation in section 7, but here we are keeping track only of the phase.

The result of substituting equation (19.54) into (19.53), recalling that $\mathbf{p} = -i\hbar\nabla$, and keeping only the largest derivatives, which come from differentiating the exponential, is

$$(\hbar\mathbf{k} + \hbar\nabla\chi - q\mathbf{A}/c)^2 = \text{constant}. \tag{19.55}$$

On multiplying this out, and keeping only the first powers of the terms χ and $\mathbf{A}$, which we are treating as small perturbations, we get

$$\mathbf{k}\cdot\nabla\chi - \frac{q}{\hbar c}\mathbf{k}\cdot\mathbf{A} = \text{constant}. \tag{19.56}$$

The net change in phase of the wave function between initial and final positions along one of the paths is

$$\chi = \int_i^f \nabla\chi\cdot d\mathbf{r}. \tag{19.57}$$

We are taking the path segment $d\mathbf{r}$ to be parallel to the momentum $\hbar\mathbf{k}$, so we have from equation (19.56)

$$\chi = \frac{q}{\hbar c}\int_i^f \mathbf{A}\cdot d\mathbf{r} + \text{constant}. \tag{19.58}$$

The difference in phases for paths 1 and 2 is

$$\chi_1 - \chi_2 = \frac{q}{\hbar c} \oint \mathbf{A} \cdot d\mathbf{r} = \frac{q\Phi}{\hbar c}. \tag{19.59}$$

The constant part of equation (19.58) is eliminated from the difference in equation (19.59), and what is left is the path integral of $\mathbf{A}$ running out along path 1 and back along path 2. This is equivalent to a closed path integral. As noted above, this closed path integral is independent of path, as long as it stays outside the cylinder, and amounts to the magnetic flux in the cylinder (eq. [19.52]).

The phase difference $\chi_1 - \chi_2$ in equation (19.59) is measurable by observing the interference of the waves from the two sides of the cylinder. That is, we have the result that although the wave function is excluded from the region where $\mathbf{B}$ is nonzero, there is an observable effect of the magnetic field on the wave function through the vector potential. This is a distinctly nonclassical effect.

It is an interesting exercise to work out the analogous effect for the scalar potential field. Imagine two long parallel metal cylinders that have an electric potential difference that is a function of time. A wave packet for a charged particle approaches the cylinders, is divided so part passes through each cylinder, and the parts then merge after leaving the cylinder. If the potential difference between the cylinders is nonzero only when the wave packet is well inside them, show that in classical physics the applied potential has no effect on the particle motion, while quantum mechanics does predict an effect that is observable in principle.

Yet another point is worth noting. If all the wave functions describing a system are multiplied by a phase factor that is independent of position, it has no effect on any of the inner products $(\psi, Q\phi)$ used in computations, so it can have no effect on any experimental prediction. This trivial symmetry of the theory is called a global phase symmetry. The theory has a local phase symmetry if we can multiply all of the wave functions for the physical system by a phase factor that is some freely chosen function of position and time. This is not trivial, because the momentum operator and the time derivative in Schrödinger's equation differentiate the phase function. To get local phase symmetry, we have to introduce scalar and vector fields, ϕ and $\mathbf{A}$, that absorb these gradients by a gauge transformation of the form of (19.18). As indicated in equations (19.45) and (19.47), the derivatives of the phase function are eliminated in the gauge transformation by requiring that space derivatives appear only in the combination $\mathbf{p} - q\mathbf{A}/c$, where $\mathbf{p} = -i\hbar\nabla$ is the momentum operator, and time derivatives appear only in the combination $i\hbar\partial/\partial t - q\phi$. The simplest Hamiltonian consistent with this condition is equation (19.32). That is, if Nature wanted to have this elegant local phase symmetry, Nature would have to (as it did!) introduce scalar and vector fields with the gauge symmetry (19.18) of electromagnetism, the simplest choice for the Hamiltonian being that of electromagnetism.

Problems

II.1) Suppose the operators P and Q satisfy the commutation relation

$$[P, Q] = a, \tag{II.1}$$

where a is a constant (a number, not an operator).

a) Reduce the commutator

$$[P, Q^n] \tag{II.2}$$

where Q^n means the product of n Qs, to the simplest possible form.

b) Reduce the commutator

$$[P, e^{iQ}] \tag{II.3}$$

to the simplest form. (The operator function e^{iQ} is defined in eq. [13.10].)

II.2) The operators P and Q are self-adjoint and satisfy the commutation relation

$$[P, Q] = ic, \tag{II.4}$$

where c is a real number. Show that the operator $[P, Q]$ is anti-self-adjoint, that is, that the adjoint of the operator is the negative of the operator, consistent with the right-hand side of the equation.

II.3) Let x be the position coordinate for a particle that moves in one dimension, and let $p = -i\hbar\partial/\partial x$ be the usual momentum operator. State whether each of the following operators is self-adjoint, anti-self-adjoint ($A^\dagger = -A$), unitary ($A^\dagger A = 1$), or, if none of the above, what the adjoint is:

$$xxp, \quad xpx, \quad xpp + ppx, \quad \frac{d^2}{dx^2}, \quad \frac{d^3}{dx^3}, \quad e^p. \tag{II.5}$$

II.4) The operator Q satisfies the two equations

$$
\begin{aligned}
Q^\dagger Q^\dagger &= 0, \\
QQ^\dagger + Q^\dagger Q &= 1.
\end{aligned}
\tag{II.6}
$$

The Hamiltonian for a system is

$$H = \alpha QQ^\dagger, \tag{II.7}$$

where α is a real constant.

a) Show H is self adjoint.
b) Find an expression for H^2, the square of H, in terms of H.
c) Find the eigenvalues of H allowed by the result from part (b).

II.5) Two operators, A and B, satisfy the equations

$$
\begin{aligned}
A &= B^\dagger B + 3, \\
A &= BB^\dagger + 1.
\end{aligned}
\tag{II.8}
$$

a) Show A is self-adjoint, $A^\dagger = A$.
b) Find the commutator $[B^\dagger, B]$.
c) Find the commutator $[A, B]$.
d) Suppose ψ is an eigenfunction of A with eigenvalue a:

$$A\psi = a\psi. \tag{II.9}$$

Show that if $B\psi \neq 0$ then $B\psi$ is an eigenfunction of A, and find the eigenvalue.

II.6) Show that if the operator Q satisfies

$$(\phi, Q\phi) = (Q\phi, \phi) \tag{II.10}$$

for all ϕ, then Q is self-adjoint, that is,

$$(\psi, Q\chi) = (Q\psi, \chi) \tag{II.11}$$

for all ψ and χ. Consider the functions

$$\phi_1 = \psi + \chi, \qquad \phi_2 = \psi + i\chi. \tag{II.12}$$

II.7) The Legendre polynomials $P_l(x)$ are a set of real polynomials orthogonal in the interval $-1 < x < 1$:

$$\int_{-1}^{1} dx\, P_l(x) P_{l'}(x) = 0, \qquad l \neq l'. \tag{II.13}$$

The polynomial $P_l(x)$ is of order l, that is, the highest power of x is x^l. It is normalized to $P_l(1) = 1$.

Starting with the set of functions

$$\psi_l(x) = x^l, \quad l = 0, 1, 2, \ldots, \tag{II.14}$$

use the orthogonalization procedure described in section 13 to derive the polynomials P_0, P_1, P_2, and P_3.

II.8) Use Ehrenfest's theorem (eq. [14.33]) to show that the expectation value of the position of a particle that moves in three dimensions with the Hamiltonian

$$H = \frac{p^2}{2m} + V(\mathbf{r}) \tag{II.15}$$

satisfies

$$\frac{d\langle \mathbf{r} \rangle}{dt} = \frac{\langle \mathbf{p} \rangle}{m}. \tag{II.16}$$

II.9) Consider a particle that moves in one dimension. Two of its normalized energy eigenfunctions are $\psi_1(x)$ and $\psi_2(x)$, with energy eigenvalues E_1 and E_2.

At time $t = 0$ the wave function for the particle is

$$\phi = c_1 \psi_1(x) + c_2 \psi_2(x) \quad \text{at} \quad t = 0, \tag{II.17}$$

where c_1 and c_2 are constants.

a) Find the wave function, $\phi(x, t)$, as a function of time, in terms of the given constants and initial condition.

b) Find, and reduce to the simplest possible form, an expression for the expectation value of the particle position, $\langle x \rangle = (\phi, x\phi)$, as a function of time, for the state $\phi(x, t)$ from part (a).

II.10) A particle that moves in three dimensions has the Hamiltonian

$$H = \frac{p^2}{2m} + \alpha(x^2 + y^2 + z^2) + \gamma z, \tag{II.18}$$

where α and γ are real nonzero constant numbers.

a) For each of the following observables, state whether and why the observable is conserved: parity, Π; energy, H; the z component of orbital angular momentum, L_z; the x component of orbital angular momentum, L_x; the z component of linear momentum, p_z.

b) Reduce the expression for the time rate of change of the expectation value of the y component of orbital angular momentum, $d\langle L_y\rangle/dt$, to the simplest possible form. Find the classical analog to the result.

II.11) The ground state wave function of a one-dimensional simple harmonic oscillator is

$$\psi_o(x) \propto e^{-x^2/x_o^2}, \tag{II.19}$$

where x_o is a constant. Given that the wave function of this system at a fixed instant of time is

$$\phi(x) \propto e^{-x^2/y^2}, \tag{II.20}$$

where y is another constant, find the probability that, if the energy is measured, the system will be found to be in the ground state.

II.12) Tritium is a radioactive isotope of hydrogen. The nucleus decays (by emitting an electron and antineutrino), changing from a triton (one proton and two neutrons) to a ^{3}He nucleus (two protons and one neutron). This changes the charge of the nucleus from e (the magnitude of the electron charge) to $2e$. For this problem, it can be assumed that the change is instantaneous. That is, the orbital electron in a tritium atom sees the charge at the nucleus suddenly change from e to $2e$. Finally, since an electron is much less massive than a proton, it can be assumed here that the reduced mass of the atom is unaffected by the decay.

If the orbital electron in the original tritium atom is in the ground state, what is the probability that the electron is in the ground state of the ^{3}He atom immediately after the decay? Note that the electron wave function immediately after the decay is the ground state wave function of a hydrogen atom, because that is what the wave function was immediately before the decay.

II.13) As discussed in problem I.11, in an ammonia molecule the three hydrogen atoms form an equilateral triangle and the nitrogen atom usually is found just above or just below the triangle. The behavior of the position of the nitrogen atom is usefully approximated as the motion of a particle moving in one dimension in the fixed potential in figure I.1, with minima on each side of the triangle and a relatively high barrier at the plane of the triangle. The ground state and first excited state energies, E_0 and E_1, are very nearly equal to each other.

a) To get the wave function for a situation in which the energy is close to E_0 and the atom is almost certainly in one of the minima of the potential energy, consider the functions

$$\psi_t(x) = [(\psi_0(x) + \psi_1(x))]/2^{1/2},$$
$$\psi_b(x) = [(\psi_0(x) - \psi_1(x))]/2^{1/2}. \tag{II.21}$$

Using the fact that E_0 and E_1 are very nearly equal, show that, with appropriate choices of phases for the ground state and first excited state wave functions ψ_0 and ψ_1, the function $\psi_t(x)$ is appreciably different from zero on only one side of the triangle, and $\psi_b(x)$ is appreciably different from zero only on the other side.

b) Show that ψ_t and ψ_b are properly normalized if ψ_0 and ψ_1 are normalized.

c) If the wave function for the molecule is $\psi_t(x)$, what is the probability that the result of the measurement of the energy will be the ground state value, E_0?

d) Given the initial condition that the wave function is $\psi_t(x)$ at time $t = 0$, find the wave function as a function of time in terms of the energy eigenfunctions and eigenvalues.

e) Show that the wave function from part (d) describes oscillations, or inversions, from the initial state, in which the molecule almost certainly is above the triangle, to a state at a later time in which it is below the triangle, and at a later time above the triangle again.

f) The frequency of the inversion oscillation is $\nu = 23000\,\mathrm{MHz} = 2.3 \times 10^{10}\,\mathrm{Hz}$. Find the energy difference, $E_1 - E_0$, in eV, between the ground state and the first excited state.

II.14) Consider a particle that moves in three dimensions with wave function ψ. Use the operator methods in section 17 to show that if ψ has total angular momentum quantum number $l = 0$, then ψ satisfies

$$L_\alpha \psi = 0, \tag{II.22}$$

for all three components L_α of the total angular momentum operator $\mathbf{L}$.

II.15) A particle of mass m is confined to a sphere of radius R by a spherically symmetric potential well of the form

$$\begin{aligned} V(r) &= 0, \quad \text{for} \quad r < r_o, \\ &= V_o, \quad \text{for} \quad r_o \leq r < R, \\ &\to \infty, \quad \text{for} \quad r \geq R. \end{aligned} \tag{II.23}$$

Here V_o, r_o, and R are positive constants.

The particle is in an energy eigenstate with definite orbital angular momentum, with angular momentum quantum number $l > 0$. (That is, the wave function is an eigenfunction of L^2 with eigenvalue $\hbar^2 l(l+1)$.)

a) Sketch the effective potential and the radial wave function for the lowest energy state with the given value of l. As usual, high art is not expected, but you should indicate reasonable curvature and reasonable positions of the zeros of the wave function.

b) At small distance r from the center of the potential well, the radial wave function varies as a power of the radius, $u \propto r^n$. Find n for the given l.

II.16) Use the results in section 17 to find the spherical harmonics Y_1^1, Y_1^0, and Y_1^{-1} as functions of the polar angles θ and ϕ, and as functions of the cartesian position coordinates x, y and z.

II.17) Let the functions $\phi_n(x)$ be the one-dimensional simple harmonic oscillator energy eigenfunctions, with energy $E_n = (n + 1/2)\hbar\omega$. We know from problem (I.9) that the ground state wave function for a spherically symmetric three-dimensional simple harmonic oscillator is

$$\psi = \phi_0(x)\phi_0(y)\phi_0(z), \tag{II.24}$$

and the three degenerate lowest-lying excited states are

$$\begin{aligned} \chi_1 &= \phi_1(x)\phi_0(y)\phi_0(z), \\ \chi_2 &= \phi_0(x)\phi_1(y)\phi_0(z), \\ \chi_3 &= \phi_0(x)\phi_0(y)\phi_1(z). \end{aligned} \tag{II.25}$$

It will be recalled from problem (I.8) that ϕ_0 is a gaussian, $\phi_0(x) \propto e^{-x^2/x_o^2}$, and that $\phi_1(x)$ is proportional to $x\phi_0(x)$.

Use the results from problems (II.14) and (II.16) to show that the ground state, ψ, has angular momentum quantum numbers $l = m = 0$, that χ_3 has quantum numbers $l = 1$, $m = 0$, and that χ_1 and χ_2 have $l = 1$ and are linear combinations of the eigenfunctions with $m = \pm 1$.

II.18) As discussed in section 17, a simultaneous eigenfunction of the angular momentum operators L^2 and L_z can be labeled as ψ_l^m. In the standard sign convention, the raising angular momentum ladder operator satisfies the equation

$$L_+\psi_l^m = \hbar(l^2 + l - m^2 - m)^{1/2}\psi_l^{m+1}. \tag{II.26}$$

The numerical factor on the right-hand side preserves the normalization, $(\psi_l^m, \psi_l^m) = 1$ and $(\psi_l^{m+1}, \psi_l^{m+1}) = 1$ if $m < l$. The sign of the right-hand side is a convention; we could multiply it by a phase factor $e^{i\phi}$.

a) Use the expression for L_-L_+ (eq. [17.32]) and the known results of operating on ψ_l^m with L^2 and L_z to derive the numerical factor on the right side of equation (II.26).

b) Use L_- operating on equation (II.26) to find the normalizing constant c in the equation

$$L_-\psi_l^m = c\psi_l^{m-1}. \tag{II.27}$$

It will be noted that the phase of the constant c in this equation is fixed by the choice of phase in equation (II.26).

II.19) Consider a two-particle system, with angular momentum observables $\mathbf{L}(a)$ and $\mathbf{L}(b)$, and total angular momentum

$$\mathbf{L} = \mathbf{L}(a) + \mathbf{L}(b). \tag{II.28}$$

As discussed in section 17, two complete sets of commuting observables for this system are

$$L(a)^2,\ L_z(a),\ L(b)^2,\ L_z(b), \tag{II.29}$$

with eigenfunctions

$$\phi(l_a, m_a, l_b, m_b), \tag{II.30}$$

and

$$L^2,\ L_z,\ L(a)^2,\ L(b)^2, \tag{II.31}$$

with eigenfunctions

$$\psi(l, m, l_a, l_b). \tag{II.32}$$

There may be other observables that commute with all components of the angular momentum operators. In that case the wave functions in equations (II.30) and (II.32) are completely specified by the angular momentum quantum numbers displayed in the equations along with other eigenvalues of other observables. We will suppose these other eigenvalues are held fixed.

a) By using the equations $L_z = L_z(a) + L_z(b)$ and $L_\pm = L_\pm(a) + L_\pm(b)$, show that the state with $l_a = m_a = 2$ and $l_b = m_b = 1$ is an eigenfunction of the set of observables in equation (II.31) with $l = m = 3$:

$$\phi(2, 2, 1, 1) = \psi(3, 3, 2, 1). \tag{II.33}$$

The indices are placed as in equations (II.30) and (II.32).

b) Using equation (II.33), the ladder operators $L_\pm$, and the results from problem (II.18), find an expression for the state $\psi(3, 2, 2, 1)$ with $l = 3$ and $m = 2$ as a linear combination of the $\phi(2, m_a, 1, m_b)$.

c) Find an expression for the state $\psi(2, 2, 2, 1)$ with $l = 2$ and $m = 2$ as a linear combination of the $\phi(2, m_a, 1, m_b)$. One approach is to seek that linear combination of the ϕ functions with the correct total m value that is orthogonal to the wave function from part (b). The other is to observe that the wanted linear combination has to be annihilated by L_+.

d) Suppose a system has state vector $\chi = \psi(2, 2, 2, 1)$ with $l = 2$ and $m = 2$ and, as before, $l_a = 2$ and $l_b = 1$. Find the probabilities for all possible results of a measurement of the value of the quantum number m_a belonging to the z component of the angular momentum of particle a. Note that the probabilities ought to add up to unity.

II.20) Consider a particle of mass m in a spherically symmetric potential well, $V(r)$, and an applied uniform static magnetic field. The Hamiltonian (eq. [19.32]) is

$$H = \frac{1}{2m}(\mathbf{p} - q\mathbf{A}/c)^2 + V(r), \tag{II.34}$$

where q is the charge of the particle. The magnetic field points along the z axis. As you can check by evaluating $\nabla \times \mathbf{A}$, a vector potential for this field is

$$\begin{aligned} A_x &= -By/2, \\ A_y &= Bx/2. \end{aligned} \tag{II.35}$$

a) Multiply out the square in equation (II.34) and use equation (II.35) to get an expression for H in which the interaction between the particle and the magnetic field is expressed in terms of the angular momentum operator.

b) Show that L_z is conserved in this Hamiltonian.

c) If the magnetic field B is not too large, the term in H that is proportional to B^2 can be ignored. In this approximation, find expressions for the time rates of change of the expectation values $\langle L_x \rangle$ and $\langle L_y \rangle$ of the x and y components of angular momentum. By combining these two equations, find the general solution for $\langle L_x \rangle$ as a function of time.

II.21) A particle with mass m and charge q moves in a uniform static magnetic field B that points along the z axis in a cartesian z, y, z coordinate system. A suitable vector potential (different from the one in eq. [II.35]) is $A_x = 0$, $A_y = Bx$, $A_z = 0$, and the corresponding Hamiltonian is

$$H = \frac{1}{2m}\left[p_x^2 + \left(p_y - \frac{q}{c}Bx\right)^2 + p_z^2\right]. \tag{II.36}$$

a) Find the components of the momenta p_α that are conserved, and use the results to reduce the Schrödinger energy eigenvalue problem to a one-dimensional equation.

b) By noting that the result from part (a) is a simple harmonic oscillator Schrödinger equation, find the energy eigenvalues. Show how the results compare to the classical cyclotron frequency (the frequency of rotation of a classical charged particle moving in a circle at speed v in a magnetic field B).

II.22) Fun with a one-dimensional simple harmonic oscillator.

a) Show that the oscillator raising operator $a^\dagger$ defined in equation (6.34) is the adjoint of the lowering operator a.

b) The Hamiltonian for the oscillator is given by the equation (eqs. [6.37] and [6.38]):

$$H = a^\dagger a + \hbar\omega/2 = aa^\dagger - \hbar\omega/2. \tag{II.37}$$

Using the result from part (a), show that these expressions are self-adjoint.

c) Let the state vector for the mth energy level of the oscillator be $\psi_m(x)$, and suppose the state vectors are normalized, $(\psi_m, \psi_m) = 1$. We saw that the raising operator gives (eq. [6.46])

$$a^\dagger \psi_n = N\psi_{n+1}. \tag{II.38}$$

Use operator methods to find the normalizing constant, N.

d) Use the definitions of a and $a^\dagger$ (eqs. [6.34]) to write the position and momentum observables x and p as linear combinations of a and $a^\dagger$. Use operator methods and the above results to evaluate, for arbitrary (nonnegative) integers m, n, the inner products

$$(\psi_n, x\psi_m), \qquad (\psi_n, p\psi_m), \tag{II.39}$$

where as usual the momentum operator is

$$p = -i\hbar\frac{d}{dx}. \tag{II.40}$$

e) Use the completeness of the ψ_n to show that

$$(\xi, AB\phi) = \sum_n (\xi, A\psi_n)(\psi_n, B\phi), \tag{II.41}$$

where A and B are operators. The trick is to use equation (13.58) to write the function $B\phi$ as a sum over the ψ_n, operate on this sum with A, and then take the inner product of the result with ξ.

f) Use the results from parts (d) and (e) to compute

$$\langle x^2 \rangle = (\psi_n, x^2\psi_n), \qquad \langle p^2 \rangle = (\psi_n, p^2\psi_n). \tag{II.42}$$

Find from this the standard deviations σ_x and σ_p (eq. [I.15]) in the position and momentum in the nth energy level. The result for $n = 0$ ought to agree with equation (I.16).

II.23) More fun: coherent states of an oscillator.

Part (d) from the last problem shows that in an energy eigenstate the expectation value of the position x of the oscillator vanishes: the displacement from equilibrium is as likely to be found to be negative as positive. States in which the predicted value of x varies with time like a classical oscillator are constructed as follows.

a) Use the commutation relation $[a, a^\dagger] = \hbar\omega$ (eq. [6.39]) and the results from the last problem with equation (II.3) to reduce to the simplest possible form the commutator

$$[a, e^{\lambda a^\dagger}], \tag{II.43}$$

where λ is a complex constant (not an operator).

b) Consider the function

$$\phi(\lambda, x) = e^{\lambda a^\dagger}\psi_0(x), \tag{II.44}$$

where $\psi_0(x)$ is the ground state wave function, so $a\psi_0 = 0$. Use the result from part (a) to show that ϕ is an eigenfunction of a with eigenvalue $\lambda\hbar\omega$.

c) Use the result from part (b) to show

$$e^{\lambda^* a}\phi = K\phi, \tag{II.45}$$

where K is a constant, and find K.

d) Evaluate (ϕ, ϕ), where ϕ is defined by equation (II.44), and the function ψ_0 is normalized to $(\psi_0, \psi_0) = 1$.

e) Use operator methods to find the expectation values of position and momentum in the state with wave function $\phi(\lambda, x)$ in equation (II.44),

$$\langle x \rangle = \frac{(\phi, x\phi)}{(\phi, \phi)}, \qquad \langle p \rangle = \frac{(\phi, p\phi)}{(\phi, \phi)}, \tag{II.46}$$

for given complex number λ.

f) As discussed in problem (I.16), a measure of the uncertainty in the predicted position is the standard deviation (root mean square deviation) δx, defined by the equation

$$\delta x^2 = \langle (x - \langle x \rangle)^2 \rangle = \langle x^2 \rangle - \langle x \rangle^2. \tag{II.47}$$

Use operator methods to find the standard deviations δx and δp in position and momentum in the wave function $\phi(\lambda, x)$. You ought to find that the product is the same as the result from problem (II.22f) for $n = 0$, and the minimum allowed by the uncertainty principle (to be derived below in problem [III.3]).

g) If λ is large enough ($|\lambda| \gg (\hbar\omega)^{-1/2}$), the function ϕ in equation (II.44), with ψ_0 and $a^\dagger$ independent of time and λ a suitably chosen function of time, is an approximate solution to Schrödinger's equation

$$i\hbar\frac{\partial \phi}{\partial t} = H\phi, \qquad \text{(II.48)}$$

with the Hamiltonian H given in equation (II.37). Find λ as a function of time in this limit, and use $\lambda(t)$ with the result from part (e) to find $\langle x \rangle$ as a function of time.

The state represented by the wave function ϕ in equation (II.44) is called a coherent state. In the limit $|\lambda| \gg (\hbar\omega)^{-1/2}$, the predicted value of the position in this coherent state has a small relative uncertainty and the predicted position varies with time as a classical simple harmonic oscillator. It will be noted that to get this classical behavior we had to make the wave function a linear combination of states with very different values of the quantum number n (because the series expansion of $e^{\lambda a^\dagger}$ in eq. [II.44] contains all powers of the raising operator $a^\dagger$). That is, in the classical limit, where the position and momentum of the oscillator are predicted to high relative accuracy, the energy level quantum number n has a very broad range of possible values.

II.24) Still more fun: squeezed states of an oscillator.

Another interesting wave function for a simple harmonic oscillator is obtained as follows. Consider the operator

$$U(\eta) = \exp\{\eta(a^2 - a^{\dagger 2})\}. \qquad \text{(II.49)}$$

Here η is a real variable (not an operator, and not complex, as is λ in eq. [II.44]).

a) Show U is unitary, $U^\dagger U = 1$.

b) The operator $g_+(\eta)$ is defined by the equation

$$g_+(\eta) = U(\eta)(a + a^\dagger)U(\eta)^\dagger = e^{\eta(a^2 - a^{\dagger 2})}(a + a^\dagger)e^{-\eta(a^2 - a^{\dagger 2})}. \qquad \text{(II.50)}$$

Show g_+ satisfies

$$\frac{dg_+}{d\eta} = 2\hbar\omega g_+(\eta). \tag{II.51}$$

c) Find $dg_-/d\eta$ for the operator

$$g_-(\eta) = U(\eta)(a - a^\dagger)U(\eta)^\dagger. \tag{II.52}$$

d) Use the results from parts (b) and (c) to show

$$U(a \pm a^\dagger)U^\dagger = e^{\pm 2\eta\hbar\omega}(a \pm a^\dagger). \tag{II.53}$$

e) The wave function of a squeezed state is

$$\chi(\eta, x) = U^\dagger\psi_0(x) = e^{-\eta(a^2 - a^{\dagger 2})}\psi_0(x). \tag{II.54}$$

As in equation (II.44), the wave function for the ground state of the oscillator is $\psi_0(x)$.

By writing x and p in terms of a and $a^\dagger$, as in problems (II.22) and (II.23), and using the above results for $g_\pm(\eta)$, find the expectation values of x, x^2, p, and p^2 in the state χ (eq. [II.54]). If all goes well, you ought to find that the standard deviations (eq. [II.47]) in position and momentum satisfy $\delta x \delta p = \hbar/2$, and that in the squeezed state at large enough η the uncertainty δp in the momentum can be made arbitrarily small. The point of the second result is that one can find a state in which the momentum is arbitrarily well predicted, at the expense of increasing the uncertainty in the position. Other squeezed states predict some chosen linear combination of position and momentum arbitrarily well. The term "squeezed" refers to a reduction of uncertainty of the predicted value of the momentum (or the predicted position, or the predicted value of some linear combination of position and momentum) relative to the uncertainty in a coherent state with the same amplitude of oscillation.

II.25) Consider a particle of mass m moving in three dimensions with potential energy $V(\mathbf{r})$. In classical mechanics, the Lagrangian for the motion of the particle expressed in cartesian x, y, z coordinates is

$$L = \frac{1}{2}m(\dot{x}^2 + \dot{y}^2 + \dot{z}^2) - V. \tag{II.55}$$

The canonical momenta (eq. [19.8]) are

$$p_x = \frac{\partial L}{\partial \dot{x}} = m\dot{x}, \tag{II.56}$$

and so on, and the Hamiltonian (eq. [19.10]) is

$$H = \frac{1}{2m}(p_x^2 + p_y^2 + p_z^2) + V. \tag{II.57}$$

In the canonical quantization procedure discussed in section 19, the components of the momentum in equation (II.57) are replaced with operators,

$$p_x = -i\hbar\frac{\partial}{x}, \tag{II.58}$$

and so on, to get the quantum mechanical Hamiltonian

$$H = -\frac{\hbar^2}{2m}\nabla^2 + V. \tag{II.59}$$

It is often convenient to compute with polar coordinates (eq. [17.42]). To get the quantum mechanical Hamiltonian expressed in polar coordinates, one might have been tempted to apply the canonical procedure using r, θ, and ϕ in place of x, y, and z. However, this can cause confusion, as follows. In polar coordinates, the classical Lagragian is

$$L = \frac{m}{2}(\dot{r}^2 + r^2\dot{\theta}^2 + r^2 \sin\theta^2 \dot{\phi}^2) - V, \tag{II.60}$$

and the canonical momenta are

$$\begin{aligned} p_r &= \frac{\partial L}{\partial \dot{r}} = m\dot{r}, \\ p_\theta &= \frac{\partial L}{\partial \dot{\theta}} = mr^2\dot{\theta}, \\ p_\phi &= \frac{\partial L}{\partial \dot{\phi}} = mr^2 \sin^2\theta\,\dot{\phi}. \end{aligned} \tag{II.61}$$

By the usual calculation, this gives the classical Hamiltonian

$$H = \frac{1}{2m}\left(p_r^2 + \frac{p_\theta^2}{r^2} + \frac{p_\phi^2}{r^2\sin^2\theta}\right) + V. \tag{II.62}$$

To get the quantum mechanical Hamiltonian in polar coordinates, one might have thought one could follow equation (II.58), and simply replace the momenta with derivative operators,

$$\begin{aligned} p_r &= -i\hbar\frac{\partial}{\partial r}, \\ p_\theta &= -i\hbar\frac{\partial}{\partial \theta}, \\ p_\phi &= -i\hbar\frac{\partial}{\partial \phi}, \end{aligned} \tag{II.63}$$

and then substitute these in equation (II.62) to get

$$H = -\frac{\hbar^2}{2m}\left(\frac{\partial^2}{\partial r^2} + \frac{1}{r^2}\frac{\partial^2}{\partial \theta^2} + \frac{1}{r^2 \sin^2\theta}\frac{\partial^2}{\partial \phi^2}\right) + V. \tag{II.64}$$

On the other hand, the Laplacian operator expressed in polar coordinates works out to

$$\begin{aligned} \nabla^2 &= \frac{\partial^2}{\partial x^2} + \frac{\partial^2}{\partial y^2} + \frac{\partial^2}{\partial z^2} \\ &= \frac{1}{r}\frac{\partial^2}{\partial r^2}r + \frac{1}{r^2}\left[\frac{1}{\sin\theta}\frac{\partial}{\partial\theta}\left(\sin\theta\frac{\partial}{\partial\theta}\right) + \frac{1}{\sin^2\theta}\frac{\partial^2}{\partial\phi^2}\right]. \end{aligned} \tag{II.65}$$

Thus, if one starts with the correct quantum Hamiltonian in equation (II.59) in cartesian coordinates, and then one uses calculus to express ∇^2 in polar coordinates, one finds

$$H = -\frac{\hbar^2}{2m}\left(\frac{1}{r}\frac{\partial^2}{\partial r^2}r + \frac{1}{r^2\sin\theta}\frac{\partial}{\partial\theta}\sin\theta\frac{\partial}{\partial\theta} + \frac{1}{r^2\sin^2\theta}\frac{\partial^2}{\partial\phi^2}\right) + V, \tag{II.66}$$

which is different from equation (II.64).

To see why equation (II.64) is wrong, note that the inner product of two wave functions is

$$(\psi, \chi) \equiv \int d^3r\ \psi^*\chi = \int r^2 dr \sin\theta d\theta d\phi\, \psi^*\chi. \tag{II.67}$$

In the last expression, the volume element has been written in polar coordinates.

Using the last expression in equation (II.67) for the inner product, show that the operator

$$\frac{\partial^2}{\partial r^2} \tag{II.68}$$

that appears in equation (II.64) is not self-adjoint, while the corresponding operator

$$\frac{1}{r}\frac{\partial^2}{\partial r^2}r = \frac{1}{r^2}\frac{\partial}{\partial r}r^2\frac{\partial}{\partial r} \tag{II.69}$$

in equation (II.66) is self-adjoint. Do the same for the derivatives with respect to θ in equations (II.64) and (II.66). The conclusion is that equation (II.64) is unacceptable because it is not self-adjoint. (The self-adjoint form [II.66] is obtained in eq. [18.16].)

Finally, use these methods to show that the adjoint of the operator expression (17.48) for L_+ in polar coordinates is indeed L_-.

CHAPTER 3

ABSTRACT LINEAR SPACE OF STATE VECTORS

The wave mechanics presented in chapter 2 is easily generalized for use in all the applications of quantum mechanics to be presented in this book. In particular, to take account of spin one just replaces the wave function with a set of functions, one for each possible choice of the quantum numbers of the z components of the spins of the particles. However, as will be seen here, it is easy to adapt the wave mechanics formalism in sections 13 and 14 to the more general scheme that represents the states of a system as elements of an abstract linear space rather than a space of wave functions. This approach has the virtue that one can explicitly see the logic of the generalization of the wave function to take account of account of spin, and of course this is the road to other generalizations, like quantum field theory.

20 Bras, Kets, and Brackets

This final step in writing down the principles of quantum mechanics is based on a generalization of the concept of the space of states of a system. An analogy is the treatment of vectors in ordinary three-dimensional space. One can regard a vector $\mathbf{r}$ in three dimensions as an entity on which operations can be performed: $\mathbf{r}$ can be multiplied by constants and added to other vectors, and $\mathbf{r}$ has a scalar or dot product $\mathbf{r}\cdot\mathbf{s}$ with all other vectors $\mathbf{s}$. The vector $\mathbf{r}$ can be specified by its components x, y, and z in some coordinate system. These components are not unique: a rotated coordinate system would give a different set of components that also represent the vector. In a similar way, the set of coefficients $c_n = (\phi_n, \psi)$ in the basis ϕ_n tells everything about the wave function ψ, because we can use completeness to reconstruct ψ as the linear combination $\psi = \sum c_n \phi_n$ (eq. [13.58]). The set of expansion coefficients c_n can be regarded as the coordinates of a vector. The vector is to be considered an element, $|\psi\rangle$, in an abstract linear space. (This linear space of course has nothing to do with the space we live in; the word "space" is only meant to be a useful analogy.) The components c_n are evaluated in the "coordinate system," or basis, ϕ_n. The expansion of the same vector $|\psi\rangle$ in another set of orthogonal functions would give another set of coefficients, that is, another representation of $|\psi\rangle$. As will be discussed, we can regard the set of values of the wave function $\psi(\mathbf{r})$ for each value of $\mathbf{r}$ as the set of components of $|\psi\rangle$ in yet another basis, the position representation.

The analogy to vectors in three dimensions can be extended to inner products. In wave function notation, the expansions of the functions χ and ψ in the basis ϕ_n are

$$\psi = \sum c_m \phi_m, \qquad \chi = \sum d_m \phi_m. \tag{20.1}$$

Using the orthonormality of the basis functions, $(\phi_m, \phi_n) = \delta_{mn}$, we can write the inner product (eq. [13.3]) of χ and ψ as

$$\langle \chi | \psi \rangle = (\chi, \psi) = \sum_n d_n^* c_n. \tag{20.2}$$

The sum looks like a conventional dot product expressed in cartesian coordinates (in a space that may have infinite dimension, and complex expansion coefficients). Like a conventional dot product, the value of the scalar product (χ, ψ) is independent of the coordinate system, or basis of functions, in which it is computed.

In the abstract linear space, we will associate with each pair of elements $|\chi\rangle$ and $|\psi\rangle$ a complex number, $\langle \chi | \psi \rangle$, that plays the role of a dot product or inner product. The first step in this section is to set up the procedures for dealing with the inner products $\langle \chi | \psi \rangle$. This involves the introduction of a dual space of elements $\langle \psi |$ that stand in one-to-one correspondence to the elements $|\psi\rangle$ in the linear space (and play the role of the complex conjugate ψ^* of the wave function ψ). Since nothing new is added one might consider this inelegant, but it is hoped that this chapter will illustrate the convenience of the notation. And one should not underestimate the importance of a powerful notation as an aid to understanding.

The Linear Space

In quantum physics, the states of a system are represented by elements or state vectors in a linear space. These elements play the role of the functions of wave mechanics. An element of the space is written as $|\psi\rangle$, where the symbol ψ inside $|\quad\rangle$ is a label. Thus, we might write the state vector for a hydrogen atom in the nth energy level, with angular momentum quantum numbers l and m, as

$$|\psi\rangle = |n, l, m\rangle. \tag{20.3}$$

The space is linear. That means that if $|\psi\rangle$ and $|\phi\rangle$ are elements of the space so is the linear combination

$$|\chi\rangle = c|\psi\rangle + d|\phi\rangle, \tag{20.4}$$

where c and d are constants (complex numbers).

The analog of the inner product (ϕ, ψ) is defined with the help of a dual space of elements $\langle\psi|$ that correspond one-to-one to the elements $|\psi\rangle$ (and that play the role of ψ^* in wave mechanics). For every pair of elements $|\psi\rangle$ from the space and $\langle\phi|$ from the dual space there is assigned a complex number, the scalar product or inner product $\langle\phi|\psi\rangle$. The rules for assigning this number follow those of (ϕ, ψ).

1) The complex conjugate of an inner product is

$$\langle\psi|\phi\rangle^* = \langle\phi|\psi\rangle. \tag{20.5}$$

This means that $\langle\psi|\psi\rangle$ is real; we will see that $\langle\psi|\psi\rangle$ is not negative.

2) The inner product is linear. If

$$|\chi\rangle = c|\psi\rangle + d|\phi\rangle, \tag{20.6}$$

then

$$\langle\eta|\chi\rangle = c\langle\eta|\psi\rangle + d\langle\eta|\phi\rangle, \tag{20.7}$$

for all $\langle\eta|$. Another way to write this is

$$\langle\eta|\{c|\psi\rangle + d|\phi\rangle\} = c\langle\eta|\psi\rangle + d\langle\eta|\phi\rangle. \tag{20.8}$$

3) The dual space also is linear, so the linear combination $e\langle\psi| + f\langle\phi|$ of elements is an element of the dual space, with

$$\{e\langle\psi| + f\langle\phi|\}|\eta\rangle = e\langle\psi|\eta\rangle + f\langle\phi|\eta\rangle. \tag{20.9}$$

4) A vector $|\psi\rangle$ is uniquely specified by the set of numbers $\langle\phi|\psi\rangle$ for all $\langle\phi|$, and a $\langle\psi|$ is uniquely specified by the set of numbers $\langle\psi|\phi\rangle$ for all $|\phi\rangle$.

In a dazzling burst of wit, Dirac called the elements $|\psi\rangle$ ket vectors, and the elements $\langle\psi|$ of the dual space bra vectors, so he could call the inner products $\langle\psi|\phi\rangle$ brackets. The word bracket refers to the notation for the expectation value of a random variable or for a quantum observable. The bracket notation for an expectation value is given in equation (21.11) below.

These rules fix some relations among the dual vectors. If

$$|\psi'\rangle = c|\psi\rangle, \tag{20.10}$$

where c is a constant, then for any dual vector $\langle\phi|$

$$\langle\phi|\psi'\rangle = c\langle\phi|\psi\rangle, \tag{20.11}$$

by linearity (eq. [20.7]). By equation (20.5), the complex conjugate of this expression is

$$\langle\psi'|\phi\rangle = c^*\langle\psi|\phi\rangle = \{c^*\langle\psi\}|\phi\rangle, \tag{20.12}$$

by linearity in the dual space (eq. [20.9]). Since this relation is true for any $|\phi\rangle$, rule 4 says the two dual vectors are the same. That is, the dual of $c|\psi\rangle$ is $c^*\langle\psi|$.

If

$$|\psi\rangle = |\psi_1\rangle + |\psi_2\rangle, \tag{20.13}$$

then for any $\langle\phi|$ linearity says

$$\langle\phi|\psi\rangle = \langle\phi|\psi_1\rangle + \langle\phi|\psi_2\rangle. \tag{20.14}$$

By equations (20.5) and (20.9), the complex conjugate of this expression is

$$\begin{aligned}\langle\psi|\phi\rangle &= \langle\psi_1|\phi\rangle + \langle\psi_2|\phi\rangle \\ &= \{\langle\psi_1| + \langle\psi_2|\}|\phi\rangle.\end{aligned} \tag{20.15}$$

As for equation (20.12), we see that since $\langle\psi|$ and $\langle\psi_1| + \langle\psi_2|$ have the same inner product with every ket vector the two dual vectors are the same, $\langle\psi| = \langle\psi_1| + \langle\psi_2|$. Since $\langle\psi|$ is the dual of $|\psi_1\rangle + |\psi_2\rangle$ (eq. [20.13]), we see that the dual of $|\psi_1\rangle + |\psi_2\rangle$ is $\langle\psi_1| + \langle\psi_2|$. Collecting this and the result from equation (20.12), we see that the dual satisfies

$$c|\psi_1\rangle + d|\psi_2\rangle \rightarrow c^*\langle\psi_1| + d^*\langle\psi_2|. \tag{20.16}$$

Operators

A linear operator Q maps each element $|\psi\rangle$ into some other element of the space:

$$Q|\psi\rangle = |\psi'\rangle. \tag{20.17}$$

Linearity means

$$Q\{c|\psi\rangle + d|\phi\rangle\} = cQ|\psi\rangle + dQ|\phi\rangle, \tag{20.18}$$

for a linear combination of elements.

Sums and products of operators are defined by the equations

$$\begin{aligned} (P+Q)|\psi\rangle &= P|\psi\rangle + Q|\psi\rangle, \\ PQ|\psi\rangle &= P\{Q|\psi\rangle\}. \end{aligned} \tag{20.19}$$

This defines any operator function that can be expanded in a power series (as in eq. [13.10]).

The action of the operator Q on elements of the dual space,

$$\langle\psi|Q = \langle\psi'|, \tag{20.20}$$

is defined by the equation

$$\langle\psi'|\phi\rangle = \{\langle\psi|Q\}|\phi\rangle \equiv \langle\psi|\{Q|\phi\rangle\} \equiv \langle\psi|Q|\phi\rangle, \tag{20.21}$$

for all $|\phi\rangle$. The fourth rule for inner products says the vector $\langle\psi'| = \langle\psi|Q$ is uniquely identified as an element of the dual space, because we have specified its inner product with all elements $|\phi\rangle$ of the linear space.

It is left as an exercise to use the linearity relation (20.9) with the definition (20.21) to show

$$\{c_1\langle\psi_1| + c_2\langle\psi_2|\}Q = c_1\langle\psi_1|Q + c_2\langle\psi_2|Q, \tag{20.22}$$

that is, the operator is linear in the dual space. A similar calculation gives

$$\langle\phi|(P+Q) = \langle\phi|P + \langle\phi|Q. \tag{20.23}$$

We need the equivalent of the adjoint of an operator in wave mechanics. Following wave mechanics, where we had (eq. [13.12]),

$$(\psi, Q\phi)^* = (\phi, Q^\dagger\psi), \tag{20.24}$$

the adjoint $Q^\dagger$ of the linear operator Q in the linear space is defined by the equation

$$\langle\psi|Q|\phi\rangle^* \equiv \langle\phi|\{Q^\dagger|\psi\rangle\} \equiv \langle\phi|Q^\dagger|\psi\rangle. \tag{20.25}$$

This defines the inner product of $Q^\dagger|\psi\rangle$ with all elements $\langle\phi|$ of the linear space, which uniquely identifies the vector $Q^\dagger|\psi\rangle$. That is, equation (20.25) defines the operation of $Q^\dagger$ on all $|\psi\rangle$.

Equation (20.25) says the dual of the vector $Q|\psi\rangle$ is

$$|\chi\rangle = Q|\psi\rangle \rightarrow \langle\chi| = \langle\psi|Q^\dagger, \tag{20.26}$$

because, for all vectors $|\phi\rangle$, equations (20.5) and (20.25) give

$$\begin{aligned}\langle\chi|\phi\rangle &= \langle\phi|\chi\rangle^* \\ &= \langle\phi|Q|\psi\rangle^* \\ &= \langle\psi|Q^\dagger|\phi\rangle.\end{aligned} \tag{20.27}$$

To see that the operator $Q^\dagger$ is linear, let us work out the operation of $Q^\dagger$ on the ket vector

$$|\chi\rangle = a|\alpha\rangle + b|\beta\rangle. \tag{20.28}$$

The dual of this vector is (eq. [20.16])

$$\langle\chi| = a^*\langle\alpha| + b^*\langle\beta|. \tag{20.29}$$

By the linearity rule (eq. [20.9]), the inner product of this with the vector $Q|\phi\rangle$ is

$$\langle\chi|Q|\phi\rangle = a^*\langle\alpha|Q|\phi\rangle + b^*\langle\beta|Q|\phi\rangle. \tag{20.30}$$

The complex conjugate of this equation is, by equation (20.25) with the linearity relation (20.8),

$$\begin{aligned}\langle\phi|Q^\dagger|\chi\rangle &= a\langle\phi|Q^\dagger|\alpha\rangle + b\langle\phi|Q^\dagger|\beta\rangle \\ &= \langle\phi|\{aQ^\dagger|\alpha\rangle + bQ^\dagger|\beta\rangle\}.\end{aligned} \tag{20.31}$$

This says

$$Q^\dagger\{a|\alpha\rangle + b|\beta\rangle\} = aQ^\dagger|\alpha\rangle + bQ^\dagger|\beta\rangle, \tag{20.32}$$

because the vectors on both sides of this equation have the same inner products with all $\langle\phi|$, so $Q^\dagger$ is linear.

To find the adjoint of the sum of two operators, consider $(P + Q)|\phi\rangle = P|\phi\rangle + Q|\phi\rangle$. Equation (20.26) says the dual of this expression is

$$\langle\phi|(P+Q)^\dagger = \langle\phi|P^\dagger + \langle\phi|Q^\dagger, \tag{20.33}$$

because the dual of the sum of the ket vectors is the sum of the duals (eq. [20.16]). This with equation (20.23) gives

$$(P+Q)^\dagger = P^\dagger + Q^\dagger. \tag{20.34}$$

To find the adjoint of the product of two operators, let $|\phi'\rangle = P|\phi\rangle$ and $|\phi''\rangle = Q|\phi'\rangle = QP|\phi\rangle$. The adjoint of the last vector is (with eq. [20.26])

$$\langle\phi''| = \langle\phi'|Q^\dagger = \langle\phi|P^\dagger Q^\dagger. \tag{20.35}$$

Since we have also $\langle\phi''| = \langle\phi|(QP)^\dagger$, it follows that

$$(PQ)^\dagger = Q^\dagger P^\dagger. \tag{20.36}$$

This is equation (13.16) for operators in the space of wave functions. It is left as exercises to show that the adjoint of a complex number is the complex conjugate, $c^\dagger = c^*$, and that

$$Q^{\dagger\dagger} = Q. \tag{20.37}$$

Eigenvectors and Eigenvalues

As in wave mechanics, observables are represented by self-adjoint operators, such that

$$Q^\dagger = Q, \tag{20.38}$$

that have real eigenvalues that are the possible values of the result of measuring the observable.

The discussion of the eigenvectors and eigenvalues of self-adjoint operators closely parallels that of wave mechanics in section 13. The eigenvalue equation is

$$Q|\phi\rangle = q|\phi\rangle, \tag{20.39}$$

where the eigenvalue q is a number and $|\phi\rangle$ is the eigenvector of Q. Eigenvalues are real, for if $|\phi\rangle$ is an eigenvector then

$$\langle\phi|Q|\phi\rangle = q\langle\phi|\phi\rangle, \tag{20.40}$$

and the complex conjugate of this expression is (eq. [20.25])

$$q^*\langle\phi|\phi\rangle = \langle\phi|Q^\dagger|\phi\rangle = \langle\phi|Q|\phi\rangle = q\langle\phi|\phi\rangle, \tag{20.41}$$

because $Q^\dagger = Q$. Since an eigenvector is not supposed to have zero "length," $\langle\phi|\phi\rangle \neq 0$, we see that $q^* = q$.

The eigenvectors belonging to different eigenvalues are orthogonal in the sense that their inner products vanish. Suppose two eigenvectors are

$$\begin{aligned} Q|q_1\rangle &= q_1|q_1\rangle, \\ Q|q_2\rangle &= q_2|q_2\rangle. \end{aligned} \tag{20.42}$$

The dual of the second equation is (eq. [20.26])

$$\langle q_2|Q = q_2\langle q_2|, \tag{20.43}$$

because Q is self-adjoint and q_2 is real. The inner product with $|q_1\rangle$ is

$$\begin{aligned} \langle q_2|Q|q_1\rangle &= q_2\langle q_2|q_1\rangle \\ &= q_1\langle q_2|q_1\rangle, \end{aligned} \tag{20.44}$$

where the second equation follows by applying the operator Q to the right. We have then

$$\langle q_2|q_1\rangle = 0 \text{ if } q_1 \neq q_2. \tag{20.45}$$

The other properties of eigenfunctions in wave mechanics follow here in the same way. By repeating the argument in section 13, one sees that linearly independent eigenvectors belonging to the same eigenvalue can be made to be orthogonal by taking suitable linear combinations of the eigenvectors, and that if operators P and Q commute there is a complete set of simultaneous eigenvectors of P and Q.

Completeness

Consider a self-adjoint operator Q associated with an observable quantity of a physical system. It is assumed that the complete set of eigenvectors of Q is complete also in the sense that any vector $|\psi\rangle$ representing a possible state of the system can be written as a linear combination of the eigenvectors.

Suppose the eigenvectors of Q have discrete eigenvalues. As noted above, a complete set of linearly independent eigenvectors $|n\rangle$ of Q can be chosen to be orthogonal and normalized,

$$\langle n|m\rangle = \delta_{nm}. \tag{20.46}$$

The completeness assumption is that any state vector can be expanded as a linear combination in this basis $|n\rangle$:

$$|\psi\rangle = \sum c_n |n\rangle. \tag{20.47}$$

On taking the inner product of this sum with the bra vector $\langle m|$ and using the orthogonality equation (20.46), we see that the expansion coefficients are

$$c_m = \langle m|\psi\rangle, \tag{20.48}$$

so the expansion (20.47) is

$$|\psi\rangle = \sum |n\rangle\langle n|\psi\rangle. \tag{20.49}$$

The corresponding expansion of a vector in the dual space is (eq. [20.16])

$$\langle\psi| = \sum \langle\psi|n\rangle\langle n|. \tag{20.50}$$

The expansion (20.49) can be written in a compact way by means of the linear operator

$$Q = \sum |n\rangle\langle n|. \tag{20.51}$$

This means

$$Q|\psi\rangle = \left(\sum |n\rangle\langle n|\right)|\psi\rangle \equiv \sum |n\rangle\langle n|\psi\rangle \ = |\psi\rangle. \tag{20.52}$$

Thus Q is the identity operator:

$$\sum |n\rangle\langle n| = 1. \tag{20.53}$$

The sum is over the complete set of orthogonal normalized eigenvectors. This is the completeness relation, in a form that will be used many times.

The eigenvalues of an observable may have a continuous range of values. One thinks of this as the limit of the discrete eigenvalue case when the eigenvalues are arbitrarily close together. Suppose the eigenvalues are discrete but very closely spaced, and let us change the normalization of the basis vectors in equation (20.46) to

$$\langle q_n|q_m\rangle = \delta_{nm}/\Delta q, \qquad \Delta q = q_{n+1} - q_n, \tag{20.54}$$

where Δq is the difference between neighboring eigenvalues q_n. This changes the expression (20.49) for the expansion of a vector $|\psi\rangle$ to

$$|\psi\rangle = \sum \Delta q|q_n\rangle\langle q_n|\psi\rangle. \tag{20.55}$$

The completeness relation (20.53) thus changes to

$$\sum |q_n\rangle\Delta q\langle q_n| = 1. \tag{20.56}$$

In the limit $\Delta q \to 0$, equations (20.54) and (20.56) become

$$\langle q|q'\rangle = \delta(q - q'), \qquad \int |q\rangle dq\langle q| = 1. \tag{20.57}$$

The second equation says the expansion of a vector in this basis is

$$|\psi\rangle = \int |q\rangle dq\langle q|\psi\rangle. \tag{20.58}$$

An observable can have a spectrum of eigenvalues that is continuous in part, discrete in part. In this case the completeness relation is the sum of an integral and a discrete sum.

As will be discussed in the next section, quantum mechanics associates with the state of a system a state vector $|\psi\rangle$ that is normalized to $\langle\psi|\psi\rangle = 1$. We can assign this normalization to the discrete eigenvectors of an observable, so a physical system can be in an eigenstate such that the quantity represented by the observable has a definite value. Since the eigenvectors belonging to an observable with a continuous spectrum of eigenvalues are not normalized to unit length, state vectors have to be integrals over these eigenvectors.

Representations: Matrix Mechanics

Suppose an observable Q has a complete discrete set of eigenvectors $|n\rangle$. Each ket vector $|\psi\rangle$ is completely specified by the set of numbers (components) $\langle n|\psi\rangle$, because the completeness relation (20.53) can be used to write the inner product $\langle\phi|\psi\rangle$ for any vector $\langle\phi|$ as

$$\langle\phi|\psi\rangle = \sum \langle\phi|n\rangle\langle n|\psi\rangle. \tag{20.59}$$

We can similarly specify the vector $|\phi\rangle = P|\psi\rangle$ produced by the linear operator P by its components

$$\langle n|\phi\rangle = \langle n|P|\psi\rangle = \sum_m \langle n|P|m\rangle\langle m|\psi\rangle. \tag{20.60}$$

Thus the observable P is completely specified by the numbers $\langle n|P|m\rangle$.

The expression (20.60) is a matrix product. The operator is represented by the matrix with elements

$$P_{nm} = \langle n|P|m\rangle. \tag{20.61}$$

The vectors $|\psi\rangle$ and $|\phi\rangle$ are represented by the column vectors,

$$\psi_m = \langle m|\psi\rangle, \qquad \phi_n = \langle n|\phi\rangle. \tag{20.62}$$

In expanded form, equation (20.60) is the matrix product

$$\begin{vmatrix} \phi_1 \\ \phi_2 \\ \phi_3 \\ \vdots \end{vmatrix} = \begin{vmatrix} P_{11} & P_{12} & P_{13} & \cdots \\ P_{21} & P_{22} & P_{23} & \cdots \\ P_{31} & P_{32} & P_{33} & \cdots \\ \vdots & & & \end{vmatrix} \begin{vmatrix} \psi_1 \\ \psi_2 \\ \psi_3 \\ \vdots \end{vmatrix}. \tag{20.63}$$

The matrix elements of the product of two operators are

$$\langle n|PQ|m\rangle = \sum_i \langle n|P|i\rangle\langle i|Q|m\rangle, \tag{20.64}$$

by completeness (eq. [20.53]). This is the product of the matrix with elements P_{ni} with the matrix with elements Q_{im}. Finally, one recognizes the inner product of a $\langle\psi|$ with a $|\phi\rangle$ in equation (20.59) as the matrix product of a row vector with a column vector.

We see from these examples that the content of the theory can be expressed as a matrix mechanics, in which operators P are represented by matrices with elements $\langle n|P|m\rangle$, products of operators are represented by matrix products, vectors $|\psi\rangle$ are represented by column vectors on which matrices operate by the usual rules of matrix multiplication, and dual vectors are represented by row vectors. The one slightly novel feature is that the matrices and vectors can have an infinite number of dimensions.

Because $\langle n|\psi\rangle^* = \langle\psi|n\rangle$, the elements of the row vector $\langle\psi|n\rangle$ that represents a member of the dual space $\langle\psi|$ are the complex conjugates of the elements of the column vector $\langle n|\psi\rangle$ that represents the corresponding ket vector $|\psi\rangle$. The matrix elements of the adjoint of an operator satisfy (eq. [20.25])

$$\begin{aligned}(Q^\dagger)_{nm} &= \langle n|Q^\dagger|m\rangle \\ &= \langle m|Q|n\rangle^*. \\ &= Q^*_{mn}\end{aligned} \tag{20.65}$$

Some terminology from the theory of matrices might be mentioned. The conjugate transpose or Hermitian adjoint of a matrix is the complex conjugate of the transpose of the matrix. That is, the Hermitian adjoint of the matrix with elements H_{nm} has matrix elements $H^\dagger_{nm} \equiv H^*_{mn}$. Thus the row vector $\langle\psi|n\rangle$ is the Hermitian adjoint of the column vector $\langle n|\psi\rangle$, and equation (20.65) says the matrix of $Q^\dagger$ is the Hermitian adjoint of the matrix of Q. A self-adjoint operator thus has a matrix that is equal to its Hermitian adjoint. A matrix with this property is said to be Hermitian (and a self-adjoint operator thus is said to be Hermitian). We know the eigenvalues of a Hermitian matrix are real and the eigenvectors orthogonal, because the same is true of the operators and vectors the matrices are representing. Finally, the matrix elements of a unitary operator, for which $U^\dagger U = 1$, are the elements of a unitary matrix, such that the Hermitian adjoint of the matrix is equal to the inverse matrix. This matrix formalism will be used in section 24 in the treatment of spin.

Quantum theory in matrix form was invented by Heisenberg, Born, and Jordan the year before Schrödinger developed his wave equation. Only later did Schrödinger, Dirac, and others see that matrix and wave mechanics are physically equivalent representations of an abstract linear space. To see how wave mechanics can arise as another representation of the abstract theory, let us consider next a basis of eigenstates of position. These position eigenstates will be taken to have a continuous spectrum of eigenvalues.

Representations: Wave Mechanics

Suppose a physical system consists of a single particle that moves in three dimensions. The particle is assigned position observables $\hat{x}$, $\hat{y}$, and $\hat{z}$, which will be written collectively as $\hat{\mathbf{r}}$. The hats are written here to distinguish these operators from their eigenvalues, x, y, and z, or collectively, $\mathbf{r}$. The position eigenstates satisfy the usual equations

$$\hat{\mathbf{r}}|\mathbf{r}\rangle = \mathbf{r}|\mathbf{r}\rangle. \tag{20.66}$$

The eigenvalues $\mathbf{r}$ are assumed to be continuous, so the the eigenvectors are normalized to

$$\langle\mathbf{r}|\mathbf{r}'\rangle = \delta(\mathbf{r}-\mathbf{r}'), \qquad \int |\mathbf{r}\rangle d^3r\langle\mathbf{r}| = 1, \tag{20.67}$$

as in equation (20.57). The vector $|\psi\rangle$ has components

$$\psi(\mathbf{r}) = \langle\mathbf{r}|\psi\rangle. \tag{20.68}$$

In a discrete representation the components might be written $c_n = \langle n|\psi\rangle$, with a discrete label, n. Here the label is continuous, and the tradition is to write the label as an argument of a function, $\psi(\mathbf{r})$ in equation (20.68). This is Schrödinger's wave function.

The complex conjugate of the wave function is

$$\psi(\mathbf{r})^* = \langle\psi|\mathbf{r}\rangle. \tag{20.69}$$

Using the completeness relation (20.67), we can write the inner product of two vectors as

$$\begin{aligned}\langle\chi|\psi\rangle &= \int \langle\chi|\mathbf{r}\rangle d^3r \langle\mathbf{r}|\psi\rangle \\ &= \int \chi(\mathbf{r})^* \psi(\mathbf{r}) d^3r \\ &= (\chi, \psi),\end{aligned} \tag{20.70}$$

which is the definition of an inner product of the wave functions (eq. [13.3]). This is the basis for the assertion at the beginning of this section (eq. [20.2]) that the inner product of wave functions, (χ, ψ), can be equivalent to the inner product $\langle\chi|\psi\rangle$ in the abstract space.

An example of the treatment of matrix elements of operators in this continuous case is discussed in section 22, after the general principles of quantum mechanics have been reviewed.

21 Principles of Quantum Mechanics

The general assumptions of quantum mechanics have already been written down in section 14 in the terms of wave mechanics; we only have to restate them in the new language.

1) *State vectors.* The states of a physical system such as an atom are represented by the vectors $|\psi\rangle$ in a linear space. The state vector tells us all that can be predicted about the system.

2) *Observables.* Any attribute of the system that in principle can be be measured is associated with a linear self-adjoint operator Q in the space of vectors.

3) *Completeness.* The state vector $|\psi\rangle$ can be expanded as a linear combination of a complete set of eigenvectors of the observable Q. If the eigenvectors of Q are degenerate, we can introduce another operator P that commutes with Q, and whose eigenvalues may distinguish different eigenvectors with the same eigenvalue of Q. A complete compatible set of observables all commute with each other, and the complete set of their simultaneous eigenvectors are uniquely specified by the set of eigenvalues. For example, if the commuting observables P and Q completely specify all eigenvectors, we can label the simultaneous eigenvectors as $|p_m, q_n\rangle$, where the eigenvalues of P and Q are p_m and q_n. If the eigenvalues are

discrete, we can use the normalization

$$\langle p_m, q_n | p_{m'}, q_{n'} \rangle = \delta_{mm'} \delta_{nn'}. \tag{21.1}$$

Then the expansion (20.49) of the state vector $|\psi\rangle$ is

$$|\psi\rangle = \sum_{m,n} |p_m, q_n\rangle \langle p_m, q_n | \psi\rangle. \tag{21.2}$$

If one of the eigenvalues were continuous and the other discrete, the eigenstates would be normalized as in equation (20.57), giving

$$|\psi\rangle = \sum_m \int |p_m, q\rangle dq \langle p_m, q | \psi\rangle. \tag{21.3}$$

4) *Probability interpretation.* Suppose P and Q are compatible observables with discrete eigenvalues that uniquely label the basis $|p_m, q_n\rangle$ with the normalization of equation (21.1). Then in the state represented by the normalized state vector $|\psi\rangle$, the joint probability that the results of a measurement of the attributes represented by P and Q are p_m and q_n is

$$P_{mn} = |\langle p_m, q_n | \psi\rangle|^2 = \langle \psi | p_m, q_n\rangle \langle p_m, q_n | \psi\rangle. \tag{21.4}$$

The expansion coefficients $\langle p_m, q_n | \psi\rangle$ are called the probability amplitudes. The probabilities summed over all possible results of the measurements of P and Q are, by the completeness relation (20.53),

$$\sum_{mn} P_{mn} = \sum_{m,n} \langle \psi | p_m, q_n\rangle \langle p_m, q_n | \psi\rangle = \langle \psi | \psi\rangle = 1. \tag{21.5}$$

The probabilities thus sum to unity, as required.

If an observable has a continuous spectrum of eigenvalues, one arrives at the probability distribution in the value of the observable by considering the limit of a discrete spectrum when the distance Δq between neighboring eigenvalues approaches zero, as in section 20. For example, suppose a basis is labeled by the one discrete closely spaced eigenvalue q_n, and write the normalization as in equation (20.54),

$$\langle q_n | q_m \rangle = \delta_{nm} / \Delta q, \tag{21.6}$$

with $\Delta q = q_{n+1} - q_n$. Because the normalization of the $|q_n\rangle$ differs from that assumed in equation (21.4) by the factor Δq, the expression for the probability for observing the value q_n in the state $|\psi\rangle$ changes to

$$P_n = \Delta q |\langle q_n|\psi\rangle|^2. \tag{21.7}$$

The number of eigenvectors in the small range of values q to $q+\delta q$ is $\delta n = \delta q/\Delta q$. The probability of observing that the result of the measurement is in the range q to $q + \delta q$ is the product of the probability P_n per eigenvector with the number δn of eigenvectors in the wanted range of q:

$$\delta P = P_n \delta n = \delta q |\langle q|\psi\rangle|^2. \tag{21.8}$$

In the limit $\Delta q \to 0$, this is a differential probability distribution in q, $dP/dq = |\langle q|\psi\rangle|^2$. This is the form we assumed for the probability distribution in position in wave mechanics (eq. [8.2]). This approach to a continuous probability distribution was used in the derivation of the momentum distribution in wave mechanics (eq. [14.13]).

For another example, suppose the basis is completely labeled by the discrete eigenvalues, p_n, of the observable P and the continuous eigenvalues, q, of the compatible observable Q. The joint probability that P is found to have the value p_n and Q is found to have the value q in the range dq is

$$dP_n = |\langle p_n, q|\psi\rangle|^2 dq, \tag{21.9}$$

where the simultaneous eigenvectors of the commuting observables P and Q are $|p_n, q\rangle$, and the normalization is $\langle p_n, q|p_m, q'\rangle = \delta_{nm}\delta(q - q')$. Completeness gives the normalization condition

$$\sum_n \int dq \frac{dP_n}{dq} = \sum_n \int \langle\psi|q, p_n\rangle dq \langle q, p_n|\psi\rangle = \langle\psi|\psi\rangle, \tag{21.10}$$

so the probabilities are normalized if the state vector $|\psi\rangle$ is normalized to $\langle\psi|\psi\rangle = 1$.

It is left as an exercise to check that the probability assumption implies that the expectation value of the observable Q in the state $|\psi\rangle$ is

$$\langle Q\rangle = \langle\psi|Q|\psi\rangle. \tag{21.11}$$

This is the same as equation (14.14) in wave mechanics. The elegant form of this equation explains Dirac's choice of symbols for the vectors and their duals.

5) *Dynamics.* In the Schrödinger representation (section 14), the state vector representing a physical system evolves with time, while operators representing observables are constant (in the absence of a time-variable applied field). The starting assumption for dynamics is that the state vector at time t_2 is obtained from the state vector at time t_1 by a linear mapping. That is, it is assumed that there is a linear time translation operator $U(t_2, t_1)$ such that

$$|t_2\rangle = U(t_1, t_2)|t_1\rangle. \tag{21.12}$$

Because the dual of equation (21.12) is $\langle t_2| = \langle t_1|U^\dagger$ (eq. [20.26]), the length of the time translated state vector is

$$\langle t_2|t_2\rangle = \langle t_1|U^\dagger U|t_1\rangle = \langle t_1|t_1\rangle = 1. \tag{21.13}$$

That is, the normalization condition $\langle\psi|\psi\rangle = 1$ is preserved by the requirement that the time translation operator is unitary,

$$U^\dagger U = 1. \tag{21.14}$$

The operator U is a continuous function of time. Thus with $t_2 = t_1 + \delta t$, we can expand U as a power series in δt,

$$U = 1 - iH\delta t/\hbar + \ldots. \tag{21.15}$$

The leading term has to be unity. The constant $\hbar$ in the first order term is introduced to make the operator H have units of energy; it also makes the differential equation (21.19) below look like Schrödinger's equation. The factor i is introduced to make the operator H self-adjoint, for the product of U with its adjoint is

$$U^\dagger U = 1 + \frac{i}{\hbar}\delta t(H^\dagger - H) + \ldots. \tag{21.16}$$

Since U is unitary, the term linear in δt has to vanish, so $H^\dagger = H$, so H is self-adjoint. Finally, if physics is unchanged by a translation in time, H is independent of time.

For an infinitesimal time step dt, equations (21.12) and (21.15) are

$$|t + dt\rangle = (1 - iHdt/\hbar)|t\rangle. \tag{21.17}$$

We can define the time derivative of $|t\rangle$ in the usual way:

$$\frac{\partial|t\rangle}{\partial t} = \frac{|t+dt\rangle - |t\rangle}{dt}, \tag{21.18}$$

so we have

$$i\hbar\frac{\partial|t\rangle}{\partial t} = H|t\rangle. \tag{21.19}$$

This is Schrödinger's equation, with Hamiltonian H.

We can get the time shift operator for a finite time interval, t, in terms of the operator H by applying N times the operator U (eq. [21.15]) for a small time shift $\delta t = t/N$ (as was done for a position shift in eqs. [16.5] and [16.6]). In the limit $N \to \infty$ we have

$$\begin{aligned} U(t_1, t_1 + t) &= \left(1 - \frac{iHt}{\hbar N}\right)^N \\ &= e^{-iHt/\hbar}. \end{aligned} \tag{21.20}$$

This is equation (14.27) in wave mechanics.

As in wave mechanics (eq. [14.26]), the general solution for the time evolution of the state vector can be expressed as a sum over the eigenvectors $|n\rangle$ of the Hamiltonian with eigenvalues E_n:

$$|t\rangle = \sum c_n|n\rangle e^{-iE_nt/\hbar}, \tag{21.21}$$

where the constants of integration c_n are determined by the initial conditions.

6) *The "details."* For a specific physical system, one must decide what are the observables P, Q, and so on, and what is their algebra, and one must decide by analogy or intuition how the Hamiltonian is constructed out of the observables. The next section shows an example of how all this is done, in the construction of wave mechanics out of the general formalism.

22 Recovering Wave Mechanics

Let us set up a quantum mechanical description of a single particle that moves in one dimension. We will assign just two observables, or linear self-adjoint operators, $\hat{x}$ to represent the position of the particle, and $\hat{p}$ to represent the momentum. Following wave mechanics, or the canonical quantization rule (eq. [19.33]), we will assume that the commutation relation between these two observables is

$$[\hat{x}, \hat{p}] = i\hbar. \tag{22.1}$$

The factor $\hbar$ is arbitrary; it reflects the choice of units. The factor i is required because $\hat{x}$ and $\hat{p}$ are self-adjoint, so the adjoint of $[\hat{x}, \hat{p}]$ is $-[\hat{x}, \hat{p}]$. Finally, we will assume that the eigenvalues of $\hat{x}$ are continuous: the particle can be observed to have any position.

Since $\hat{x}$ and $\hat{p}$ are the only observables, and they do not commute, a complete basis is the set $|x\rangle$ of position eigenvectors,

$$\hat{x}|x\rangle = x|x\rangle. \tag{22.2}$$

Since the eigenvalues x are supposed to be continuous, the normalization and completeness relations for this basis are (eq. [20.57])

$$\langle x|x'\rangle = \delta(x - x'), \qquad \int |x\rangle dx \langle x| = 1. \tag{22.3}$$

As discussed in section 20, the vector $|\psi\rangle$ is represented by its components,

$$\langle x|\psi\rangle \equiv \psi(x). \tag{22.4}$$

This defines the wave function $\psi(x)$, with the usual form for the inner product (eq. [20.70]).

The operator $\hat{p}$ is represented by its matrix elements between bra vectors $\langle x|$ and ket vectors $|x'\rangle$, just as in equation (20.61). To evaluate the matrix elements of $\hat{p}$, consider the matrix elements of the commutation relation (22.1) between $\langle x|$ and $|x'\rangle$:

$$\langle x|\hat{x}\hat{p} - \hat{p}\hat{x}|x'\rangle = i\hbar\langle x|x'\rangle. \tag{22.5}$$

In the first term, $\hat{x}$ operating to the left gives x (eq. [22.2]), and in the second term $\hat{x}$ operating to the right gives x'. The term on the right-hand side is proportional to the Dirac delta function $\delta(x-x')$ (eq. [22.3]). Thus we have

$$(x - x')\langle x|\hat{p}|x'\rangle = i\hbar\delta(x - x'). \tag{22.6}$$

This determines what $\hat{p}$ does when operating on a vector $|\psi\rangle$. The components of $|\phi\rangle = \hat{p}|\psi\rangle$ are

$$\langle x|\hat{p}|\psi\rangle = \int \langle x|\hat{p}|x'\rangle dx' \langle x'|\psi\rangle. \tag{22.7}$$

On expanding the wave function $\langle x'|\psi\rangle = \psi(x')$ in the integral as a Taylor series around the position x, we have

$$\langle x|\hat{p}|\psi\rangle = \int dx' \langle x|\hat{p}|x'\rangle [\psi(x) + (x' - x) d\psi(x)/dx + \ldots]. \tag{22.8}$$

The first term in the brackets integrates to zero because the integrand is odd around $x' = x$. In the second term we can replace the matrix element of $\hat{p}$ with equation (22.6). The higher order terms give zero because of the higher powers of $x' - x$. The result is

$$\langle x|\hat{p}|\psi\rangle = -i\hbar \frac{d\psi(x)}{dx}. \tag{22.9}$$

This calculation shows that the result of operating on the vector $|\psi\rangle$ with $\hat{p}$ is, in the position representation, the function

$$\phi(x) = \langle x|\hat{p}|\psi\rangle = -i\hbar \frac{d\psi}{dx}, \tag{22.10}$$

where $\psi(x) = \langle x|\psi\rangle$ is the position representation of $|\psi\rangle$. That is, the momentum operator in the position representation acts as the derivative operator

$$\hat{p} = -i\hbar \frac{d}{dx}, \tag{22.11}$$

which is what is used in wave mechanics (eq. [19.34]).

To complete the theory we must specify the Hamiltonian. The standard and successful model for a particle of mass m in a potential well is the familiar form

$$\hat{H} = \frac{\hat{p}^2}{2m} + V(\hat{x}). \tag{22.12}$$

Schrödinger's equation (21.19) is

$$i\hbar \frac{\partial}{\partial t} |\psi\rangle = \left(\frac{\hat{p}^2}{2m} + V(\hat{x}) \right) |\psi\rangle. \tag{22.13}$$

We get Schrödinger's equation in the position representation by taking the inner product of this equation with the bra vector $\langle x|$. The potential energy operator $\hat{V}$ is some given function of $\hat{x}$, so

$$\langle x|V(\hat{x}) = V(x)\langle x|. \tag{22.14}$$

The calculation in equation (22.8) carried through a second time gives

$$\langle x|\hat{p}^2|\psi\rangle = -\hbar^2 \frac{d^2\psi(x)}{dx^2}. \tag{22.15}$$

Thus we have from equation (22.13)

$$i\hbar \frac{\partial\psi(x,t)}{\partial t} = -\frac{\hbar^2}{2m}\frac{\partial^2\psi}{\partial x^2} + V(x)\psi(x,t). \tag{22.16}$$

This is the usual Schrödinger equation in wave mechanics for a particle in a potential well.

We could equally well have used as basis vectors the eigenvectors of $\hat{p}$. In this momentum representation, $\hat{x}$ acts as a derivative operator with respect to p, of the form $\hat{x} = +i\hbar\partial/\partial p$, as equation (22.11), but with the opposite sign. This symmetry of position and momentum representations was seen in section 10 (eqs. [10.42] to [10.44]).

To describe the motion of a single particle in three dimensions, one uses three position observables, $\hat{r}_\alpha$, with $\alpha = 1$, 2, and 3 meaning $\hat{x}$, $\hat{y}$, and $\hat{z}$, and three momentum observables, $\hat{p}_\alpha$. The position observables commute among themselves, the momentum observables commute among themselves, and position and momentum are assumed to have the cannonical commutation relations (19.33),

$$[\hat{r}_\alpha, \hat{p}_\beta] = i\hbar\delta_{\alpha\beta}. \tag{22.17}$$

Since the three position observables commute with each other, and do not commute with the components of momentum, the $\hat{r}_\alpha$ are a complete set of compatible observables. Their simultaneous eigenvectors are $|x, y, z\rangle = |\mathbf{r}\rangle$. These eigenvectors are the basis for the position representation, in which the components of the vector $|\psi\rangle$ are the wave function

$$\langle x, y, z|\psi\rangle = \langle \mathbf{r}|\psi\rangle = \psi(\mathbf{r}). \tag{22.18}$$

The matrix elements of the momentum operators are computed using the generalization of equation (22.6),

$$(r_\beta - r'_\beta)\langle \mathbf{r}|\hat{p}_\alpha|\mathbf{r}'\rangle = i\hbar\delta_{\alpha\beta}\delta(\mathbf{r}-\mathbf{r}'). \tag{22.19}$$

The generalization of the calculation in equation (22.8) gives

$$\langle \mathbf{r}|\hat{\mathbf{p}}|\psi\rangle = -i\hbar\nabla\psi(\mathbf{r}). \tag{22.20}$$

The usual Schrödinger equation for the wave function $\langle \mathbf{r}|\psi\rangle = \psi(\mathbf{r}, t)$ for a single particle in the potential $V(\mathbf{r})$ follows from the Hamiltonian

$$H = \frac{\hat{\mathbf{p}}^2}{2m} + V(\hat{\mathbf{r}}), \tag{22.21}$$

as in the one-dimensional case.

The generalization to an N-particle system follows in the same way. There are $3N$ position observables, $\hat{\mathbf{r}}_j$, which commute with each other and whose eigenvectors $|\mathbf{r}_1, \mathbf{r}_2 \ldots\rangle$ are the basis for the position representation. The representation of the vector $|\psi\rangle$ is the wave function

$$\langle \mathbf{r}_1, \mathbf{r}_2, \ldots |\psi\rangle = \psi(\mathbf{r}_1, \mathbf{r}_2, \ldots). \tag{22.22}$$

This is a function of $3N$ variables, plus time in the Schrödinger representation. There are $3N$ momentum observables, which commute among themselves. The momenta belonging to particle j satisfy the commutation relations (22.17) for the position observables of particle j, and commute with all the other position observables. The momentum matrix elements are computed as in equation (22.20). This leads to the treatment described in section 12 for a many-particle system in the position representation.

23 Spin

Spin and Magnetic Moments

In the discussion so far, a particle such as an electron or proton moving in three dimensions has been assigned six observables, the three components each of position and momentum. A fuller description requires another triplet of observables, the three components of the intrinsic or spin angular momentum, $\mathbf{s} = s_x,\ s_y,\ s_z$. This was introduced by Goudsmit and Uhlenbeck in 1925, to account for features such as the fine-structure in the lines of atomic spectra (section 42).

The total angular momentum operator, $\mathbf{J}$, for a particle with spin is the sum of the orbital angular momentum part $\mathbf{L}$ discussed in section 17 and the intrinsic spin angular momentum part:

$$\mathbf{J} = \mathbf{L} + \mathbf{s}. \tag{23.1}$$

This equation means components add, $J_x = L_x + s_x$ and so on. The components of the observable $\mathbf{s}$ are assumed to commute with the position and momentum observables, so $\mathbf{s}$ commutes with the components of $\mathbf{L}$. One arrives at the commutation relations for the components of $\mathbf{s}$ among themselves by the following argument.

We want to preserve the law of conservation of angular momentum, where total angular momentum is represented by the sum in equation (23.1). It will be recalled from the discussion in section 17 that this conservation law follows in quantum mechanics from the fact that the components of the angular momentum operator generate rotations of the system, as in equation (17.60). If the Hamiltonian H of a system is symmetric under rotations, then H commutes with the angular momentum. That means energy eigenstates can be labeled with definite angular momentum quantum numbers (for the square and one component of the angular momentum), and these quantum numbers are conserved.

This argument requires that rotations of the full system, positions plus spin orientations, be generated by the components of the total angular momentum operator $\mathbf{J}$. Thus the operator that rotates the state $|\psi\rangle$ by angle θ around the axis with unit normal $\mathbf{n}$ is is generalized from equation (17.60) to

$$|\psi\rangle \to |\psi'\rangle = e^{-i\theta\mathbf{n}\cdot\mathbf{J}/\hbar}|\psi\rangle. \tag{23.2}$$

For a rotation through angle χ around the x axis, this equation is

$$|\psi\rangle \to |\psi'\rangle = e^{-i\chi J_x/\hbar}|\psi\rangle. \tag{23.3}$$

For a single-particle system, the total angular momentum operator $\mathbf{J}$ is given by equation (23.1). For an N-body system, $\mathbf{J}$ is the sum of the vector operators $\mathbf{J}_i$ for each particle.

As discussed in section 17, the commutation relations (17.7) for the orbital angular momentum operators L_x and L_y reflect the difference between the results of applying a rotation about the x axis followed by a rotation about the y axis, and the result of applying the rotations in the opposite order. Since this difference is a general property of rotations, the commutation relations (17.7) must apply also to the components of $\mathbf{J}$. We must therefore adopt the same commutation relations for the components of $\mathbf{s}$:

$$\begin{aligned}[s_x, s_y] &= i\hbar s_z, \\ [s_y, s_z] &= i\hbar s_x, \\ [s_z, s_x] &= i\hbar s_y.\end{aligned} \tag{23.4}$$

The operator arguments in section 17 for the angular momentum eigenvalues apply to $\mathbf{s}$ as well as $\mathbf{L}$, because the algebra is the same, so equation (17.41) gives the possible values of the quantum numbers of the square and one component of $\mathbf{s}$. For an electron or proton or neutron the latter is observed to have only two values, $\pm\hbar/2$. By equation (17.41), this means the eigenvalue of $\mathbf{s}^2$ for such a particle always is

$$\mathbf{s}^2|\psi\rangle = \hbar^2 s(s+1)|\psi\rangle = \hbar^2 \frac{1}{2}\frac{3}{2}|\psi\rangle. \tag{23.5}$$

These particles are said to have spin one-half, because that is the value of the total spin quantum number s (though of course the square of $\mathbf{s}$ is $3\hbar^2/4$), and it is the maximum value of the spin component along a given direction.

The word spin naturally leads one to think of an electron as behaving like a spinning top. The picture is helpful, but one must remember that it is only an analogy. An electron behaves as a pointlike particle in a position measurement, or in interactions with other particles, and one cannot really imagine that a point spins. One indication of the failure of the spinning top picture is the fact that the angular momentum quantum number for the spin of an electron is $s = 1/2$, which the algebra allows but is not possible for the orbital angular momentum of a particle (eq. [17.47]). Spin must be considered an intrinsic attribute of a particle. It is associated with linear operators, the triplet of observables $\mathbf{s}$, whose algebra can be used to (successfully) predict the results of experiments.

The spinning top model does allow one to understand the order of magnitude of the intrinsic magnetic moment of an electron (though dimensional analysis would do just as well). Consider the classical circular

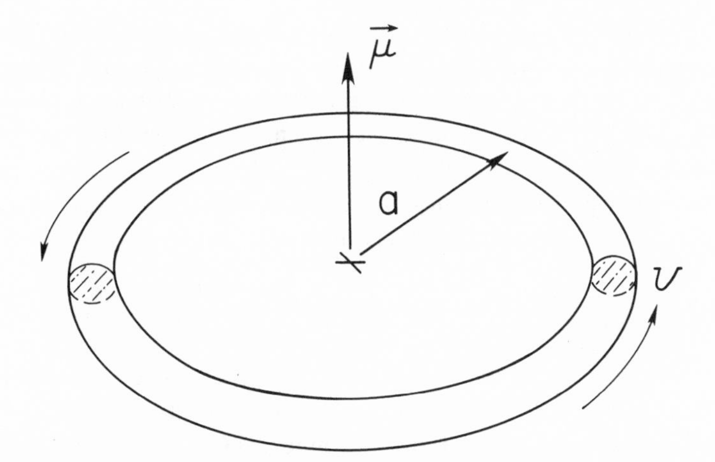

Fig. 23.1 Orbital model for a magnetic dipole. The ring has mass m, charge q, and radius a. The mass and charge are rotating in the plane of the ring at speed v, producing a current loop with angular momentum.

ring of material of mass m and charge q shown in figure 23.1. The charge and mass are uniformly distributed around a circle of radius a. The ring is rotating in its plane at speed $v \ll c$. This produces a current loop. To find the current, note that the total charge q passes a fixed point during the rotation period $2\pi a/v$, so the current is

$$i = \frac{qv}{2\pi a}. \tag{23.6}$$

This current makes a magnetic field that at large distances from the ring is the dipole field of a magnetic dipole moment $\vec{\mu}$ with direction along the normal to the current loop and magnitude equal to the product of the current and area of the current loop and divided by the velocity of light. Thus the magnetic dipole moment of this current loop is

$$\mu = \frac{\pi a^2 i}{c} = \frac{qva}{2c}. \tag{23.7}$$

The magnetic field produced by this current loop at distances large compared to the ring radius a is given in equation (34.4).

The angular momentum of the ring is $l = mav$, so we can write the magnetic dipole moment as

$$\mu = \frac{q}{2mc} l. \tag{23.8}$$

The ratio of μ to l is the gyromagnetic ratio; this spinning ring has gyromagnetic ratio $q/2mc$. The same value applies to any nonrelativistic spinning object whose charge and mass have the same distribution and motion, because the mean charge and mass distributions can be decomposed into a sum of rings.

In classical electromagnetism a current loop with magnetic dipole moment $\vec{\mu}$ in an externally applied magnetic field $\mathbf{B}$ has energy

$$U = -\vec{\mu} \cdot \mathbf{B}. \tag{23.9}$$

The energy U is minimized when the magnetic moment is parallel to $\mathbf{B}$. (This describes the tendency of a bar magnet to point along the direction of the applied magnetic field, so as to minimize U.) In an inhomogeneous magnetic field, there is a force on the magnetic dipole tending to move it in the direction that reduces U. (This is the familiar attraction of one bar magnet to another.) The force is the gradient of the potential energy:

$$\mathbf{F} = -\nabla U. \tag{23.10}$$

Problem (II.20) shows that in quantum mechanics an applied static homogeneous magnetic field $\mathbf{B}$ adds to the Hamiltonian for a particle of charge q the term $-q\mathbf{B} \cdot \mathbf{L}/(2mc)$, where $\mathbf{L}$ is the orbital angular momentum operator. This is equation (23.9), with the gyromagnetic ratio of equation (23.8).

To account for the behavior of an electron in a magnetic field, Goudsmit and Uhlenbeck were led to assume that an electron has a magnetic dipole moment with twice the gyromagnetic ratio of equation (23.8). That is, an electron has a magnetic dipole moment

$$\vec{\mu} = -g\frac{e}{2mc}\mathbf{s}, \qquad g = 2.0022\ldots, \tag{23.11}$$

In the presence of a magnetic field $\mathbf{B}$, this contributes to the Hamiltonian a term $U = -\vec{\mu} \cdot \mathbf{B}$. In the expression for $\vec{\mu}$ in equation (23.11), $\mathbf{s}$ is the spin vector observable in equation (23.1). The form of the expression follows equation (23.8), where m is the electron mass and the charge e is a positive quantity, the magnitude of the electron charge. The minus sign indicates that the magnetic moment is antiparallel to the spin, as one might expect from the orbital motion model with negative charge. As indicated by the dimensionless factor g, the magnitude of the dipole moment is slightly more than twice what one would expect from a naive application of the orbital motion model. As shown in chapter 8, Dirac's relativistic electron theory predicts $g = 2$. The small correction to this value (which is known to considerably more decimal places than are shown here), is successfully predicted by the quantum field theory of electromagnetism.

An expression like (23.11) applies to the proton, with the proton mass replacing m and $g = 5.59$. Here $\vec{\mu}$ is parallel to $\mathbf{s}$, again in the direction one would expect from the orbital motion model. The magnetic moment of the neutron is $g = -3.83$, the minus sign indicating that $\vec{\mu}$ is antiparallel to $\mathbf{s}$.

Single-Particle Wave Function

To see how the spin observable is incorporated into a position representation, let us write down the wave function for a system consisting of a single particle with spin 1/2. A complete set of commuting observables is the three position coordinates $\mathbf{r}$ and one component of the spin $\mathbf{s}$. (We could include $\mathbf{s}^2$, but that always has the same value so it adds nothing.) It is conventional to use the z-component of $\mathbf{s}$.

A complete set of simultaneous eigenvectors of $\mathbf{r}$ and s_z is

$$|\mathbf{r}, +\rangle, \qquad |\mathbf{r}, -\rangle, \tag{23.12}$$

where

$$s_z|\mathbf{r}, \pm\rangle = \pm\frac{\hbar}{2}|\mathbf{r}, \pm\rangle. \tag{23.13}$$

The position and spin representation of the state vector $|\psi\rangle$ is

$$\psi_a(\mathbf{r}) = \langle\mathbf{r}, a|\psi\rangle. \tag{23.14}$$

The index a has two values, $a = \pm$, for the two possible values of the spin component along the z axis, $\pm\hbar/2$. The particle thus is described by two wave functions, $\psi_+(\mathbf{r})$ and $\psi_-(\mathbf{r})$. They can be written as the column vector

$$\psi = \begin{bmatrix} \psi_+(\mathbf{r}) \\ \psi_-(\mathbf{r}) \end{bmatrix} = \begin{bmatrix} \langle\mathbf{r}, +|\psi\rangle \\ \langle\mathbf{r}, -|\psi\rangle \end{bmatrix}. \tag{23.15}$$

The inner product of two state vectors can be written in terms of the wave functions by using completeness, as in equation (21.3):

$$
\begin{aligned}
\langle\psi|\phi\rangle &= \sum_{a=\pm}\int \langle\psi|\mathbf{r},a\rangle d^3r\langle\mathbf{r},a|\phi\rangle \\
&= \sum_{a}\int d^3r\ \psi_a(\mathbf{r})^*\phi_a(\mathbf{r}) \qquad\qquad (23.16)\\
&= \int d^3r\{\psi_+(\mathbf{r})^*\phi_+(\mathbf{r}) + \psi_-(\mathbf{r})^*\phi_-(\mathbf{r})\}.
\end{aligned}
$$

The probability that the particle with state vector $|\psi\rangle$ is found at position $\mathbf{r}$ in the volume element d^3r with its z component of spin up is, according to the rules in section 21 (eqs. [21.4] and [21.8]),

$$dP = |\langle\mathbf{r},+|\psi\rangle|^2 d^3r = |\psi_+(\mathbf{r})|^2 d^3r, \qquad (23.17)$$

and similarly for spin down. As usual, one can use completeness to check that this expression is correctly normalized if $\langle\psi|\psi\rangle = 1$. If the state vector is an eigenvector of s_z with eigenvalue $+\hbar/2$, then there is no chance the particle will be found to have spin down: $\langle\mathbf{r},-|\psi\rangle = 0$ because the eigenvectors are orthogonal.

The Hamiltonians we have considered so far do not contain the spin, so the two components of the wave function obey the same Schrödinger equation. An example where that is not so is discussed next.

Stern-Gerlach Effect

This refers to the influence on the motion of a particle in an inhomogeneous magnetic field due to the gradient of the $\vec{\mu}\cdot\mathbf{B}$ term in the Hamiltonian. The classical expression for the force due to the gradient of the magnetic field is given in equation (23.10). In this discussion of the quantum version we will ignore the presence of the magnetic field in the kinetic energy term in the Hamiltonian for a charged particle (as discussed in section 19), because that has the same effect on each spin component. Also, we will assume the electric field is negligible, so the Hamiltonian is

$$H = \frac{p^2}{2m} - \vec{\mu}\cdot\mathbf{B}. \qquad (23.18)$$

To avoid unnecessary complications with the matrix elements of the spin operators (to be discussed in the next section), let us imagine that the magnetic field is everywhere parallel to the z axis. (Since we want $\mathbf{B}$ to be a function of z that is not possible, but a full treatment of the components of $\mathbf{B}$ in the x and y directions does not add anything interesting.) With this simplification, and using equation (23.11) for $\vec{\mu}$, the Hamiltonian becomes

$$H = \frac{p^2}{2m} + \frac{ge}{2mc} B(z) s_z. \tag{23.19}$$

Schrödinger's equation is

$$i\hbar \frac{\partial}{\partial t} |\psi\rangle = H|\psi\rangle. \tag{23.20}$$

The inner product of this equation with $\langle \mathbf{r}, a|$ gives two differential equations, for $a = \pm$ in the wave functions $\langle \mathbf{r}, a|\psi\rangle = \psi_a(\mathbf{r})$. The treatment of the position part is the same as in section 22 (eq. [22.19]). The operator s_z on $|\mathbf{r}, \pm\rangle$ gives $\pm\hbar/2$, so we see that the two differential equations are

$$\begin{aligned} i\hbar \frac{\partial \psi_+}{\partial t} &= -\frac{\hbar^2}{2m} \nabla^2 \psi_+ + \frac{ge\hbar}{4mc} B(z) \psi_+, \\ i\hbar \frac{\partial \psi_-}{\partial t} &= -\frac{\hbar^2}{2m} \nabla^2 \psi_- - \frac{ge\hbar}{4mc} B(z) \psi_-. \end{aligned} \tag{23.21}$$

We can use Ehrenfest's theorem (section 14) to see how the difference in signs in the last terms affects the motion of a wave packet. The time rate of change of the z component of momentum of the particle is (eq. [14.33])

$$i\hbar \frac{\partial}{\partial t} \langle p_z \rangle = \langle [p_z, H] \rangle. \tag{23.22}$$

The commutator of p_z with the Hamiltonian in equation (23.19) is

$$[p_z, H] = -i\hbar \frac{ge}{2mc} \frac{\partial B}{\partial z} s_z. \tag{23.23}$$

We have then

$$\frac{\partial \langle p_z \rangle}{\partial t} = -\frac{ge}{2mc} \frac{\partial B}{\partial z} \langle s_z \rangle. \tag{23.24}$$

The expectation value of p_z then fixes the time rate of change of the expectation value of the z component of position, as in equation (14.41).

Suppose the spin of the particle is up, that is, the state vector is an eigenvector of s_z with eigenvalue $+\hbar/2$. Because the Hamiltonian (23.19) commutes with s_z, the spin stays up, so $\langle s_z \rangle = +\hbar/2$. Thus the factor multiplying $\partial B/\partial z$ in the right-hand side of equation (23.24) is negative, and the wave packet is driven away from the direction of the gradient of $\mathbf{B}$. If the spin is down, $\langle s_z \rangle = -\hbar/2$, the factor is positive, and the wave packet is driven the opposite way, toward the direction of the gradient of the magnetic field. The magnetic field gradient therefore separates particles in the beam with spin parallel to the gradient from those with spin antiparallel.

What happens if the system is not an eigenstate of s_z? In this case, the state vector $|\psi\rangle$ is a linear combination of eigenvectors of s_z with spin up and spin down, so ψ has nonzero components $\psi_+(\mathbf{r})$ and $\psi_-(\mathbf{r})$. As we have seen, the magnetic field gradient drives the wave packets $\psi_+(\mathbf{r})$ and $\psi_-(\mathbf{r})$ in opposite directions. Thus, in the general case the wave packet splits into two parts in space, each having a definite spin along the direction of the gradient of the magnetic field.

If we chose a basis $|\mathbf{r}, \pm\rangle$ of eigenvectors of position and s_z with the z axis not along the gradient of $\mathbf{B}$, we would of course still predict that the splitting is along the gradient of $\mathbf{B}$, but the calculation would be more difficult because the differential equations (23.21) for ψ_+ and ψ_- would not be independent. The method of computing the matrix elements of the spin observables that would be needed in this case is discussed in the next section.

The splitting of a beam of atoms into discrete beams by a magnetic field gradient was observed in 1921, by Stern and Gerlach. The effect cannot be understood in classical physics because the force $\nabla\vec{\mu} \cdot \mathbf{B}$ in equation (23.10) could assume any value between maximum and minimum, depending on the orientation of the dipole moment $\vec{\mu}$. In quantum mechanics the wave packet is split into discrete parts because the component of the spin along the magnetic field gradient is quantized.

24 Single Spin 1/2 System

Pauli Spin Matrices

The simplest of all physical systems in quantum mechanics consists of a single spin 1/2 particle whose motion can be ignored. In this case, the only interesting dynamical variables are the three components of the spin $\mathbf{s}$, with the commutation relations in equation (23.4). A complete set of commuting observables is one component of $\mathbf{s}$, conventionally taken to be s_z, and the square of $\mathbf{s}$. The latter is uninteresting because it always has the same value, so a complete set of eigenvectors can be written as

$$|+\rangle, \qquad |-\rangle. \tag{24.1}$$

These are the eigenvectors of s_z, with eigenvalues $\pm\hbar/2$.

The completeness assumption in section 20 is that any state vector $|\psi\rangle$ can be written as a linear combination of these basis vectors:

$$|\psi\rangle = |+\rangle\langle+|\psi\rangle \; + \; |-\rangle\langle-|\psi\rangle. \tag{24.2}$$

That is, any state vector is completely specified by its two components $\psi_\pm = \langle\pm|\psi\rangle$. In the matrix notation of section 20, the representation of the state vector is the column vector

$$\psi = \begin{bmatrix} \psi_+ \\ \psi_- \end{bmatrix}. \tag{24.3}$$

The inner product of two state vectors is the matrix product of row and column vectors:

$$\langle\psi|\phi\rangle = \sum_{a=\pm} \langle\psi|a\rangle\langle a|\phi\rangle = \psi_+^*\phi_+ + \psi_-^*\phi_- = [\psi_+^* \; \psi_-^*] \begin{bmatrix} \phi_+ \\ \phi_- \end{bmatrix}. \tag{24.4}$$

The last expression is multiplied out by the usual rules for matrix multiplication. The spin operators likewise are completely specified in this representation by their matrix elements in the basis $|\pm\rangle$, which we will now compute.

The $|\pm\rangle$ are eigenvectors of s_z:

$$s_z|+\rangle = \frac{\hbar}{2}|+\rangle, \qquad s_z|-\rangle = -\frac{\hbar}{2}|-\rangle. \tag{24.5}$$

Thus we immediately see that the matrix elements of the operator s_z are

$$\begin{aligned} \langle+|s_z|+\rangle &= \frac{\hbar}{2}\langle+|+\rangle = \frac{\hbar}{2}, \\ \langle-|s_z|+\rangle &= \frac{\hbar}{2}\langle-|+\rangle = 0, \\ \langle-|s_z|-\rangle &= -\frac{\hbar}{2}\langle-|-\rangle = -\frac{\hbar}{2}. \end{aligned} \tag{24.6}$$

The middle line vanishes by orthogonality. The results can be collected in the matrix

$$\langle a|s_z|b\rangle = \frac{\hbar}{2}\begin{bmatrix}1 & 0\\ 0 & -1\end{bmatrix}. \tag{24.7}$$

To get the matrix elements of s_x and s_y, recall from section 17 the ladder operators (eq. [17.20]),

$$\begin{aligned} s_+ &= s_x + is_y, \\ s_- &= s_x - is_y. \end{aligned} \tag{24.8}$$

Since s_+ raises the eigenvalue of s_z by the amount $\hbar$, and s_- lowers it by the same amount, s_+ must annihilate $|+\rangle$ and s_- must bring $|+\rangle$ to a vector proportional to $|-\rangle$:

$$s_+|+\rangle = 0, \qquad s_-|+\rangle = c|-\rangle. \tag{24.9}$$

To find the normalizing constant c, note that the dual of the second of these equations is (eq. [20.26])

$$\langle +|s_-^\dagger = \langle +|s_+ = c^*\langle -|, \tag{24.10}$$

because the adjoint of s_- is s_+. Therefore,

$$\langle +|s_+s_-|+\rangle = c^*c\langle -|-\rangle = |c|^2. \tag{24.11}$$

The product s_+s_- can be written in terms of s^2 and s_z (as in eq. [17.33]):

$$\begin{aligned} s_+s_- &= (s_x + is_y)(s_x - is_y) \\ &= s_x^2 + s_y^2 - i[s_x, s_y] \\ &= s_x^2 + s_y^2 + \hbar s_z \\ &= s^2 - s_z^2 + \hbar s_z. \end{aligned} \tag{24.12}$$

This in equation (24.11) gives

$$\begin{aligned} |c|^2 &= \langle +|s^2 - s_z^2 + \hbar s_z|+\rangle \\ &= \hbar^2\left(\frac{1}{2}\frac{3}{2} - \frac{1}{4} + \frac{1}{2}\right) \\ &= \hbar^2. \end{aligned} \tag{24.13}$$

Thus the normalizing constant c in equation (24.9) is equal to $\hbar$ up to an arbitrary phase factor (that is, a number $e^{i\phi}$ with unit modulus). The standard convention is to take the phase factor to be unity, so $c = \hbar$, and equation (24.9) is

$$s_-|+\rangle = \hbar|-\rangle. \tag{24.14}$$

This is a special case of the relations found in problem (II.18).

The result of operating on equation (24.14) with s_+ and using equation (24.12) is

$$(s^2 - s_z^2 + \hbar s_z)|+\rangle = \hbar s_+|-\rangle. \tag{24.15}$$

As in equation (24.13), the left side can be simplified because $|+\rangle$ is an eigenvector of s^2 and s_z. This yields

$$s_+|-\rangle = \hbar|+\rangle. \tag{24.16}$$

The phase in this expression is fixed by the phase choice in equation (24.14).

With equations (24.14) and (24.16) we can work out the matrix elements of s_x and s_y. In terms of the ladder operators (24.8), these spin components are

$$s_x = \frac{s_+ + s_-}{2}, \qquad s_y = \frac{s_+ - s_-}{2i}. \tag{24.17}$$

For example,

$$\begin{aligned} \langle +|s_x|+\rangle &= 0, \\ \langle +|s_x|-\rangle &= \langle +|s_+|-\rangle/2 = \hbar/2, \end{aligned} \tag{24.18}$$

and so on.

The final results are that the matrix elements of the components of the spin operator in the representation of equation (24.1) are

$$\mathbf{s} = \frac{\hbar}{2}\vec{\sigma}, \tag{24.19}$$

where the three components of $\vec{\sigma}$ are the Pauli spin matrices,

$$\sigma_x = \begin{bmatrix} 0 & 1 \\ 1 & 0 \end{bmatrix}, \quad \sigma_y = \begin{bmatrix} 0 & -i \\ i & 0 \end{bmatrix}, \quad \sigma_z = \begin{bmatrix} 1 & 0 \\ 0 & -1 \end{bmatrix}. \tag{24.20}$$

In this matrix representation, the eigenvectors of the spin operators are eigenvectors of the Pauli spin matrices. One readily checks that the eigenvector of σ_x with eigenvalue $+1$ is

$$\psi_{x+} = \frac{1}{2^{1/2}} \begin{bmatrix} 1 \\ 1 \end{bmatrix}. \tag{24.21}$$

The factor $2^{1/2}$ normalizes the state vector (eq. [24.4]). This is the representation of the eigenvector of s_x with eigenvalue $+\hbar/2$. The eigenvector of σ_x with eigenvalue -1 is

$$\psi_{x-} = \frac{1}{2^{1/2}} \begin{bmatrix} 1 \\ -1 \end{bmatrix}. \tag{24.22}$$

The eigenvectors of σ_y and σ_z are

$$\begin{aligned} \psi_{y+} &= \frac{1}{2^{1/2}} \begin{bmatrix} 1 \\ i \end{bmatrix}, \qquad \psi_{y-} = \frac{1}{2^{1/2}} \begin{bmatrix} 1 \\ -i \end{bmatrix}, \\ \psi_{z+} &= \begin{bmatrix} 1 \\ 0 \end{bmatrix}, \qquad \psi_{z-} = \begin{bmatrix} 0 \\ 1 \end{bmatrix}. \end{aligned} \tag{24.23}$$

If the system is in the state $\phi = \psi_{x+}$, the result of the measurement of s_x certainly is $+\hbar/2$. If instead the y component of the spin is measured in the state $\phi = \psi_{x+}$, what is the probability of finding the value $+\hbar/2$? The rules of quantum mechanics (eq. [21.4]) say the probability is the absolute value squared of the probability amplitude, which is the inner product of the state vector $|\phi\rangle = |\psi_{x+}\rangle$ with the eigenvector $|\psi_{y+}\rangle$ of s_y with eigenvalue $+\hbar/2$. The probability amplitude is

$$\langle\psi_{y+}|\phi\rangle = \langle\psi_{y+}|\psi_{x+}\rangle = \frac{1}{2}\begin{bmatrix} 1 & -i \end{bmatrix} \begin{bmatrix} 1 \\ 1 \end{bmatrix} = (1-i)/2. \tag{24.24}$$

The elements in the row vector in the third expression are the complex conjugate of the elements in the column vector in equation (24.23), because $\langle\psi|+\rangle = \langle+|\psi\rangle^*$. The probability is then

$$P = |\langle\psi_{y+}|\phi\rangle|^2 = 1/2. \tag{24.25}$$

The Pauli spin matrices satisfy two useful identities. First, the square of any of the matrices is the unit matrix,

$$\sigma_\alpha^2 = 1, \tag{24.26}$$

as can be checked by multiplying out the matrix products. Second, it is easy to check that the product of two different spin matrices is the third:

$$\sigma_x \sigma_y = i\sigma_z,$$
$$\sigma_y \sigma_z = i\sigma_x, \tag{24.27}$$
$$\sigma_z \sigma_x = i\sigma_y.$$

We see from these equations that the commutation relations for the spin matrices are

$$[\sigma_x, \sigma_y] = 2i\sigma_z, \tag{24.28}$$

and so on. This agrees with the angular momentum commutation relations in equation (23.4) (allowing for the factor $\hbar/2$ in eq. [24.19]). These identities will be used in the discussion of the angular momentum operators as generators of rotations.

Rotations

As discussed in sections 17 and 23, the components of angular momentum are generators of rotation. The unitary operator that rotates a spin 1/2 system by angle χ around the x axis is (eq. [23.3])

$$U_x = \exp -i\chi s_x/\hbar = \exp -i\chi\sigma_x/2. \tag{24.29}$$

The last expression replaces the spin operators with the matrix representation in equation (24.19). It will be recalled that the exponential function U_x of the matrix σ_x is defined as the series expansion in powers of σ_x. Since $\sigma_x^2 = 1$ (eq. [24.26]), all even powers of σ_x in this expansion are the identity, and all odd powers of σ_x are just σ_x. In the series expansion of $e^{i\theta}$, the even powers add up to $\cos\theta$, and the odd powers to $\sin\theta$, so we have

$$U_x = e^{-i\chi\sigma_x/2} = \cos(\chi/2) - i\sigma_x \sin(\chi/2). \tag{24.30}$$

The same applies to rotations about the y and z axes generated by s_y and s_z.

Now let us rotate a state vector. In the Pauli spin matrix representation, the vector

$$\psi_{z+} = \begin{bmatrix} 1 \\ 0 \end{bmatrix} \tag{24.31}$$

is an eigenvector of σ_z with eigenvalue $+1$ (eq. [24.23]), so it represents a state in which the z component of the spin certainly is up. We can rotate this to an eigenvector of σ_y with eigenvalue $+1$ (which means it represents an eigenvector of s_y with eigenvalue $+\hbar/2$) by rotating the system around the x axis by $\chi = -\pi/2$ radians. (Recall that a positive rotation around the x axis swings a point along the positive y axis toward the positive z axis, so we want to rotate the opposite direction.) Equation (24.30) says the rotated state vector is

$$\psi' = \{\cos(\pi/4) + i\sin(\pi/4)\sigma_x\}\psi_{z+}. \tag{24.32}$$

The matrix product on the right is

$$\sigma_x\psi_{z+} = \begin{bmatrix} 0 & 1 \\ 1 & 0 \end{bmatrix}\begin{bmatrix} 1 \\ 0 \end{bmatrix} = \begin{bmatrix} 0 \\ 1 \end{bmatrix}. \tag{24.33}$$

Since $\sin(\pi/4) = \cos(\pi/4) = 1/2^{1/2}$, the result of adding the elements in the top row and the elements in the bottom row of the column vectors in equation (24.32) is

$$\psi' = \frac{1}{2^{1/2}}\begin{bmatrix} 1 \\ i \end{bmatrix}. \tag{24.34}$$

This is an eigenvector of σ_y with eigenvalue $+1$, properly normalized to $\langle\psi'|\psi'\rangle = 1$ (eq. [24.23]). As expected, the state vector has been rotated from spin certainly up along the z axis to spin certainly up along the y axis.

Suppose we rotate the state (24.31) with spin up along the z axis through the angle $-\chi$ around the x axis. This brings a unit vector along the z axis to a unit vector $\mathbf{n}$ in the yz plane tilted at angle χ away from the z axis toward the y axis, at polar coordinates $\theta = \chi$ and $\phi = \pi/2$. We see from equation (24.30) that the rotated state vector is

$$\begin{aligned} \psi' &= \{\cos(\chi/2) + i\sin(\chi/2)\sigma_x\}\psi_{z+} \\ &= \begin{bmatrix} \cos\chi/2 \\ i\sin\chi/2 \end{bmatrix}. \end{aligned} \tag{24.35}$$

In terms of ket vectors, this is

$$|\psi'\rangle = \cos\chi/2\,|+\rangle + i\sin\chi/2\,|-\rangle. \tag{24.36}$$

This state vector represents a system in which the component of the spin measured along the tilted unit vector $\mathbf{n}$ certainly is $+\hbar/2$. The component of the matrix vector $\vec{\sigma}$ along $\mathbf{n}$ is

$$\sigma_n = \vec{\sigma} \cdot \mathbf{n} = \sigma_z \cos\chi + \sigma_y \sin\chi. \tag{24.37}$$

It is an interesting exercise to work through the matrix products and sums to check that the rotated state vector, ψ', in equation (24.35) really is an eigenvector of σ_n (eq. [24.37]) with eigenvalue +1.

We can use the rotated eigenvectors to answer the following question. Suppose a system has been arranged so the spin measured along a given axis has a definite value. What is the predicted result of a measurement of the component of the spin along some other axis? The state ψ' in equation (24.35) has spin $+\hbar/2$ measured along the axis $\mathbf{n}$ with polar angles $\theta = \chi$ and $\phi = \pi/2$. The probability amplitude for the result $+\hbar/2$ in a measurement of the z component of the spin is the inner product of ψ' with the eigenvector ψ_{z+} (eq. [24.31]) of s_z with eigenvalue $+\hbar/2$:

$$\langle\psi_{z+}|\psi'\rangle = [1 \quad 0]\begin{bmatrix} \cos\chi/2 \\ i\sin\chi/2 \end{bmatrix} = \cos\chi/2. \tag{24.38}$$

The probability is the square of this amplitude:

$$P = \cos^2(\chi/2). \tag{24.39}$$

If $\chi = 0$, the probability is unity. If $\chi = \pi/2$, the probability is $P = 1/2$, which means there is equal probability of finding that the spin in the z direction is up or down, for we have rotated the system to an eigenvector of s_y. If $\chi = \pi$ the probability is zero: the system has been rotated to a state in which the spin along the z axis is down.

It will be noted finally that if $\chi = 2\pi$ the state in equation (24.35) is

$$\psi' = \begin{bmatrix} -1 \\ 0 \end{bmatrix}. \tag{24.40}$$

This is an eigenvector of s_z with eigenvalue $+\hbar/2$: the 2π rotation has brought the state back to its original orientation. However, the state vector has the opposite sign from what it started with (eq. [24.31]). The change in sign is observable. The wave function for a spin 1/2 neutron can be split into two parts by passing the neutron beam through a thin sheet of material that partially reflects the wave function, partially transmits it. The two parts can be brought together again by further reflections. One part of the wave function can be rotated by passing it through a magnetic field, as discussed next. The sign change caused by a full rotation of 2π radians is observable in the interference of the two parts of the recombined wave packet.

Spin Precession in a Magnetic Field

Suppose the spin 1/2 particle is placed in a static magnetic field. Because the only dynamical variable is the spin, the Hamiltonian is given by equation (23.11):

$$H = -\vec{\mu} \cdot \mathbf{B} = \frac{ge}{2mc} \mathbf{s} \cdot \mathbf{B}. \tag{24.41}$$

The time evolution of the state vector is given by Schrödinger's equation,

$$i\hbar \frac{\partial |\psi\rangle}{\partial t} = H|\psi\rangle. \tag{24.42}$$

Since H does not depend on time, we can write the formal solution to Schrödinger's equation as (eq. [21.20])

$$|\psi(t)\rangle = e^{-iHt/\hbar}|\psi\rangle. \tag{24.43}$$

With equation (24.41), this is

$$|\psi(t)\rangle = e^{-i\vec{\theta}\cdot\mathbf{s}/\hbar}|\psi\rangle, \tag{24.44}$$

where

$$\vec{\theta} = \frac{get\mathbf{B}}{2mc}. \tag{24.45}$$

As in equations (23.2) and (24.30), we recognize equation (24.44) as a rotation of the initial state around the axis defined by $\mathbf{B}$ by the angle $\theta \propto t$. That is, the state is rotating with angular velocity ω, period T:

$$\omega = \frac{geB}{2mc} \qquad T = \frac{2\pi}{\omega} = \frac{4\pi mc}{geB}. \tag{24.46}$$

The torque caused by the magnetic field is causing the spin to precess, just as a gyroscope precesses when torque is applied in a direction not parallel to the gyroscope angular momentum.

It is a useful exercise to solve Schrödinger's equation for spin precession another way, using the spin matrices to write out the differential equations for the time evolution of the two components of the state vector. To simplify the equations, let us take $\mathbf{B}$ to be along the z axis. Then the Hamiltonian (24.41) in the Pauli matrix representation of equation (24.19) is

$$H = \frac{geB\hbar}{4mc}\sigma_z = \mu_o B\sigma_z, \tag{24.47}$$

where

$$\mu_o = \frac{g}{2}\frac{e\hbar}{2mc}. \tag{24.48}$$

The second factor is called the Bohr magneton.

The state vector has two components, $\psi_+(t)$ and $\psi_-(t)$ (eq. [24.3]). The z component of the spin operating on the state vector gives

$$\sigma_z\psi = \begin{bmatrix} 1 & 0 \\ 0 & -1 \end{bmatrix}\begin{bmatrix} \psi_+ \\ \psi_- \end{bmatrix} = \begin{bmatrix} \psi_+ \\ -\psi_- \end{bmatrix}, \tag{24.49}$$

so the two components of Schrödinger's equation (24.2) are

$$\begin{aligned} i\hbar\frac{\partial\psi_+}{\partial t} &= \mu_o B\psi_+, \\ i\hbar\frac{\partial\psi_-}{\partial t} &= -\mu_o B\psi_-. \end{aligned} \tag{24.50}$$

These are simpler versions of equation (23.21). The solutions are

$$\begin{aligned} \psi_+(t) &= A_+e^{-i\mu_o Bt/\hbar}, \\ \psi_-(t) &= A_-e^{i\mu_o Bt/\hbar}. \end{aligned} \tag{24.51}$$

The normalization condition on the two constants of integration is (eq. [24.4])

$$|A_+|^2 + |A_-|^2 = 1. \tag{24.52}$$

Suppose that at time $t = 0$ the spin is up along the z axis. Then suitable initial conditions are $A_+ = 1$, $A_- = 0$. The solution (24.51) says the spin along the z axis stays up, while the phase of the state vector increases at the rate $\mu_o B/\hbar$ radians per unit of time. As expected, the system remains an eigenstate of s_z, because s_z commutes with the Hamiltonian.

Suppose that at time $t = 0$ the spin measured along the x axis is $+\hbar/2$. In this case, suitable initial conditions are $A_+ = A_- = 1/2^{1/2}$ (eq. [24.21]). The solution (24.51) is

$$\psi = \frac{1}{2^{1/2}} \begin{bmatrix} e^{-i\mu_o Bt/\hbar} \\ e^{i\mu_o Bt/\hbar} \end{bmatrix}. \tag{24.53}$$

This equation says the system is precessing about the z axis. The initial conditions have been fixed so that at time $t = 0$ the state vector is an eigenvector of σ_x with eigenvalue $+1$. A quarter of the precession period T later (eq. [24.46]), the state vector (24.53) is

$$\psi = \frac{1}{2} \begin{bmatrix} 1-i \\ 1+i \end{bmatrix} = \frac{1-i}{2} \begin{bmatrix} 1 \\ i \end{bmatrix}. \tag{24.54}$$

Up to a phase factor, this is the eigenvector of σ_y with eigenvalue $+1$ in equation (24.23). That is, the system has swung around a quarter of a revolution. It is left as an exercise to check that at after half of the period of revolution defined in equation (24.46) the system has swung through half a revolution, with the state vector (24.53) now an eigenvector of σ_x with eigenvalue -1, and so on.

25 Two Spin 1/2 Particles

Singlet and Triplet States

Now let us consider a slightly more complicated system, consisting of two spin 1/2 particles. It will be assumed that the only significant dynamical variables are the components of the two spin operators, $\mathbf{s}(1)$ and $\mathbf{s}(2)$, for the particles. Spin operators belonging to different particles commute, and the components of the spin for one of the particles satisfy the angular momentum commutation relations (eq. [23.4]). Thus we have

$$\begin{aligned} [s_x(1), s_y(1)] &= i\hbar s_z(1), \\ [s_x(2), s_y(2)] &= i\hbar s_z(2), \\ [s_x(1), s_y(2)] &= 0, \end{aligned} \tag{25.1}$$

and so on.

The total spin vector operator, $\mathbf{s}$, for the system is the sum of the operators for the two particles:

$$\mathbf{s} = \mathbf{s}(1) + \mathbf{s}(2). \tag{25.2}$$

As in equation (23.1), this equation means the components add. Because $\mathbf{s}(1)$ and $\mathbf{s}(2)$ commute with each other, the components of $\mathbf{s}$ satisfy the angular momentum commutation relations (23.4).

The square of the total spin operator is $s^2 \equiv s_x^2 + s_y^2 + s_z^2$. On multiplying out the terms $s_x^2 = (s_x(1) + s_x(2))^2$, and so on, we get

$$s^2 = s(1)^2 + s(2)^2 + 2\mathbf{s}(1) \cdot \mathbf{s}(2). \tag{25.3}$$

We can choose two different representations, based on different choices of the set of commuting observables, as in equations (17.72) and (17.73):

$$\begin{aligned} &\text{set A:}\;\; s(1)^2,\; s_z(1),\; s(2)^2,\; s_z(2)\;; \\ &\text{set B:}\;\; s^2,\; s_z,\; s(1)^2,\; s(2)^2. \end{aligned} \tag{25.4}$$

The observable s^2 cannot be added to set A because $\mathbf{s}(1)\cdot\mathbf{s}(2)$ in equation (25.3) does not commute with $s_z(1)$. The observable s_z commutes with the observables in set A, but that adds nothing because s_z is just the sum of the operators $s_z(1)$ and $s_z(2)$ in set A. Since the particles have spin 1/2 the operators $s(1)^2$ and $s(2)^2$ always are equal to $3\hbar^2/4$, so we can ignore them. The interesting parts of the two sets of commuting observables are then

$$\begin{aligned} &\text{set A:}\;\; s_z(1),\; s_z(2)\;; \\ &\text{set B:}\;\; s^2,\; s_z. \end{aligned} \tag{25.5}$$

A complete set of eigenvectors for set A is $|m_1, m_2\rangle$, with eigenvalues $\pm\hbar/2$. These states will be written as

$$|++\rangle, \quad |+-\rangle, \quad |-+\rangle, \quad |--\rangle. \tag{25.6}$$

We know from the angular momentum operator algebra worked out in section 17 that a complete set of eigenvectors of set B can be labeled as (eq. [17.41])

$$|s, m\rangle, \qquad -s \leq m \leq s. \tag{25.7}$$

These states satisfy the eigenvalue equations

$$s^2|s,m\rangle = \hbar^2 s(s+1)|s,m\rangle,$$
$$s_z|s,m\rangle = \hbar m|s,m\rangle. \tag{25.8}$$

By the triangle rule in section 17, the possible values of the quantum number s are $s = 0$ and $s = 1$ (eq. [17.74]). It is an interesting exercise to derive this result by working the matrix eigenvalue calculation outlined in section 17 for the particular case of two spin $1/2$ particles, as follows.

Completeness says that we can write each of the eigenvectors of set B as a linear combination of the eigenvectors of set A:

$$|s,m\rangle = \sum |m_1,m_2\rangle\langle m_1,m_2|s,m\rangle. \tag{25.9}$$

The expansion coefficients $\langle m_1,m_2|s,m\rangle$ in this sum are derived with the help of the ladder operators

$$s_\pm = s_x \pm i s_y = s_\pm(1) + s_\pm(2). \tag{25.10}$$

As in equation (24.12), the angular momentum commutation relations say these operators satisfy the equations

$$s^2 = s_- s_+ + s_z^2 + \hbar s_z \;\; = s_+ s_- + s_z^2 - \hbar s_z. \tag{25.11}$$

To begin the calculation, note that

$$s_z|++\rangle = \hbar|++\rangle, \qquad s_+|++\rangle = 0. \tag{25.12}$$

The first equation follows because $|++\rangle$ is an eigenvector of $s_z(1)$ and $s_z(2)$ with eigenvalues $\hbar/2$, and $s_z = s_z(1) + s_z(2)$. Thus $|++\rangle$ is an eigenvector of s_z with $m = 1$. The second of equations (25.12) follows because $s_+ = s_+(1) + s_+(2)$, and neither of the raising operators can raise the z component of the spin in this state. Since the eigenvalue of s_z cannot be raised by s_+, the state $|++\rangle$ must be an eigenvector of s^2 with $s = m = 1$:

$$s^2|++\rangle = 1\cdot 2\cdot \hbar^2|++\rangle. \tag{25.13}$$

One can check this in a more direct way by using the first of equations (25.11) and equation (25.12). We have then

$$|s=1,m=1\rangle = |++\rangle. \tag{25.14}$$

The two eigenvectors $|1,1\rangle$ and $|++\rangle$ could differ by a phase factor; in equation (25.14) the factor has been arbitrarily set equal to unity.

To get the eigenvector $s = 1$, $m = 0$, operate on equation (25.14) with the lowering operator s_-. We get (on taking the same phase convention for both particles)

$$\begin{aligned} |1,0\rangle &\propto s_-|1,1\rangle = s_-|++\rangle \\ &= \{s_-(1) + s_-(2)\}|++\rangle \\ &\propto |-+\rangle + |+-\rangle. \end{aligned} \tag{25.15}$$

The result of applying s_- to this is

$$|1,-1\rangle = |--\rangle. \tag{25.16}$$

This could have been obtained by the same argument that led to equation (25.14).

We have written down three linear combinations of the four $|m_1, m_2\rangle$. There is just one more independent combination, which we can write as

$$|0,0\rangle \propto |+-\rangle - |-+\rangle. \tag{25.17}$$

This state satisfies

$$\begin{aligned} s_z|0,0\rangle &= 0, \\ s_+|0,0\rangle &\propto |++\rangle - |++\rangle = 0. \end{aligned} \tag{25.18}$$

The first equation says the s_z quantum number is $m = 0$. The second equation, which follows as in equation (25.15), says m cannot be raised (or lowered), so this must be an eigenvector of total angular momentum with $s = 0$. This is the fourth eigenvector of the set B in equation (25.5).

To summarize, the compatible observables s^2 and s_z in set B have three eigenvectors with $s = 1$, one with $s = 0$ (consistent with the triangle rule in eq. [17.74]). The first three are called the triplet states, and $s = 0$ is the singlet state. With the standard phase factors, the states are

$$
\begin{aligned}
|1,1\rangle &= |++\rangle, \\
|1,0\rangle &= \frac{|+-\rangle + |-+\rangle}{2^{1/2}}, \\
|1,-1\rangle &= |--\rangle, \\
|0,0\rangle &= \frac{|+-\rangle - |-+\rangle}{2^{1/2}}.
\end{aligned}
\tag{25.19}
$$

It is easy to check that these linear combinations $|s, m\rangle$ of the orthonormal state vectors $|m_1, m_2\rangle$ are orthogonal and normalized. We know they are a complete set of eigenvectors of s^2 and s_z for two spin 1/2 particles, because there are only four linearly independent combinations to be made out of the four eigenvectors $|\pm, \pm\rangle$ of set A in equation (25.6).

For the discussion of Bell's theorem in measurement theory, we will need the following probability. Suppose the two particles are in the singlet spin state $|0,0\rangle$. We can imagine simultaneous measurements of one component of the spin of particle 1 and one component of the spin of particle 2, because the spin operators for the two particles commute. For particle 1, the z component of the spin will be measured. For particle 2, the measurement will be the component of the spin along the axis $\mathbf{n}$ in the yz plane tilted at angle χ to the z axis. To compute the probabilities for possible results of measurements along these axes, we need the simultaneous eigenvectors of $s_z(1)$ and $\mathbf{n}\cdot\mathbf{s}(2)$. The eigenvector with both eigenvalues equal to $+\hbar/2$ is

$$
|\phi\rangle = \cos\chi/2\,|++\rangle + i\sin\chi/2\,|+-\rangle. \tag{25.20}
$$

This is the same as equation (24.36) for particle 2. The same calculation applies here because the quantum number for $s_z(1)$ is unaffected by all the operations on the quantum numbers for particle 2 that produce an eigenvector of $\mathbf{n}\cdot\mathbf{s}(2)$.

The probability amplitude for finding that the measurement of the spin of particle 1 along the z axis is $+\hbar/2$ and that the measurement of the spin of particle 2 along the axis $\mathbf{n}$ is $+\hbar/2$ is the inner product of the eigenvector $\langle\phi|$ in equation (25.20) with the given singlet state vector $|0,0\rangle$ of the system. The singlet state is given in the last line of equation (25.19). Using the orthogonality of the $|\pm,\pm\rangle$, we see that the probability amplitude works out to

$$
\langle\phi|0,0\rangle = \frac{i}{2^{1/2}}\sin\chi/2. \tag{25.21}
$$

The probability is the absolute value squared of this amplitude:

$$P = \frac{1}{2}\sin^2(\chi/2). \tag{25.22}$$

This vanishes if $\chi = 0$: if the spin of particle 1 along the z axis is found to be up, then the spin of 2 along this axis has to be down to make the sum of the two spins vanish in the singlet state. If $\chi = \pi$, the probability is $P = 1/2$, again as expected because in the isotropic singlet state the spin of particle 1 is up with 50 percent probability, and if the spin of particle 1 is up the spin of particle 2 certainly is down.

Two Interacting Spin 1/2 Systems

The interaction of the magnetic dipole moments of the electron and proton in a hydrogen atom contributes to the Hamiltonian a term proportional to the dot product of the spin operators of the two particles. This interaction is discussed in section 34. As a preliminary computation, we consider here a system in which the only dynamical variables are the two spin vector observables, each with spin 1/2, with Hamiltonian

$$H = c\,\mathbf{s}(1)\cdot\mathbf{s}(2). \tag{25.23}$$

Since H is self-adjoint, the constant c must be real.

The total spin of the system is

$$\mathbf{s} = \mathbf{s}(1) + \mathbf{s}(2). \tag{25.24}$$

It is left as an exercise to show that the components of the total spin commute with the Hamiltonian (25.23):

$$[H, s_\alpha] = 0. \tag{25.25}$$

This form for the Hamiltonian is adopted because it conserves angular momentum.

The trick for solving this energy eigenvalue problem is to note that the square of the total angular momentum is (eq. [25.3])

$$s^2 = s(1)^2 + s(2)^2 + 2\mathbf{s}(1)\cdot\mathbf{s}(2). \tag{25.26}$$

Thus we can rewrite the Hamiltonian (25.23) as

$$\begin{aligned} H &= \frac{c}{2}(s^2 - s(1)^2 - s(2)^2) \\ &= \frac{c}{2}\left(s^2 - 2\hbar^2 \frac{1}{2}\frac{3}{2}\right). \end{aligned} \tag{25.27}$$

This means eigenvectors of s^2 are eigenvectors of H. We saw that two spin 1/2 particles can have total spin $s = 0$ or 1 (eq. [25.19]). The singlet $s = 0$ state has energy

$$E_s = \frac{c}{2}\left(0 - \frac{3}{2}\hbar^2\right) = -\frac{3}{4}c\hbar^2. \tag{25.28}$$

The triplet $s = 1$ states have energy

$$E_t = \frac{c\hbar^2}{4}. \tag{25.29}$$

For practice, let us find the energy eigenvector for the Hamiltonian (25.23) another way, as the solutions to a matrix eigenvalue problem. The matrix elements of the spin operators for each particle can be chosen with the same phase conventions as for the Pauli spin matrices (eqs. [24.14], [24.16], and [24.20]). We have for example

$$\begin{aligned} \langle + - |s_x(2)| + + \rangle &= \frac{\hbar}{2}, \\ \langle + - |s_y(2)| + + \rangle &= \frac{i\hbar}{2}, \\ \langle + - |s_z(2)| + + \rangle &= 0. \end{aligned} \tag{25.30}$$

The matrix elements of products of spin operators for the two particles are

$$\begin{aligned} &\langle + - |s_x(1)s_x(2)| - + \rangle = \hbar^2/4, \\ &\langle + - |s_y(1)s_y(2)| - + \rangle = (-i)(i)\hbar^2/4 = \hbar^2/4, \\ &\langle + - |s_z(1)s_z(2)| - + \rangle = 0, \end{aligned} \tag{25.31}$$

and so on. The matrix elements of H in the representation $|++\rangle$, $|+-\rangle$, $|-+\rangle$, and $|--\rangle$ are

$$H = \frac{c\hbar^2}{4}\begin{bmatrix} 1 & 0 & 0 & 0 \\ 0 & -1 & 2 & 0 \\ 0 & 2 & -1 & 0 \\ 0 & 0 & 0 & 1 \end{bmatrix}. \tag{25.32}$$

The energy eigenvalue problem $H\psi = E\psi$ is

$$\frac{c\hbar^2}{4}\begin{bmatrix} 1 & 0 & 0 & 0 \\ 0 & -1 & 2 & 0 \\ 0 & 2 & -1 & 0 \\ 0 & 0 & 0 & 1 \end{bmatrix}\begin{bmatrix} e \\ f \\ g \\ h \end{bmatrix} = E\begin{bmatrix} e \\ f \\ g \\ h \end{bmatrix}. \tag{25.33}$$

The column vector represents the ket vector

$$|\psi\rangle = e|++\rangle + f|+-\rangle + g|-+\rangle + h|--\rangle. \tag{25.34}$$

One solution to equation (25.33) is $e = 1$ with $f = g = h = 0$, and $E = c\hbar^2/4$. This is the first of the triplet eigenvectors in equation (25.29). Another triplet eigenvector is $h = 1$ with $e = f = g = 0$. The eigenvectors orthogonal to these two must have $e = h = 0$. In this case the result of multiplying out the matrix product in the right-hand side of equation (25.33) and then equating the elements on the left- and right-hand sides is

$$\begin{aligned} (-f + 2g)c\hbar^2/4 &= Ef, \\ (2f - g)c\hbar^2/4 &= Eg. \end{aligned} \tag{25.35}$$

The solutions to this equation are

$$E = \frac{c\hbar^2}{4}, \quad -\frac{3c\hbar^2}{4}. \tag{25.36}$$

These complete the solutions in equations (25.28) and (25.29). It is left as an exercise to check that the values for f and g for these two solutions are such that the state vector (25.34) is respectively triplet and singlet.

Finally, let us use the energy eigenvectors to work out a problem in time evolution. Suppose that at time $t = 0$ the spin of particle (1) along the z axis certainly is up, and the spin of particle (2) along the z

axis certainly is down. That means the initial value of the state vector is $|+-\rangle$. What is the probability that the spin of particle (1) measured along the z axis at a later time t is found to be up?

To find the time evolution of the state vector, write it as a sum over the energy eigenvectors, as in equation (21.21). With the Hamiltonian (25.23), energy eigenstates are eigenstates of $\mathbf{s}^2$, so using equation (25.19) for the singlet and triplet states and equations (25.28) and (25.29) for the energy eigenvalues we have

$$\begin{aligned} |\psi(t)\rangle =& \{A|++\rangle + B[|+-\rangle + |-+\rangle] + C|--\rangle\}e^{-ict\hbar/4} \\ &+ D(|+-\rangle - |-+\rangle)e^{3ict\hbar/4}. \end{aligned} \tag{25.37}$$

This is the general solution for the time evolution of the state vector. The wanted particular solution is obtained by choosing the four constants to make the state vector at time $t = 0$ agree with the given initial condition, $|\psi(0)\rangle = |+-\rangle$. This brings the solution to

$$|\psi(t)\rangle = \frac{1}{2}(|+-\rangle + |-+\rangle)e^{-ict\hbar/4} + \frac{1}{2}(|+-\rangle - |-+\rangle)e^{3ict\hbar/4}. \tag{25.38}$$

Suppose the z components of the spins of the two particles are measured at time t. The probability that the spin of particle (1) is found to be up and the spin of particle (2) is found to be up is, according to the general rules, $P_{++} = |\langle ++|\psi(t)\rangle|^2$ (eq. [21.4]). We see from equation (25.38) that this vanishes, because $\langle ++|$ is orthogonal to $|+-\rangle$ and $|-+\rangle$. The probability that at time t the spin of particle (1) is found to be up and the spin of particle (2) is found to be down along the z axis is

$$\begin{aligned} P_{+-} &= |\langle +-|\psi(t)\rangle|^2 \\ &= \left|\frac{1}{2}e^{-ict\hbar/4} + \frac{1}{2}e^{3ict\hbar/4}\right|^2 \\ &= \cos^2(ct\hbar/2). \end{aligned} \tag{25.39}$$

At $t = 0$, $P_{+-} = 1$. This is the initial condition that spin 1 is up and spin 2 is down. At time $t = \pi/(c\hbar)$, $P_{+-} = 0$: the spin of particle 1 certainly is down and the spin of 2 is up. At time $2\pi/(c\hbar)$ the state vector is back to the initial value multiplied by a phase factor, which is to say that the spins have returned to the initial conditions.

Problems

III.1) Suppose the commuting observables P and Q define a basis with continuous eigenvalues, so the completeness relation is

$$\int |p,q\rangle dp\, dq \langle p,q| = 1. \tag{III.1}$$

Use the probability assumption, as in equation (21.8), to show that the ensemble average value of the observable Q in the state $|\psi\rangle$ is

$$\langle Q\rangle = \langle\psi|Q|\psi\rangle. \tag{III.2}$$

III.2) The observable Q has the complete set of discrete, normalized eigenvectors $|n\rangle$ with eigenvalues q_n:

$$Q|n\rangle = q_n|n\rangle, \qquad \langle n|m\rangle = \delta_{n,m}. \tag{III.3}$$

The completeness relation for this basis is

$$\sum |n\rangle\langle n| = 1. \tag{III.4}$$

Find a derivation for each of the following inequalities.

a) For any element, $|\phi\rangle$, in the linear space,

$$\langle\phi|\phi\rangle \geq 0. \tag{III.5}$$

A good way to start is to insert unity, in the form of equation (III.4), between $\langle\phi|$ and $|\phi\rangle$, and then recall that $\langle n|\phi\rangle^* = \langle\phi|n\rangle$.

b) For any observable, P, the expectation value of P^2 in the state $|n\rangle$ from the above basis is not less than $|\langle n|P|m\rangle|^2$, where $|m\rangle$ is any vector from the above basis. Here a good start is to insert unity, in the form of equation (III.4), between the two operators in the matrix element $\langle n|PP|n\rangle$.

c) For any normalized state vector $|\psi\rangle$,

$$\langle\psi|Q|\psi\rangle \geq q_0\langle\psi|\psi\rangle, \tag{III.6}$$

where q_0 is the smallest of the eigenvalues q_n belonging to the eigenvectors $|n\rangle$ of Q. A good start here is to introduce unity in the form of equation (III.4) between Q and $|\psi\rangle$ in the left-hand side of equation (III.6).

d) For any two normalized state vectors $|\psi\rangle$ and $|\phi\rangle$, the real part of the inner product $\langle\phi|\psi\rangle$ satisfies

$$\text{Re}\ \langle\phi|\psi\rangle \leq 1. \tag{III.7}$$

One approach is to consider the inner product of $|\psi\rangle - |\phi\rangle$ with its dual, $\langle\psi| - \langle\phi|$.

e) For any two normalized state vectors $|\psi\rangle$ and $|\phi\rangle$, the real part of the inner product satisfies

$$\text{Re}\ \langle\phi|\psi\rangle \geq -1. \tag{III.8}$$

f) For any observable P and normalized state vector $|\psi\rangle$,

$$\langle P^2\rangle = \langle\psi|P^2|\psi\rangle \geq \langle P\rangle^2. \tag{III.9}$$

III.3) The Heisenberg uncertainty principle can be derived by operator algebra, as follows. Consider a one-dimensional system, with position and momentum observables x and p. The goal is to find the minimum possible uncertainties in the predicted values of the position and momentum in any state $|\psi\rangle$ of the system. We need the following preliminaries.

a) Suppose the self-adjoint observables q and r satisfy the commutation relation

$$[r, q] = ic, \tag{III.10}$$

where c is a constant (not an operator). Show c is real.

b) Let the system have the normalized state vector $|\psi\rangle$, and define the ket vector

$$|\phi\rangle = (\alpha r + iq)|\psi\rangle, \tag{III.11}$$

where α is a real constant (again, a number, not an operator). Use equations (III.5) and (III.10) to show

$$\alpha^2\langle r^2\rangle - \alpha c + \langle q^2\rangle \geq 0, \tag{III.12}$$

where $\langle r^2\rangle = \langle\psi|r^2|\psi\rangle$ and $\langle q^2\rangle$ are the expectation values of the squares of the observables r and q in the state $|\psi\rangle$.

d) By seeking the value of α that minimizes the left side of equation (III.12), show

$$\langle r^2 \rangle \langle q^2 \rangle \geq c^2/4. \tag{III.13}$$

e) Define the operators r and q by the equations

$$\begin{aligned} r &= x - \langle x \rangle = x - \langle \psi | x | \psi \rangle, \\ q &= p - \langle p \rangle = p - \langle \psi | p | \psi \rangle, \end{aligned} \tag{III.14}$$

where x and p are the position and momentum observables, with $[x, p] = i\hbar$. As in problem I.16 (eq. [I.15]), $\sigma_x = \langle r^2 \rangle^{1/2} = \langle \psi | (x - \langle x \rangle)^2 | \psi \rangle^{1/2}$ is the standard deviation (rms fluctuation around the mean in an ensemble of measurements) in the result of measuring the position x, and $\sigma_p = \langle q^2 \rangle^{1/2}$ is the standard deviation in the momentum. Collecting the above results, show that $\sigma_x \sigma_p \geq \hbar/2$. This is the Heisenberg uncertainty relation in position and momentum.

III.4) A spinless particle moving in one dimension has position and momentum observables $\hat{x}$ and $\hat{p}$. The position eigenvectors satisfy the normalization in equations (22.2) and (22.3). The analogous relations for the eigenvectors of momentum are

$$\hat{p}|p\rangle = p|p\rangle, \tag{III.15}$$

with the normalization

$$\langle p|p' \rangle = \delta(p - p'), \qquad \int |p\rangle dp \langle p| = 1. \tag{III.16}$$

This assumes the eigenvalues of x and p are continuous and can range from $-\infty$ to $+\infty$.

a) By considering $\langle x|\hat{p}|p\rangle$, and using the result in equation (22.9), find the differential equation satisfied by the function $\langle x|p\rangle$, and show the solution is

$$\langle x|p\rangle \propto e^{ipx/\hbar}. \tag{III.17}$$

Use the normalization in equation (III.16) (along with eq. [10.18]) to find the constant of proportionality.

b) Consider a state vector $|\psi\rangle$, with $\langle\psi|\psi\rangle = 1$. The position representation of this state vector is the wave function

$$\psi(x) = \langle x|\psi\rangle, \tag{III.18}$$

and the momentum representation is

$$f(p) = \langle p|\psi\rangle. \tag{III.19}$$

The probability distribution in the result of the measurement of the momentum in the state $|\psi\rangle$ is (eq. [21.8])

$$dP = |\langle p|\psi\rangle|^2 dp. \tag{III.20}$$

Use part (a) to express this probability distribution in terms of the position wave function $\psi(x)$ defined in equation (III.18). (The result ought to be consistent with eqs. [10.39] and [10.41].)

III.5) Show by multiplying out the matrix product of $\mathbf{s} \cdot \mathbf{n}$ (eq. [24.37]) with the column vector in equation (24.35) that the latter is in fact an eigenvector of the component $\mathbf{s} \cdot \mathbf{n}$ of the spin along the axis $\mathbf{n}$.

III.6) A single spin one-half system has Hamiltonian

$$H = \alpha s_x + \beta s_y, \tag{III.21}$$

where α and β are real numbers, and s_x and s_y are the x and y components of spin.

a) Using the representation of the spin components s_α as Pauli spin matrices, find an expression for H^2 in terms of the above parameters.

b) Use the result from part (a) to find the energy eigenvalues.

c) Find the eigenvectors of H in equation (III.21) in the Pauli spin matrix representation.

d) Suppose that at time $t = 0$ the system is an eigenstate of s_z, with eigenvalue $+\hbar/2$. Find the state vector as a function of time in the Pauli spin matrix representation.

e) Suppose the z component of the spin in the state found in part (d) is measured at time $t > 0$. Find the probability that the result is $+\hbar/2$.

III.7) Precession in a magnetic field for a spin 1/2 system was discussed in section 24. It is good practice to work the spin 1 case, as follows.

Consider a dynamical system with the three observables s_x, s_y, and s_z that satisfy the angular momentum commutation relations

$$[s_x, s_y] = i\hbar s_z, \tag{III.22}$$

and so on. The system has spin 1, so the square of the total angular momentum, $s^2 = s_x^2 + s_y^2 + s_z^2$, has the single eigenvalue $\hbar^2 \times 1 \times 2$. A convenient basis is the set of eigenvectors $|m\rangle$ of s_z with eigenvalues $m\hbar$, where $m = 1, 0, -1$. In this basis, a state vector $|\psi\rangle$ is represented by the three-component column vector $\langle m|\psi\rangle$, and the observables are represented by the 3×3 matrices

$$(s_\alpha)_{mm'} = \langle m|s_\alpha|m'\rangle. \tag{III.23}$$

a) Show that the methods in section 24 lead to the representations

$$\begin{aligned} s_x &= \frac{\hbar}{2^{1/2}} \begin{bmatrix} 0 & 1 & 0 \\ 1 & 0 & 1 \\ 0 & 1 & 0 \end{bmatrix}, \\ s_y &= \frac{\hbar}{2^{1/2}} \begin{bmatrix} 0 & -i & 0 \\ i & 0 & -i \\ 0 & i & 0 \end{bmatrix}, \\ s_z &= \hbar \begin{bmatrix} 1 & 0 & 0 \\ 0 & 0 & 0 \\ 0 & 0 & 1 \end{bmatrix}. \end{aligned} \tag{III.24}$$

b) Find the normalized eigenvector of s_x with eigenvalue $+\hbar$ in this representation.

c) Suppose the Hamiltonian is

$$H = g\mathbf{s} \cdot \mathbf{B}, \tag{III.25}$$

where g is a real constant and $\mathbf{B}$ is the constant magnetic field. Find the general solution to Schrödinger's equation,

$$i\hbar \frac{\partial|\psi\rangle}{\partial t} = H|\psi\rangle, \tag{III.26}$$

for **B** parallel to the z axis, in the representation of equation (III.24) (as in eq. [24.51]).

d) Suppose that at time $t = 0$ the state vector is an eigenstate of s_x with eigenvalue $+\hbar$. Use the solutions from parts (b) and (c) to find the spin precession frequency.

III.8) The spin operators belonging to a system consisting of two spin one-half particles are $\mathbf{s}_1$ and $\mathbf{s}_2$, and the total spin is $\mathbf{s} = \mathbf{s}_1 + \mathbf{s}_2$. Total spin projection operators are

$$P_1 = \frac{s^2}{2\hbar^2} \qquad P_0 = 1 - \frac{s^2}{2\hbar^2}. \tag{III.27}$$

By using the fact that any state vector $|\psi\rangle$ of the system can be written as a linear combination of total spin eigenvectors, show that $|\chi\rangle = P_1|\psi\rangle$ is a triplet eigenvector with $\langle\chi|\chi\rangle$ equal to the probability that the total spin of the state $|\psi\rangle$ is $s = 1$.

III.9) Consider a system consisting of three spin one-half particles, with spin vector operators $\mathbf{s}_1$, $\mathbf{s}_2$, and $\mathbf{s}_3$. The square of the total spin is

$$s^2 = (\mathbf{s}_1 + \mathbf{s}_2 + \mathbf{s}_3)^2. \tag{III.28}$$

By following the methods in sections 17 and 25 and in problems (II.18) and (II.19), find the eigenvalues of s^2, and for each eigenvalue find the number of different eigenvectors with the same eigenvalue.

III.10) Consider a system consisting of two interacting spin one-half objects, with spin operators $\mathbf{s}_1$ and $\mathbf{s}_2$, as in section 25. The Hamiltonian is

$$H = c\,\mathbf{s}(1) \cdot \mathbf{s}(2) + d\,s_z(1), \tag{III.29}$$

where c and d are real constants. This is a good model for the hyperfine structure of atomic hydrogen in an applied magnetic field, the first term representing the interaction between the electron and proton magnetic moments, the second the $\vec{\mu} \cdot \mathbf{B}$ energy of the electron in the applied field. (The proton magnetic moment is much smaller because of the larger mass.)

The energy eigenvalues for the Hamiltonian in equation (III.29) for the case where $d = 0$ were worked out in section 25 (eqs. [25.31] to [25.34]). Generalize the calculation to find the energy eigenvalues in the case $d \neq 0$. Sketch a graph of energy as a function of the parameter d (which is proportional to the applied magnetic field), showing how the energy eigenvalues vary with d, from $d = 0$ through to the case where the d term dominates the energy.

III.11) A Hamiltonian for a deuteron that improves on the one in problem I.14 is

$$H = \frac{p_p^2}{2m_p} + \frac{p_n^2}{2m_n} + V_1(r) + V_2(r)\,\mathbf{s}_n \cdot \mathbf{s}_p, \tag{III.30}$$

where m_n and m_p are the neutron and proton masses, V_1 and V_2 are functions of the neutron-proton separation $r = |\mathbf{r}_n - \mathbf{r}_p|$, and as usual the components of the spin operators $\mathbf{s}_n$ and $\mathbf{s}_p$ commute with each other and satisfy the angular momentum commutation relations among themselves.

a) What are the quantum numbers that can be assigned to an energy eigenstate, that is, what are the allowed eigenvalues of the conserved observables?

b) Write down the one-dimensional differential eigenvalue equation one would solve to find the energy of a state with definite energy and other quantum numbers, and state the boundary conditions.

III.12) The system described by the Hamiltonian H_o has just two orthogonal energy eigenstates, $|1\rangle$ and $|2\rangle$, with

$$\langle 1|1\rangle = 1, \quad \langle 1|2\rangle = 0, \quad \langle 2|2\rangle = 1. \tag{III.31}$$

The two eigenstates have the same eigenvalue, E_o:

$$H_o|i\rangle = E_o|i\rangle, \tag{III.32}$$

for $i = 1$ and 2.

Now suppose the Hamiltonian for the system is changed by the addition of the term V, giving

$$H = H_o + V. \tag{III.33}$$

The matrix elements of V are

$$\langle 1|V|1\rangle = 0, \quad \langle 1|V|2\rangle = V_{12}, \quad \langle 2|V|2\rangle = 0. \tag{III.34}$$

a) Find the eigenvalues of the new Hamiltonian, H, in terms of the above quantities.

b) Find the normalized eigenstates of H in terms of $|1\rangle$, $|2\rangle$, and the other given expressions.

CHAPTER 4

MEASUREMENT THEORY

We come now to the last element of quantum mechanics, the prescription for how the mathematical theory is to be related to the results of measurements. This is required for any physical theory, not just quantum mechanics. For example, the formalism of general relativity theory is quite misleading if one does not understand that coordinate dependent objects like the four-velocity of a particle must be distinguished from the scalars that are in principle measurable.

The measurement prescription in quantum mechanics can be stated in a few lines (as has already been done in sections 14 and 21 on the principles of quantum mechanics, and is done in a little more detail in the first section of this chapter). There is no controversy over this or the other elements of quantum physics. The theory has found an enormous range of applications, in all of which it has proved to be consistent with logic and experimental tests. Why then is this chapter so long? It is because the implications seem so bizarre that such thoughtful people as Einstein and Wigner have argued that the theory cannot be physically complete as it stands. A review of these bizarre implications, and the ways in which the reservations about the physical completeness of quantum physics have been rationalized or otherwise laid to rest to the satisfaction of many (but certainly not all) physicists gives us an excellent framework for a discussion of some of the physics of this subject.

The first section in this chapter reviews measurement theory in quantum mechanics, and extends the prescription to the case where the state vector is not known. Some of the "paradoxes" of quantum mechanics (none of which prove to be paradoxical) are discussed in section 27. Section 28 presents Bell's theorem, that shows there cannot be a local underlying deterministic theory for which quantum mechanics plays the role of a statistical approximation. Section 29 summarizes the worldview of a modern-day "standard thoughtful physicist."

26 Quantum Measurement Theory

As discussed in sections 14 and 21, in quantum physics the fullest possible description of a physical system is given by its state vector. It may be that a prior measurement has uniquely determined the state vector (up to an arbitrary phase factor) for the purpose of predicting the results of subsequent measurements. In this case one says the prior measurement has placed the system in a pure state. In other cases we may have to be content with assigning probabilities for various possible state vectors, in a mixed state. The use of the density matrix to describe mixed states is presented in part (b) of this section.

Measurement of a Pure State

Physics has two parts: initial conditions express our prior knowledge of a physical system, resulting from previous measurements of it, and the laws of physics yield predictions of the results of subsequent measurements. In classical physics there is no limit in principle to the precision of predictions from initial conditions. It makes sense to think of a particle as having an objectively real position $\mathbf{r}(t)$ at time t, because we can be perfectly sure that that is where the particle would be found if its position were measured. The initial conditions for quantum physics also take account of the knowledge of the system accumulated through previous observations, and the laws of the theory yield predictions for the results of further measurements. A quantum physics prediction may be so sharp as to be a sure thing, but we have also seen cases where the prediction is only probabilistic. In this latter case it is not so clear that one can think of a particle as having definite attributes such as an objectively real position $\mathbf{r}(t)$ at time t; in fact, we will have to argue that the existence of an attribute can depend on a decision to measure it.

Let us consider the measurement of a component of the spin of a spin 1/2 particle (such as a neutron) by means of the Stern-Gerlach effect discussed in section 23. The wave packet for the particle initially is moving to the right in figure 26.1. It passes into a region of inhomogeneous magnetic field. We saw in section 23 that the energy of the particle has the contribution $U = -\vec{\mu}\cdot\mathbf{B}$, where $\vec{\mu}$ is the magnetic moment of the

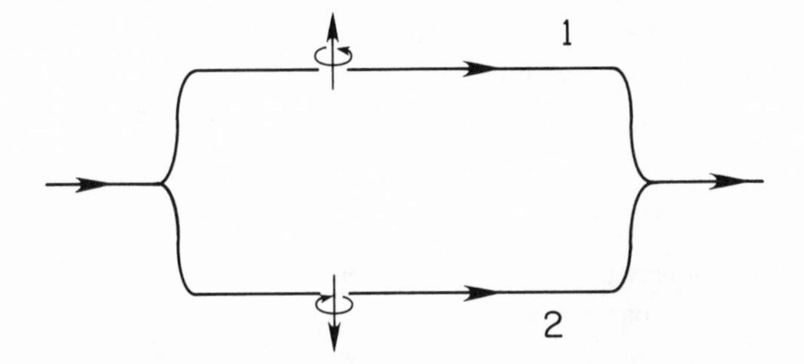

Fig. 26.1 Stern-Gerlach effect for spin 1/2 particles. A particle with spin up is deflected along path 1; a particle with spin down, along path 2.

particle and **B** is the magnetic field. The gradient of U produces a rate of change of momentum that deflects the wave packet. We can arrange that, if the z component of the spin of the particle is up, the wave packet is deflected up, along path 1 in the figure. In this case the wave packet continues moving to the right along the upper arm 1, with the spin up. If the z component of the spin is down, the packet is deflected down, to move along path 2, with the spin down.

We can imagine that a second region of inhomogeneous magnetic field is arranged so as to bring the two paths back together again, leaving the z component of spin as it was initially. We can imagine in principle that the paths 1 and 2 have identical lengths (though that would be exceedingly difficult in practice).

Let the wave function for a particle with spin up, that moves along path 1, be $\psi(\mathbf{r}, t, +)$. The last argument indicates the sign of the z component of spin. In matrix notation, the wave function is

$$\psi_{z+} = \begin{bmatrix} \psi(\mathbf{r}, t, +) \\ 0 \end{bmatrix}. \tag{26.1}$$

We can shorten this by writing the spin part of the wave function as

$$\psi_{z+} = \begin{bmatrix} 1 \\ 0 \end{bmatrix}. \tag{26.2}$$

We are assuming that things are so arranged that if equation (26.2) is the state of the spin of the particle as the wave packet enters the apparatus in figure 26.1, then equation (26.2) gives the spin state as the particle passes through and leaves the apparatus. Similarly, the wave function for a particle with spin down, that moves along path 2, is $\psi(\mathbf{r}, t, -)$, and the spin part of this wave function is and remains

$$\psi_{z-} = \begin{bmatrix} 0 \\ 1 \end{bmatrix}. \tag{26.3}$$

Now consider what happens if we pass through the apparatus the wave packet of a particle with x component of spin known to be $+1/2$ (that is, the system is an eigenstate of s_x with eigenvalue $+\hbar/2$). The spin part of this wave function is (eq. [24.21])

$$\psi = \frac{1}{2^{1/2}} \begin{bmatrix} 1 \\ 1 \end{bmatrix}. \tag{26.4}$$

We know how to write down the solution to Schrödinger's equation for this case in terms of the functions $\psi(\mathbf{r}, t, \pm)$ given above, because Schrödinger's equation is linear. That is, if $|a, t\rangle$ and $|b, t\rangle$ are solutions to Schrödinger's equation with a given Hamiltonian, then another solution is the linear combination

$$|t\rangle = \alpha|a, t\rangle + \beta|b, t\rangle, \tag{26.5}$$

where α and β are constants. In the present case, the solution with the initial conditions of equation (26.4) is

$$\psi_{x+} = \frac{1}{2^{1/2}}\{\psi(\mathbf{r}, t, +) + \psi(\mathbf{r}, t, -)\}, \tag{26.6}$$

for one sees that this makes the linear combination of spin wave functions in equations (26.2) and (26.3) agree with equation (26.4).

In equation (26.6) the wave function passes along both paths. Did the particle travel along both paths? The standard answer, following Bohr, is to say that the question is meaningless. The concept of a particle as an entity with a definite position, that would have to be either in path 1 or in path 2, is classical, derived from our experience with the macroscopic world, and Bohr emphasized that there is no reason why what we have learned from macroscopic experience need apply in the realm of atomic and subatomic physics. Indeed, we will have to argue that the concept of an objective particle position is not one we can carry

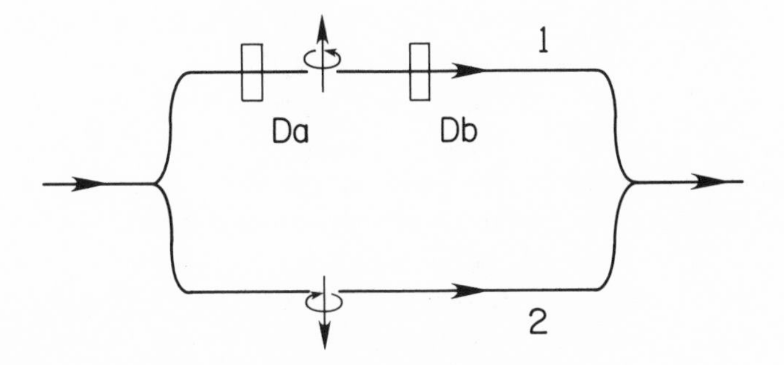

Fig. 26.2 Spin measurement by the Stern-Gerlach effect.

over from the macroscopic world to quantum physics. This is illustrated in the following examples.

Suppose we put particle detectors Da and Db in the top path in the Stern-Gerlach apparatus, as shown in figure 26.2. The detectors are supposed to register the presence of a particle with good reliability, but let it pass with relatively little energy loss. The probability that Da registers a particle is the probability that the z component of the spin is found to be up,

$$P_a = |\langle\psi_{z+}|\psi\rangle|^2. \tag{26.7}$$

As in equation (21.4), $|\psi\rangle$ is the state vector for the particle, $|\psi_{z+}\rangle$ represents a particle with spin definitely up, as in equation (26.1), and the inner product $\langle\psi_{z+}|\psi\rangle$ is the probability amplitude. For the wave function in equation (26.6), this works out to $P_a = 1/2$: there is an equal chance of finding the particle in path 1 or path 2. That is, if the experiment were repeated many times with the same initial state vector for the particle, Da would detect the particle in 50 percent of the measurements.

If Da registers the particle, what is the probability that the second detector, Db, on the same path in figure 26.2, also registers the particle? Experience says Db certainly detects it. A familiar example of this effect is the track of a particle in a bubble chamber. A particle may enter the chamber in a broad wave packet. The first bubble in the chamber localizes the particle, and subsequent bubbles show a continuous path from the first: the particle moves from where it was detected.

The observation that Db detects the particle if and only if Da detects it is interpreted to mean that after the detection of the particle in Da we have new information, so we must compute predictions for subsequent measurements using a new wave function, here ψ_{z+}. One way to put this is that the detection of the particle in Da has caused a "collapse of the wave function," from the initial one, ψ_{x+} (eq. [26.6]), to the new one, ψ_{z+}. However, this does not mean that some physical entity has discontinuously changed. The less dramatic way to put it is that the knowledge gained from the result of the measurement in Da affects the prediction for the result of the subsequent measurement in Db. Since the prediction is based on a state vector, we have to update the state vector used in the computation. A measurement of this kind in a Stern-Gerlach apparatus to the left of the one in figure 26.2 might have been used to prepare the state vector (26.6) we have been using.

Still assuming the initial condition is ψ_{x+} in equation (26.6), what is the spin state after the paths rejoin? If the apparatus is the one shown in figure 26.1, where there are no counters, then when the wave packet leaves the apparatus $\psi(\mathbf{r}, t, +)$ and $\psi(\mathbf{r}, t, -)$ again have a common space part and the spin part is the linear combination of up and down in equation (26.4). (We have assumed that the paths 1 and 2 have the same length, so the phase shifts due to motions of the wave packets along the two paths are the same.) This means a measurement of the x component of spin after the wave packet leaves the apparatus certainly yields $+\hbar/2$.

Suppose the detectors are in place, as in figure 26.2. If Da registered the particle, the wave function for subsequent measurements would be ψ_{z+}. In this case, the particle would leave the apparatus as an eigenstate of s_z with eigenvalue $+\hbar/2$, and a subsequent measurement of s_x in another Stern-Gerlach apparatus further downstream would give $\pm\hbar/2$ with equal probability. That is, if the particle is detected in path 1 we have to compute with a new state vector with eigenvalue $+\hbar/2$ for s_z, rather than the old eigenstate of s_x.

Suppose the detectors are in place, and the particle enters as an eigenstate of s_x. If the particle is not registered in detector Da, what is the final spin state when the particle leaves the system? The particle has to have passed through path 2, because it is known that it is not in path 1. This knowledge requires us to use in subsequent predictions the wave function ψ_{z-}, for which the z component of spin certainly is down. Here again, at the end of the experiment the spin is not an eigenstate of s_x: a measurement of s_x yields $\pm\hbar/2$ with equal probability. This is an example of the "collapse of the wave function" as the result of the absence of a detection. One often reads that the statistical character of the predictions of quantum mechanics is a result of the quantized nature of the interaction between system and measuring apparatus: the measurement necessarily disturbs the system. This can be a helpful concept but it certainly is too simple, as we see from this example, where the system has been disturbed from its original state as the result of the *absence* of an interaction, a point Dicke likes to emphasize.

*Measurement of a Mixed State**

Probability distributions are familiar in classical statistical mechanics, where one introduces probabilities because as a practical matter it is impossible to know the detailed microscopic state of a system containing a large number of particles. This is different from the statistical predictions of quantum physics, where the theory says that it is impossible in principle to make predictions sharper than those afforded by the state vector. But it generally is the case in quantum physics that the predictions are less sharp than is allowed in principle because the state vector is not known: one can only say the state vector is one of some statistical ensemble of possibilities. In this case the probability distribution of a prediction has two contributions, one intrinsic to the limited predictive power of the state vector, the other resulting from the practical matter of the uncertainty in what the state vector is. They are combined in the following way.

Let us imagine we have an isolated arrangement of objects with a definite state vector, $|\psi\rangle$. The arrangement will be supposed to consist of two parts. One is the system whose properties we wish to study. The remainder, to be called the surroundings of the system, will not be observed; it might be the walls of a heat reservoir, or the distant parts of the universe. The system to be measured has a complete set of commuting observables whose eigenvalues will be labeled collectively as a lower case letter, as a. The surroundings have a complete set of commuting observables whose eigenvalues will be labeled collectively as an upper case letter, as A. The observables belonging to the system and surroundings commute because they refer to different things, so a basis for the full arrangement, system plus surroundings, is the set of simultaneous eigenstates $|a, A\rangle$. For simplicity, it will be assumed that all the eigenvalues are discrete, so the completeness relation (20.53) is

$$\sum_{a,A} |a, A\rangle\langle a, A| = 1. \qquad (26.8)$$

Suppose f is some attribute of the system. This means f is a function of the observables of the system, and therefore that f commutes with all the observables of the surroundings. The expectation value of f in the state $|\psi\rangle$ is as usual (eq. [21.11])

$$\langle f \rangle = \langle \psi | f | \psi \rangle. \tag{26.9}$$

With the completeness relation (26.8), we can write this as

$$\langle f \rangle = \sum_{a,A,b,B} \langle \psi | a, A \rangle \langle a, A | f | b, B \rangle \langle b, B | \psi \rangle. \tag{26.10}$$

The matrix elements of f in this expression vanish unless $A = B$, and are independent of A. To see why this follows, let $\hat{A}$ be an observable for the surroundings, with eigenvalue A in the state $|a, A\rangle$:

$$\hat{A}|a, A\rangle = A|a, A\rangle. \tag{26.11}$$

Since f contains no observables of the surroundings it commutes with $\hat{A}$:

$$[\hat{A}, f] = 0. \tag{26.12}$$

A matrix element of this equation is

$$\langle a, A | [\hat{A}, f] | b, B \rangle = (A - B) \langle a, A | f | b, B \rangle = 0. \tag{26.13}$$

Thus $\langle a, A | f | b, B \rangle$ vanishes unless $A = B$. The computation by which one finds the value of $\langle a, A | f | b, A \rangle$ from the algebra of the observables $\hat{a}$ of the system, as in the computation of the Pauli spin matrices in section 24, goes through whatever the value of A. We have therefore

$$\langle a, A | f | b, A \rangle = f_{ab}, \tag{26.14}$$

where f_{ab} depends on the quantum numbers of the system, but not those of the surroundings.

Equations (26.10) and (26.14) give

$$\langle f \rangle = \sum_{a,b} f_{ab} \rho_{ba}, \tag{26.15}$$

where the density matrix for the system is defined to be

$$\rho_{ba} = \sum_A \langle b, A|\psi\rangle\langle\psi|a, A\rangle. \tag{26.16}$$

To summarize, the matrix f_{ab} depends only on the algebra of the observables of the system. The density matrix ρ_{ba} describes the state of the system.

Using the completeness relation (26.8), we see that the density matrix satisfies the normalization condition

$$\sum_a \rho_{aa} = 1. \tag{26.17}$$

Also, f_{ab} and ρ_{ba} are Hermitian matrices,

$$\rho^*_{ab} = \rho_{ba}, \qquad f^*_{ab} = f_{ba}. \tag{26.18}$$

If the system is in a pure state of the kind discussed in part (a) of this section, and if the basis vectors $|a, A\rangle$ are chosen so the pure state has the quantum numbers c of one of the basis vectors, then the coefficients $\langle a, A|\psi\rangle$ vanish unless $a = c$. This means the density matrix (26.16) vanishes unless $a = b = c$, and $\rho_{cc} = 1$ (as follows from eq. [26.17] when there is only one nonzero diagonal component of the density matrix). Equation (26.15) says the expectation value of f in this case is f_{cc}, which is the usual expression for a pure state.

The system is in a mixed state if we can only say that there is probability P_a that the state vector of the system has the quantum numbers a of one of the basis vectors $|a, A\rangle$. If the system had quantum numbers c, the expectation value of the observable f would be f_{cc}. The result of averaging the expectation value of f over the ensemble of possible states of the system is then

$$\langle f\rangle = \sum P_c f_{cc}. \tag{26.19}$$

This agrees with equation (26.15) with the diagonal density matrix $\rho_{cd} = P_c\delta_{cd}$.

As an example of this formalism, consider the spin part of the wave function in the Stern-Gerlach experiment discussed above. The density matrix for a particle in a pure state with spin up (along the $+z$ axis) is

$$\rho_{z+} = \begin{bmatrix} 1 & 0 \\ 0 & 0 \end{bmatrix}. \tag{26.20}$$

To find the density matrix in the basis of eigenstates of s_z for the eigenstate of s_x with eigenvalue $+\hbar/2$, compute the inner products of the latter with the former, using equations (24.21) and (24.23), and substitute into equation (26.16). The result is

$$\rho_{x+} = \frac{1}{2}\begin{bmatrix} 1 & 1 \\ 1 & 1 \end{bmatrix}. \qquad (26.21)$$

A density matrix for an unpolarized beam, which is a mixed state, is

$$\rho = \frac{1}{2}\begin{bmatrix} 1 & 0 \\ 0 & 1 \end{bmatrix}. \qquad (26.22)$$

This is a statistical mixture with equal probabilities of states in which the spin is up along the z and states in which the spin is down. It is an interesting exercise to check that equation (26.22) also represents a mixture with equal probabilities of states with s_x equal to $+\hbar/2$ and states with s_x equal to $-\hbar/2$. This is done by writing the basis vectors $|a, A\rangle$, where a represents the eigenvalues of s_z, as linear combinations of eigenstates $|i, A\rangle$ of s_x.

The time evolution of the density matrix may be written down in a convenient approximation if the basis vectors $|a, A\rangle$ are energy eigenstates. Suppose the interaction between system and surroundings is weak, so we can approximate the Hamiltonian as the sum of operators for the two parts, $H = H_a + H_A$. Let the bases be eigenstates of H_a and H_A, so

$$H|a, A\rangle = (E_a + E_A)|a, A\rangle, \qquad (26.23)$$

where E_a and E_A are eigenvalues of the Hamiltonians for system and surroundings. Then equation (21.20) for the time evolution of the state vector is

$$\langle a, A|\psi\rangle \sim e^{-i(E_a+E_A)t/\hbar}. \qquad (26.24)$$

The terms in the matrix element ρ_{ba} in equation (26.16) thus oscillate with time as

$$\langle a, A|\psi\rangle\langle\psi|b, A\rangle \sim e^{i(E_b-E_a)t/\hbar}. \qquad (26.25)$$

The energy E_A of the surroundings cancels in this approximation. It will be recalled, however, that this assumes the energy is the sum of energies of the system and of the surroundings. As we now discuss, a more detailed description would include a term representing the interaction between the two.

In section 1, we considered a system in thermal contact with a heat reservoir. The reservoir is supposed to be macroscopic, so its energy levels E_A are very closely spaced. This means the density matrix ρ_{ab} for the system, with the reservoir as the surroundings, is a sum over an exceedingly large number of terms $\langle a, A|\psi\rangle\langle\psi|b, A\rangle$. The time evolution of each term with $a \neq b$ is approximated by equation (26.25). However, because there is some coupling between reservoir and walls, this is only an approximation: the phase in the exponential function in equation (26.25) that describes the oscillation of $\langle a, A|\psi\rangle\langle\psi|b, A\rangle$ must evolve with time, at a rate that depends on the quantum numbers A for the reservoir. Therefore, after a sufficiently long time the off-diagonal matrix elements ρ_{ab} (with $a \neq b$) are sums of an enormous number of terms each with a phase that has wandered to some random value. These sums are much smaller than the diagonal terms ρ_{aa}, because the latter are sums of absolute values squared of the $\langle a, A|\psi\rangle$. That is, the density matrix becomes very nearly diagonal in this energy eigenvalue representation. As remarked above, in a diagonal density matrix the system has the state vector labeled by quantum numbers a with probability ρ_{aa} (eq. [26.19]). Thus we see that the thermal (statistical) equilibrium of a system loosely coupled to a heat reservoir is described by the probabilities ρ_{aa} that the system is in each of its energy levels E_a. In section 1 we found that these probabilities are given by the Boltzmann distribution,

$$\rho_{aa} \propto e^{-E_a/kT}. \tag{26.26}$$

This is the thermal equilibrium density matrix.

Summary

There are situations in which a system can be imagined to be in a pure state, described by a state vector $|\psi\rangle$. An example of the preparation of a pure state is shown in figure 26.2: if a detector Da in the upper arm shows the particle is not there then the particle is in the lower arm, and the wave function has been placed in the pure state ψ_{z-}.

The state vector of an isolated system changes with time according to Schrödinger's equation. This is deterministic in the sense that a definite initial condition, fixed by a prior measurement, yields a definite time evolution of the state vector. The result of an observation of a system is to introduce a new state vector that is consistent with what has been found: new information is gained and the results must be incorporated in new initial conditions for the state vector for use in the prediction of subsequent measurements.

The more general description of a system in quantum mechanics is the density matrix, ρ_{ab}, which determines the expectation value of any observable of the system (eq. [26.15]). The system is in a pure state if there is a basis $|a, A\rangle$ such that the only nonzero component of the density matrix is the one element $\rho_{cc} = 1$. In this case, we can say the system is in the state $|c\rangle$, ignoring the quantum numbers of the surroundings, because the latter do not enter the calculations of expectation values of all functions of the observables of the system. In a mixed state, there is no basis in which the density matrix reduces to a single term, but we can use familiar methods to find a basis in which the density matrix is diagonal. In this diagonal representation, ρ_{cc} is the probability that the system is in the pure state c. A measurement of the mixed state may reveal that the state is d, and may leave it there. In this case, we now compute with a new density matrix, the pure state with $\rho_{dd} = 1$.

27 "Paradoxes" of Quantum Physics

Because some features of quantum physics are so contrary to our instincts, it is natural to ask whether the theory really could be a complete description of physical reality. Three famous examples, in the forms of paradoxes, are presented here. The discussion of why these effects are not generally considered to be paradoxical is well worth following as a way to sharpen understanding of the physics.

Complementarity

In the late 1920s and early 1930s there was a famous series of discussions between Bohr and Einstein on the physical and logical consistency of the quantum picture, with Einstein proposing ingenious thought experiments that seem paradoxical in quantum physics, Bohr working through resolutions within his complementarity picture. Bohr's beautiful description of these debates is reprinted in *Quantum Theory and Measurement*, edited by Wheeler and Zurek, where you will also find many other classical papers on the subject. Here is an example from the Bohr-Einstein debates.

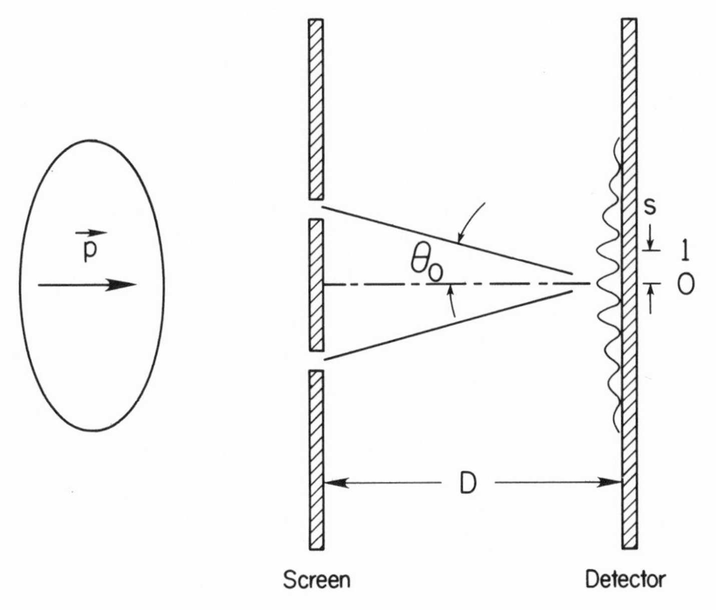

Fig. 27.1 Double slit experiment.

Suppose a sequence of single particle wave packets, each with fairly sharply defined momentum p, is incident on the screen with two slits shown in figure 27.1. If the particle does not hit the screen then we have a measurement, that says the particle got past the screen. Therefore, for the purpose of further computation, the wave function is nonzero only in the narrow regions of the two slits. The two parts of the wave function then propagate to the right to the detector. Because the two parts of the wave function that get through the slit are very narrow they spread as they move to the right, so the two waves overlap at the detector. If the path lengths along the two waves from the screen to a spot on the detector differ by an integral multiple of the particle de Broglie wavelength, λ, then the two waves add in phase; where the path difference is $\lambda/2$, the two parts of the wave function cancel. This produces a fringe pattern in the wave function at the detector. That is, the probability distribution in where the particle is detected is alternately large and small, in bands or fringes, at the detector.

In quantum mechanics, a particle that arrived at the right-hand detector passed through both slits, for how else could the particle know to avoid the regions on the detector where the two waves destructively interfere? Indeed, if one slit is blocked, so the particle is forced to pass through only one slit, the open one, the interference pattern disappears. What happens if a thin detector (that allows the particle to pass with only a small energy loss) is placed at one of the two slits so we can discover which slit the particle really passed through? One finds that the interference pattern disappears. The interpretation is the same as in the last section for the experiment in figure 26.2. The particle has been localized at one slit, so the wave function must vanish at the other.[1] Thus only one wave packet leaves the screen, and there is no interference pattern at the detector.

Einstein asked whether we might determine which slit the particle went through in a way that would not be expected to affect the interference pattern in particle positions at the detector, as follows. If the particle went through the top slit and reached the central part of the pattern at the detector, it had to have been deflected down by the angle $\sim \theta_o$ in figure 27.1. This means the particle received a downward momentum transfer $\sim p \sin\theta_o$ for incident particle momentum p. If the particle passed through the bottom slit and reached the central part of the pattern, there was an upward momentum transfer of the same amount. This momentum must have come from the screen, for momentum is conserved in quantum physics. Therefore, we would know which slit the particle passed through if we could measure the change in vertical momentum of the screen with sufficient accuracy. We need not measure the final momentum of the screen until after the particle has been detected, so this momentum measurement could have no effect on the probability distribution of the position of the particle at the detector. And we can take as long as we like to measure the final momentum of the screen with arbitrary accuracy, by the time of flight method discussed in section 10. So it would appear that we could in principle see the interference pattern as the statistics of particle detections builds up, and also know which slit each particle passed through. But if each particle passed through a definite slit, how could the particles know that there is a second slit that causes an interference pattern at the particle positions at the detector?

[1] The thin detector has shifted the phase of the wave by a large and uncontrolled amount, and one might be tempted to say that this is what has eliminated the interference pattern between the waves coming from the two slits: since the waves are incoherent, their mean interference vanishes. But we know that that is not right, because if the thin detector detects the particle near the upper slit, there is zero probability that the particle will be found near the lower slit, so we have to set the wave function from the lower slit to zero.

Bohr pointed out that the uncertainty principle prevents this from happening. To detect the momentum transfer, one must know the initial momentum of the screen with some precision, and that implies an uncertainty in the vertical position of the slits. As we will now see, this uncertainty is large enough to erase the interference pattern.

At the point 0 in figure 27.1, the paths from the slits are at angles $\pm\theta_o$ to the normal of the detector, and the path lengths from the two slits are the same. This is a point of constructive interference of the two waves, that is, a point of maximum probability of detection. At the point 1, the next point of maximum probability up from point 0, the path lengths differ by one wavelength, $\lambda = h/p$, and the angles of the beams to the normal of the detector are

$$\begin{aligned} \theta_a &= \theta_o + \delta\theta, \\ \theta_b &= \theta_o - \delta\theta. \end{aligned} \tag{27.1}$$

To simplify the calculation, let us assume that $\delta\theta \ll \theta_o$.

At point 1 on the detector, the path length difference is

$$\lambda = D\left(\frac{1}{\cos\theta_a} - \frac{1}{\cos\theta_b}\right). \tag{27.2}$$

where D is the perpendicular distance from screen to detector. The result of expanding this to first order in $\delta\theta$, and dropping factors of order unity, is

$$\lambda \sim D\sin\theta_o\delta\theta/\cos^2\theta_o \sim \hbar/p, \tag{27.3}$$

where the last line is the de Broglie relation. The difference of momenta of particles deflected down and those deflected up is

$$\delta p \sim 2p\sin\theta_o. \tag{27.4}$$

Finally, the vertical distance from the bottom slit to fringe 1 is $\sim D\tan\theta_a$, and the derivative of this gives the vertical distance, s, between successive fringes in the interference pattern at the detector:

$$s \sim D\delta\theta/\cos^2\theta_o. \tag{27.5}$$

To detect the momentum transfer δp, we would have to make the uncertainty δP in the initial momentum of the screen less than the needed precision in the measurement in the momentum transfer from the particle to the screen, $\delta P < \delta p$. By the uncertainty principle (section 10, problem III.3), the vertical position of the screen therefore must be uncertain by at least

$$\delta X \gtrsim \hbar/\delta P > \hbar/\delta p. \tag{27.6}$$

The result of combining equations (27.3) to (27.6) and dropping factors of order unity is

$$\delta X > s. \tag{27.7}$$

This means the vertical uncertainty in the slit position is larger than the distance between fringes, so the uncertainty in the position of the slits has smeared out the interference pattern.

The conclusion is that if the initial momentum of the screen were well enough known that its momentum change could be used to determine which slit the particle passed through, the resulting uncertainty in the vertical position of the slits would eliminate the interference pattern on the detector. That is, if the experiment is set up so the momentum transfer measurement can be used to determine which slit the particle passed through, so we can see that the object acts like a classical particle that really passes through only one slit, the arrangement removes the possibility of seeing the interference pattern that signals the wave nature of an object that passes through both slits. If on the other hand the vertical positions of the slits were well enough fixed that we could see the development of the interference pattern in particle positions at the detector, so as to see the wave character of the particle, then the momentum of the screen would not be well enough known to allow us to observe the behavior of a particle with a definite orbit that passes through a single slit. This is an example of Bohr's complementarity principle. The attributes of position and momentum (wavelength) are incompatible, represented by observables that do not commute. Any experimental arrangement designed to observe one attribute precludes the possibility of simultaneous observation of the other. This is not an intuitive concept of reality, but then our intuition is based on the macroscopic world of experience, which need not apply in the microscopic world.

The EPR Effect

Einstein's most subtle puzzle is said to have been first mentioned by him in 1933, and was later described in detail in a paper with Podolsky and Rosen. The paper begins with the statement, "in a complete theory there is an element corresponding to each element of reality. A sufficient condition for the reality of a physical quantity is the possibility of predicting it with certainty, without disturbing the system." The authors then present an example showing that this does not apply to the observables of quantum physics, and conclude that quantum physics is an incomplete theory. A more recent and by now standard example of how quantum physics conflicts with the EPR definition of reality was introduced by Bohm, as follows.

Consider an electron and positron that were produced by the annihilation of a system with zero net angular momentum. We can assume the particles fly apart with zero orbital angular momentum. Since the total angular momentum vanishes, the particles must have zero total spin, which is to say that they are in the singlet spin state

$$|\psi\rangle = \frac{1}{2^{1/2}}(|+-\rangle - |-+\rangle). \tag{27.8}$$

As in equation (25.19), the ket vectors are eigenstates of the z components of the particle spin operators. In the first vector, the eigenvalues are $+\hbar/2$ and $-\hbar/2$ for the two particles, and the eigenvalues are reversed in the second vector.

After the particles have become well separated, the z component of the spin of the electron is measured. If the electron spin is found to be up, the state vector becomes the first term in equation (27.8), $|+-\rangle$. Then without disturbing the positron, we know that if its z component of spin were measured the result certainly would be down. Thus, according to the EPR criterion a complete theory ought to grant that s_z for this particular positron exists and it is down (has the definite value $-\hbar/2$). But we could equally well have decided to measure the x component of the electron spin, and if we found that the value is $+\hbar/2$ we would similarly conclude that s_x for this particular positron exists and it is $-\hbar/2$. Since the positron has no way of knowing which component we decided to measure, the EPR criterion says both s_z and s_x must exist for this positron. But that certainly is wrong: we know that if the positron had a definite value of s_x, here $-\hbar/2$, then its state vector must be written as the linear combination $(|+\rangle - |-\rangle)/2^{1/2}$, which does not have a definite value of s_z.

Bohr's response was that the EPR definition of reality conflicts with his complementarity principle. He argued that here, as in the Stern-Gerlach and double slit experiments discussed above, "the procedure of measurements has an essential influence on the conditions on which the very definition of the physical quantities in question rests." That is, the complementarity principle denies the existence of an element of reality corresponding to s_x in advance of a decision to measure this component of the spin, either directly, or, in the present case, indirectly by measuring a particular component of the spin of the partner particle. We saw another application of this principle in the double slit experiment, which can be arranged either to reveal the particle nature of matter, or the wave nature, but not both at the same time.

Wigner's Friend

If we accept the proposition that an electron can be in two places at once, as in the double slit experiment in figure 27.1, or that its spin state can depend on a decision to observe another distant particle, as in the EPR effect, can we believe the same of a sentient being? The question has been discussed at least since 1935, when Schrödinger proposed that we consider what it could mean for a cat. The version presented here is based on a discussion by Wigner.

Suppose a physicist who acts as our assistant observer is standing near a system consisting of a particle with spin 1/2. We will suppose the physicist can be put in a pure state, with quantum numbers A representing a state of readiness to observe the particle, and that the particle can be put in a pure state, with quantum number a representing the z component of the spin. Then we can write the state vector for particle plus observer at the starting time for the experiment as

$$|\psi\rangle = |A, a\rangle. \tag{27.9}$$

As in equation (26.14), the expectation value of any function F of observables of the physicist is independent of the quantum numbers a of the system to be observed:

$$\langle F\rangle = \langle A, a|F|A, a\rangle = \langle A, b|F|A, b\rangle. \tag{27.10}$$

This is the wanted situation: the expectation values of all functions F of the observer are independent of the quantum number a of the system; the physicist does not know what the state of the system is.

Next we need to specify time evolution *via* the Hamiltonian. It will be assumed that, before the physicist looks at the system, physicist and particle are not coupled. That is to say that the Hamiltonian is the sum of two terms, one containing only observables of the physicist, the other only observables of the system: $H = H_p + H_s$. Observables for physicist and system commute, so H_p and H_s commute. Therefore, the time translation operator $U = \exp -iHt/\hbar$ (eq. [21.20]) can be written as the product

$$U = U_p U_s, \tag{27.11}$$

where $U_p = \exp -iH_p t/\hbar$ contains only observables of the physicist, U_s only observables of the system. At time t after the start of the experiment, and before the physicist looks at the system, the state vector is $U|A, a\rangle$, so the expectation value of the function F of the physicist is

$$\langle F \rangle_t = \langle A, a|U^\dagger F U|A, a\rangle = \langle A, a|U_p^\dagger F U_p|A, a\rangle. \tag{27.12}$$

The last expression follows because U_s commutes with F. This last expression is the expectation value of the function $U_p^\dagger F U_p$ of observables of the physicist, and so is independent of a, as in equation (27.10). That is, the expectation value $\langle F \rangle_t$ at time t of any function of the observer is independent of a: the physicist still does not know the state of the system.

When the physicist observes the system, physicist and particle are coupled, in the sense that the Hamiltonian is no longer a sum of separate functions of the observables of the physicist and of the system. This means there develops a correlation between physicist and system, so subsequently $\langle F \rangle$ depends on a. It can be arranged that the observation does not affect the state of the system. (An example is the Stern-Gerlach experiment in figure 26.2: a particle with spin down is not observed in the upper arm and leaves the apparatus with spin down.) In the present case, this means that if the spin of the particle in the system initially were up, $a = +$, a careful physicist would be able to discover that and leave the spin up, $a = +$, at the end of the observation.

In the final step, the physicist moves away from the system and the coupling is removed. The physicist tells us the result of the observation.

If the initial state of the system were an eigenstate of the spin operator s_z with eigenvalue $+\hbar/2$, the initial state vector would be $|A, +\rangle$. The final state vector would be $|A_u, +\rangle$, where A_u represents the state of the physicist after having discovered the spin is up. If the initial spin were down, the physicist would discover that and leave the final state of the spin as it was. Here the initial state would be $|A, -\rangle$ and the final state would be $|A_d, -\rangle$, where A_d represents the state of the physicist on learning that the spin is down.

Now suppose the particle initially is in a state that is a linear combination of spin up and spin down,

$$|\psi_i\rangle = \alpha|A, +\rangle + \beta|A, -\rangle. \tag{27.13}$$

The physicist initially is in the state A, as before, with no knowledge of the system. By the linearity of Schrödinger's equation, we know the final state after physicist and particle are again separated, at time t_f, is

$$|\psi_f\rangle = U(t_f)|\psi\rangle = \alpha|A_u, +\rangle + \beta|A_d, -\rangle, \tag{27.14}$$

This follows from the linearity of the time translation operator, $U(t_f)$, just as in the discussion of the Stern-Gerlach effect in section 23 and in equation (26.6). But can we imagine that the physicist has been placed in a linear superposition of two states? Surely the physicist knows for certain whether the spin is up or down!

This sequence of events, with the physicist replaced by a macroscopic measuring device, is the basis of von Neumann's description of a measurement in quantum physics. The idea is that we can arrange to make the final state of a macroscopic measuring device a function of the initial state of the system, and we can observe the state of the measuring device at our leisure. Having discovered the state of the device we have a measurement of the system. But what are we to make of the measurement procedure when an animate object is used in the chain of measurement? One might be tempted to argue that quantum physics cannot be extrapolated from the scales of length and energy where it was invented to a macroscopic object like a cat or a physicist, but we do have examples from superconductors and superfluids of macroscopic quantum effects.

There is no generally accepted answer to this question, but we can express the situation in a less dramatic way, by using the density matrix formalism of section 26. The density matrix (eq. [26.16]) for the physicist in the state vector of equation (27.14), with the particle considered as the surroundings, is[2]

$$\rho = \begin{bmatrix} |\alpha|^2 & 0 & 0 \\ 0 & 0 & 0 \\ 0 & 0 & |\beta|^2 \end{bmatrix} \tag{27.15}$$

The first column and the first row represent states of a physicist who is sure the spin is up, the second row and second column represent states of a physicist who does not know, and the last row and column represent states of a physicist who is sure the spin is down. To see that the off-diagonal matrix elements vanish, note that the matrix elements connecting up and down states of mind of the physicist look like

$$\sum_{a=\pm} \langle A_u, a|\psi\rangle\langle\psi|A_d, a\rangle. \tag{27.16}$$

This vanishes in the state (27.14), because orthogonal states of mind of the physicist always appear in connection with orthogonal spin states. As discussed in section 26, equation (27.15) is equivalent to a mixed state, in which pure states of mind appear in the ensemble with probabilities $|\alpha|^2$ and $|\beta|^2$. This means we fully describe the predictions of quantum mechanics for this experiment by using a classical statistical ensemble, in which the probability is $|\alpha|^2$ that the physicist really knows that the spin is up, and $|\beta|^2 = |\alpha|^2 - 1$ that the physicist really knows the spin is down.

Under this density matrix picture, we can use the standard rules of measurement theory, treating our physicist assistant as part of the quantum system, and at the same time the physicist can apply measurement theory treating the particle system alone in quantum mechanics, with no contradiction in the results. One cannot always separate subsystems this way; in general there is the possibility of interference of different states of a subsystem such as the physicist. That is avoided in the present example because the orthogonality of the spin states in the system makes different states of mind of the physicist orthogonal, which in turn makes the density matrix (27.15) diagonal. It is conjectured that anything large

[2] This equation is shorthand for a matrix that is block-diagonal. The element $|\alpha|^2$ in the upper left hand corner stands for the matrix of elements among states of a physicist who is sure the spin is up, the element in the first row and third column stands for the matrix of elements between a physicist who knows the spin is up and a physicist who knows the spin is down, and so on.

enough to talk to is complicated enough that it always requires such a diagonal density matrix.

28 Hidden Variables

Two aspects of quantum physics bothered Einstein and still bother many people. The more serious to Einstein was the point discussed in the last two sections, that physical attributes exist only as the result of a measurement. The other aspect is the statistical character of the predictions of the quantum theory. Einstein's famous remark on this point usually is translated as "The Lord does not play dice with the world."

One recalls that thermodynamics was invented before the discovery of the underlying physics of statistical mechanics. Could it be that quantum physics plays the role of thermodynamics for some deeper and maybe nonstatistical theory? Perhaps in an ensemble of physical systems with identical state vectors, $|\psi\rangle$, the "hidden variables" of the deeper theory uniquely determine the results of all possible measurements on any particular one of the systems, so that different results are obtained from apparently identical members of the ensemble because the hidden variables are different for each member. If there were enough hidden variables to permit a unique correspondence of values of the variables with all possible results of all possible measurements, the deterministic hidden variables theory is trivially made equivalent to quantum theory, by arranging that the distribution of values of the hidden variables for the different members of the ensemble makes the distribution of results of measurements of members of the ensemble agree with the probability distributions of quantum mechanics. However, if the hidden variables of a system do not know what measurement we are going to decide to make on the system, then the construction can fail. The point was made explicit in Bell's theorem. The version presented here follows Wigner.

Consider a system consisting of two spin 1/2 particles, 1 and 2, in the isotropic singlet (zero net spin) state (eq. [27.8])

$$|\psi\rangle = \frac{1}{2^{1/2}}(|+-\rangle - |-+\rangle). \tag{28.1}$$

Suppose the component of the spin of particle 1 is measured along axis a, and the component of the spin of particle 2 is measured along axis b tilted relative to a by the angle θ. It will be recalled from section 25 that the quantum mechanical probability that the result for particle 1 along a is $+1/2$ and that the result for 2 along b is $+1/2$ is (eq. [25.22])

$$P = \frac{1}{2}\sin^2\theta/2. \tag{28.2}$$

Now imagine we have a statistical ensemble of systems, each with the state vector of equation (28.1). For each member of the ensemble we will allow three possible choices for the axis along which a particle spin is to be measured: along axis a parallel to the z axis, along axis b tilted at 45 degrees to the z axis, or along axis c tilted at 90 degrees to the z axis and in the plane of a and b, so c is tilted at 45 degrees to b. Thus, for example, equation (28.2) says the spin of particle 1 is found to be up along a and the spin of 2 is found to be up along c in 25 percent of the cases where one chooses to measure the spins along these two axes; and 1 is up along a and 2 up along b in 7.3 percent of the cases where the spins are measured along these particular axes.

We can choose two quantities to be measured: the spin of particle 1 could be measured along the axis a, yielding the value $s_a = \pm 1/2$, or along axis b, yielding $s_b = \pm 1/2$, or along c, yielding s_c; and the spin of particle 2 could be measured along a, yielding $t_a = \pm 1/2$, or along b, yielding t_b, or along c, yielding t_c. In the hidden variables program, the particular values of the hidden variables belonging to a specific one of the systems in the ensemble uniquely determine the results of measuring the values of any particular choice of two of these six quantities, one for particle 1, one for 2. Note that just one quantity can be measured for each particle in the initial state: if spin 1 is measured along a the act of measurement can affect the hidden variables that determine the result of a subsequent measurement of spin 1 along b or c. But in a hidden variables picture it seems reasonable to assume that a measurement of spin 1 cannot affect the outcome of a measurement of spin 2, because we can imagine the particles are well separated, as in the EPR effect discussed in section 27.

This last point is the locality assumption, that the values of the hidden variables belonging to a particular member of the ensemble do not know what measurement we are going to decide to make on that member. How could the system anticipate what observable we are going to single out for a measurement? In the present example, the locality assumption means that the hidden variables belonging to a specific member of the ensemble of systems have to fix ahead of time the results of all nine of the possible experiments. That is, a specific system in the ensemble has to yield a definite value s_a if we choose to measure spin 1 along axis a, it must yield a definite value s_b if instead we choose to measure spin 1 along b, and so on to t_c if we choose to measure spin 2 along axis c. For example, the hidden variables for a particular system in the ensemble might be rigged to yield $(+--;-++)$. This would mean that if we decided to measure spin 1 along axis a we certainly would get $+1/2$; if instead we decided to measure spin 1 along axis b we certainly would get $-1/2$, and so on to $+1/2$ if we choose to measure spin 2 along axis c.

In the ensemble of systems, a specific combination of predetermined results, s_a to t_c, appears with some frequency, or probability,

$$P = P(s_a, s_b, s_c; t_a, t_b, t_c). \tag{28.3}$$

Many of these probabilities vanish. For example, in a measurement of the components of spin 1 and spin 2 both along axis a, the only possible results are $+-$ and $-+$, because the total spin along any axis vanishes (as we see from eq. [28.2] with $\theta = 0$, or from eq. [28.1] for axis a).

Now suppose we claim to have chosen these functions P to match the quantum predictions. The probability that spin 1 measured along axis a gives $+$ and spin 2 measured along b gives $+$ is

$$\frac{1}{2}\sin^2\frac{\pi}{8} = P(+-+;-+-) + P(+--;-++). \tag{28.4}$$

The left side is the quantum prediction in equation (28.2). The right side is the sum over all the possible contributions of predetermined arrangements in the local hidden variables theory. The first and fifth arguments have to be "+" because that is what we stipulated is the result of the measurement. Therefore, the second and fourth arguments have to be "−" because, as we observed, measurements of the two spins along the same axis have to give opposing results. This means that the sum is only over the two possible choices for arguments three and six. In the same way, one finds that the probability that spin 1 measured along b and 2 along c both give $+$ is

$$\frac{1}{2}\sin^2\frac{\pi}{8} = P(++-;--+) + P(-+-;+-+), \tag{28.5}$$

and the probability that spin 1 measured along a and 2 along c both give + is

$$\frac{1}{2}\sin^2\frac{\pi}{4} = P(++-;--+) + P(+--;-++). \tag{28.6}$$

The result of adding equations (28.4) and (28.5) and subtracting equation (28.6) is

$$\sin^2\frac{\pi}{8} - \frac{1}{2}\sin^2\frac{\pi}{4} = P(+-+;-+-) + P(-+-;+-+). \tag{28.7}$$

The right-hand side cannot be negative, because probabilities are not negative. But one readily checks that the left side is negative. This contradiction means the local hidden variables scheme cannot reproduce the quantum probability distributions.

Since the quantum probability distribution in equation (28.2) agrees with the experiments, we have to conclude that a local hidden variables theory is wrong. It still could be that the hidden variables of the system are coupled to our choice of what measurement to make, in violation of the locality assumption, but that seems unacceptably contrived. The simpler and generally accepted interpretation is that there is not a non-statistical theory underlying quantum physics.

29 Summary

The basic elements of the quantum theory are the linear operators Q in a space of state vectors $|\psi\rangle$. What is the physical meaning of these objects? It is curious that there is no generally accepted answer, despite the fact that the application of the theory in laboratory physics is well defined, and leads to no known operational[3] or experimental inconsistency. We arrive at the generally accepted accommodation to the question by recalling what is meant by a physical theory.

In the beginning, physics was the mechanics of the motions of the moons and planets, and of the motions of projectiles and pendulums. Given initial positions, mechanics successfully predicts orbits. It is natural to assign physical reality to these orbits; almost anyone can see

[3] It must be admitted that there has been little discussion of our firm belief that a good theory must be logically consistent, because the belief is not known to be challenged.

the planets move through the sky. A more careful description would be that we are following the motions of the centers of mass of the many mass elements in each planet. A center of mass is not so visible, but a belief in its objective presence is an easy extrapolation from our experience. Another extrapolation was the development of the kinetic theory of gasses. In the early days of kinetic theory a particularly thoughtful physicist, Ernst Mach, objected to the interpretation of atoms as real objects rather than a mathematically convenient model or mental artifice, but the many successes of the theory have made atoms seem very real to us.

One might have thought that, if quantum theory were complete, its elements, the state vectors $|\psi\rangle$ and observables Q, also could be interpreted in terms of some sort of objective physical reality. We have seen that that cannot be done within a classical version of reality. This may mean the theory is incomplete, or it may simply mean we are asking too much of a physical theory. A more modest requirement is that a successful theory correlate results of measurements. In the example of Wigner's friend, the theory presents us with the density matrix in equation (27.15), in which the probability is $|\alpha|^2$ that the physicist discovers that the spin of the particle is up and tells us about it. The theory presents the physicist with a density matrix in which the probability is $|\alpha|^2$ that the spin of the particle is up. Both descriptions agree with experience. Of course, agreement with experience is not a complete criterion for a good theory, for that would include a straight record of experimental results, which may be of great practical use but by itself would be of no fundamental interest. A physically interesting theory makes successful experimental predictions that substantially outnumber the experimental elements that were used as guides or constraints to its construction. This the quantum theory has done in spectacular abundance (only an exceedingly small part of which is recorded in this book).

Quantum physics does predict bizarre things. The EPR effect in section 27 shows that the choice of state vector for a system can depend on a causally unconnected operation on another system. As another example, a cosmic ray proton arriving from a distant galaxy has a wave function with the local protons that is antisymmetric (as discussed in section 41 below), even though the proton has not interacted with the local matter for an exceedingly long time, if ever. However, it would be illogical, even ungrateful, to fault quantum physics for such bizarre predictions, because to the accuracy of available experimental tests the predictions agree with what is observed. The successes of the quantum principles are convincing evidence that they are showing us a physical reality deeper than classical physics. On the other hand, there is no reason to believe the quantum paradigm is the ultimate truth, and there certainly will continue to be great interest in experimental tests of the standard theory and possible variants. And the search for variations on the standard theory undoubtedly will continue to be motivated by the question, what does the state vector $|\psi\rangle$ *really* mean?

CHAPTER 5

PERTURBATION THEORY

The remainder of this book deals with applications of quantum mechanics. This is coupled to a presentation of approximation methods, because the list of interesting applications that have analytic solutions is nearly exhausted by the simple harmonic oscillator and hydrogen atom considered in section 6. For the next simplest atom, helium, the wave function of the two electrons is a function of six variables. A search for analytic solutions to Schrödinger's equation is not promising here, and a straightforward numerical computation, with the wave function sampled at ~ 100 points along each of the six coordinate axes, for a reasonably accurate approximation to the wave function, would require storage of $\sim 100^6 = 10^{12}$ numbers, a challenge for the largest computers. Yet before there were electronic computers people had quite an accurate theoretical understanding of the energy levels in helium and more complicated systems. The trick was (and is) to find approximation schemes that treat unimportant parts of a physical system in quite crude approximations while reducing the interesting parts to a problem simple enough that it is feasible to compute but yet detailed enough to yield accurate results.

The approximation methods in this chapter deal with the effects of small changes in the Hamiltonian, resulting for example from the application of a static or time variable electric or magnetic field. This may cause small changes in energy levels, and it may induce transitions among eigenstates of the original Hamiltonian.

30 Time-Independent Perturbation Theory

Suppose the Hamiltonian for a system is written as the sum of two terms,

$$H = H_o + V. \tag{30.1}$$

In the case to be considered here, the operators H_o and V both are independent of time. The eigenvalues and eigenvectors of H_o are E_n and $|n\rangle$:

$$H_o|n\rangle = E_n|n\rangle. \tag{30.2}$$

It may be easy to find these eigenvalues and eigenvectors, while the addition of V to the Hamiltonian complicates the computation. The perturbation theory to be presented here gives a systematic way to approximate the solutions to the eigenvalue problem for H. An example in section 34 below is the hyperfine structure of atomic hydrogen. Even when the eigenvectors of H_o are not accurately known, we can use perturbation theory to analyze the effect of introducing V, as in the theory of the shifts in the energy levels of an atom when an electric or magnetic field is applied.

In this section it will be supposed that the eigenvectors of H_o are not degenerate, that is, each eigenvector $|n\rangle$ has a different energy E_n. The degenerate case is considered in section 33.

A convenient trick for keeping track of the order of the approximation in perturbation theory is to rewrite the full Hamiltonian as

$$H = H_o + \epsilon V, \tag{30.3}$$

where ϵ is a parameter that measures the strength of the perturbation of H from H_o. One assumes the solutions to the energy eigenvalue problem,

$$H|\psi\rangle = E|\psi\rangle, \tag{30.4}$$

can be expanded as power series in ϵ,

$$\begin{aligned} E &= E_n + \epsilon\delta E(1) + \epsilon^2\delta E(2) + \dots, \\ |\psi\rangle &= |n\rangle + \epsilon|\delta_1\rangle + \epsilon^2|\delta_2\rangle + \dots. \end{aligned} \tag{30.5}$$

The vector $\epsilon|\delta_1\rangle$ is the first-order correction to the state vector due to the perturbation ϵV, the vector $\epsilon^2|\delta_2\rangle$ is the second-order correction, and so on.

Since the eigenvectors $|n\rangle$ of H_o are a complete set, we can write these corrections to the state vector as linear combinations of the $|n\rangle$. This brings the second of equations (30.5) to the form

$$|\psi\rangle = |n\rangle + \epsilon \sum_i C_i|i\rangle + \epsilon^2 \sum_j D_j|j\rangle + \ldots . \tag{30.6}$$

The coefficients C_i, D_j, and so on in the expansion of the vectors $|\delta_1\rangle, |\delta_2\rangle, \ldots$, are independent of ϵ.

On substituting the expansions (30.5) and (30.6) into the energy eigenvalue equation (30.4), using $H_o|n\rangle = E_n|n\rangle$, and collecting powers of ϵ, we get

$$\begin{aligned}
E_n|n\rangle + \epsilon &\left[V|n\rangle + \sum C_i E_i|i\rangle\right] \\
&+ \epsilon^2 \left[\sum C_i V|i\rangle + \sum D_i E_i|i\rangle\right] \\
&+ \ldots \\
&= E_n|n\rangle \\
&+ \epsilon \left[\delta E(1)|n\rangle + E_n \sum C_i|i\rangle\right] \\
&+ \epsilon^2 \left[\delta E(2)|n\rangle + \delta E(1) \sum C_i|i\rangle + E_n \sum D_i|i\rangle\right] \\
&+ \ldots .
\end{aligned} \tag{30.7}$$

Now equate coefficients of each power of ϵ on each side of this equation. The coefficients of the first power of ϵ are

$$V|n\rangle + \sum C_i E_i|i\rangle = \delta E(1)|n\rangle + E_n \sum C_i|i\rangle. \tag{30.8}$$

The inner product of this with $\langle n|$, with the usual orthonormality condition $\langle n|i\rangle = \delta_{ni}$, gives

$$\delta E(1) = \langle n|V|n\rangle. \tag{30.9}$$

The parameter ϵ was used to keep track of the order of perturbation theory. Now that we have this result we can set ϵ to unity, so equation (30.9) is the energy shift due to the perturbation V in first-order, or linear, perturbation theory.

This result is worth remembering. It says that, to first order, the perturbation $\delta E(1)$ to the energy is the expectation value of the perturbation V to the Hamiltonian in the unperturbed state vector $|n\rangle$. For example, consider a particle that moves in one dimension and is confined to a box. The potential energy term in H_o vanishes at $0 < x < L$ and is very big elsewhere, making the wave function negligibly small at $x = 0$ and $x = L$. With these boundary conditions, the eigenfunctions of H_o are (as in eq. [8.5])

$$\psi_n(x) = (2/L)^{1/2} \sin(n\pi x/L), \qquad n = 1, 2, 3, \ldots. \tag{30.10}$$

Suppose the perturbation is

$$V(x) = a\delta(x - L/2), \tag{30.11}$$

where a is a constant. This is supposed to represent a narrow bump in the potential energy at the center of the potential well. In first-order linear perturbation theory, V shifts the energy of the ground state (that is represented to zero order as $\psi_1(x)$ in eq. [30.10]) by the amount

$$\begin{aligned}\delta E(1) &= \langle\psi_1|V|\psi_1\rangle = (\psi_1, V\psi_1) \\ &= \int dx\, \psi_1^*(x) V(x) \psi_1(x) \\ &= 2a/L.\end{aligned} \tag{30.12}$$

The position representation has been used to write the expectation value as a wave function integral, as in equation (20.70). One similarly finds that the first-order perturbation to the energy of the first excited state ($n = 2$) vanishes, because the wave function vanishes at $x = L/2$.

So far we have used the component of equation (30.8) along the unperturbed state $|n\rangle$. The components along the other basis vectors $|m\rangle$ are used to get the state vector in first-order perturbation theory, as follows.

The inner product of equation (30.8) with $\langle m|$ for $m \neq n$ gives

$$C_m = \frac{\langle m|V|n\rangle}{E_n - E_m}, \qquad m \neq n. \tag{30.13}$$

The energy eigenstate (30.6) to first order in ϵ is then

$$|\psi\rangle = |n\rangle(1 + \epsilon C_n) + \epsilon \sum_{m \neq n} \frac{|m\rangle\langle m|V|n\rangle}{E_n - E_m}. \tag{30.14}$$

The coefficient C_n is not determined, but note that we can rewrite this expression correct to terms of first order in ϵ as

$$|\psi\rangle = (1 + \epsilon C_n)\left[|n\rangle + \epsilon \sum_{m \neq n} \frac{|m\rangle\langle m|V|n\rangle}{E_n - E_m}\right] + \text{order}\,\epsilon^2, \qquad (30.15)$$

for this only differs from equation (30.14) in terms of second order in the expansion (30.5) in powers of ϵ. Thus we see that in first-order perturbation theory the coefficient C_n represents a normalizing factor. We know the normalizing factor is arbitrary because we are solving a linear equation $H|\psi\rangle = E|\psi\rangle$. The normalization condition $\langle\psi|\psi\rangle = 1$ requires that $(1 + \epsilon C_n)$ have unit modulus to first order in ϵ, so C_n has to be a pure imaginary number. In higher-order perturbation theory, that is, computing the state vector to higher powers of ϵ, these undetermined coefficients build up an arbitrary phase factor in $|\psi\rangle$, as one would expect.

Now let us consider the energy in second-order perturbation theory. The terms proportional to ϵ^2 in equation (30.7) are

$$\sum_i C_i V|i\rangle + \sum_i D_i E_i|i\rangle = \delta E(2)|n\rangle + \delta E(1)\sum_i C_i|i\rangle + E_n \sum D_i|i\rangle. \qquad (30.16)$$

The inner product of this expression with the state vector $\langle n|$ we started with gives an expression for $\delta E(2)$. Using again the orthogonality relation $\langle n|m\rangle = \delta_{nm}$, equation (30.9) for $\delta E(1)$, and equation (30.13) for the expansion coefficients C_m, we get

$$E = E_n + \epsilon\langle n|V|n\rangle + \epsilon^2 \sum_{m \neq n} \frac{\langle n|V|m\rangle\langle m|V|n\rangle}{E_n - E_m} + \text{order}\ \epsilon^3. \qquad (30.17)$$

This is the expression for the energy in second-order perturbation theory.

We see from equations (30.14) and (30.17) why it was stipulated that the energy levels are not degenerate: the expansion can be meaningless if the sums contain eigenvectors $|m\rangle$ of H_o different from $|n\rangle$ but with energy $E_m = E_n$, because the denominators vanish when $E_m = E_n$. If the matrix elements $\langle n|V|m\rangle$ vanish when $E_m = E_n$, then the C_m for the states degenerate with $|n\rangle$ vanish, and we have no problem. This is the case in the examples of first-order perturbation theory in the following sections. Section 33 formalizes the treatment of degenerate perturbation theory.

31 Zeeman Effect

For an example that is easy to compute, let us consider the effect of a static homogeneous magnetic field on the energy levels of an atom. We will imagine the atom consists of a single electron moving in a fixed spherically symmetric potential well. The effect of the magnetic field on the Hamiltonian was discussed in section 19: the momentum operator is replaced with the combination $\mathbf{p}-q\mathbf{A}/c$, where $\mathbf{A}$ is the magnetic vector potential (eqs. [19.32] and [19.34]). On writing the electron charge as $q=-e$, where $e>0$ is the magnitude of the charge, we have

$$H=\frac{1}{2m}(\mathbf{p}+e\mathbf{A}/c)^2-e\phi(r). \tag{31.1}$$

The last term is the potential energy of the electron.

A vector potential that represents a uniform magnetic field $\mathbf{B}$ is

$$\mathbf{A}=\frac{1}{2}\mathbf{B}\times\mathbf{r}. \tag{31.2}$$

It is left as an exercise to check that for homogeneous (uniform through space) $\mathbf{B}$ this agrees with $\mathbf{B}=\nabla\times\mathbf{A}$.

We will compute to first order in the magnetic field strength B, so we can multiply out equation (31.1) and discard the term $\propto A^2\propto B^2$, to get

$$H=H_o+\frac{e}{2mc}(\mathbf{p}\cdot\mathbf{A}+\mathbf{A}\cdot\mathbf{p}),\qquad H_o=\frac{\mathbf{p}^2}{2m}-e\phi(r). \tag{31.3}$$

The terms in H proportional to $\mathbf{A}$ can be simplified as follows. With equation (31.2) for the vector potential, we have

$$\begin{aligned}2\mathbf{p}\cdot\mathbf{A}&=\mathbf{p}\cdot\mathbf{B}\times\mathbf{r}\\&=\mathbf{B}\times\mathbf{r}\cdot\mathbf{p}\\&=\mathbf{B}\cdot\mathbf{r}\times\mathbf{p}=\mathbf{B}\cdot\mathbf{L}.\end{aligned} \tag{31.4}$$

We can change the order of the observables $\mathbf{r}$ and $\mathbf{p}$ in going from the first to the second line because in the cross product $\mathbf{B}\times\mathbf{r}$ the component p_x sees only y and z, and so on. The last step brings us to the usual expression for the orbital angular momentum vector operator, $\mathbf{L}$ (eq. [17.1]). The term $\mathbf{A}\cdot\mathbf{p}$ gives the same result, so the magnetic field part of the Hamiltonian in equation (31.3) is, to first order in the magnetic field strength,

$$V = \frac{e}{2mc}\mathbf{B}\cdot\mathbf{L}. \tag{31.5}$$

This is the same as the classical model in equations (23.8) and (23.9) (with charge $q = -e$).

We must also take account of the energy of interaction of the magnetic field with the magnetic dipole moment of the electron (eq. [23.11]). To simplify the expression, let us approximate the gyromagnetic ratio as $g = 2$. Then with $\mathbf{B}$ along the z axis the sum of equation (31.5) for the orbital energy of interaction with the magnetic field and equations (23.9) and (23.11) for the spin interaction energy is

$$V = \frac{eB}{2mc}(L_z + 2s_z). \tag{31.6}$$

Here s_z represents the z component of the electron spin angular momentum operator.

Since the Hamiltonian H_o in the absence of the magnetic field (eq. [31.3]) is supposed to be spherically symmetric, it commutes with the components of $\mathbf{L}$, and H_o commutes with the components of $\mathbf{s}$ because H_o does not contain $\mathbf{s}$. Thus H_o, L^2, L_z, and s_z all commute with each other, so we can find a complete set of simultaneous eigenvectors of these observables, the eigenvectors being labeled as

$$|\psi\rangle = |n, l, m_l, m_s\rangle. \tag{31.7}$$

The eigenvalue of H_o is E_n, the eigenvalue of L^2 is $\hbar^2 l(l+1)$, the eigenvalue of L_z is $m_l\hbar$, with $-l \le m_l \le l$, and the eigenvalue of s_z is $m_s\hbar$, with $m_s = \pm 1/2$.

Now it is easy to find the expectation value of the magnetic energy perturbation V in equation (31.6) in an eigenstate $|\psi\rangle$ of H_o, because as indicated in equation (31.7) we can choose simultaneous eigenstates of H_o and $L_z + 2s_z$. We have then

$$\delta E = \langle\psi|V|\psi\rangle = \frac{eB\hbar}{2mc}(m_l + 2m_s). \tag{31.8}$$

This is the Zeeman effect to first order in the applied magnetic field B.

It will be noted that we did not have to worry about the fact that the eigenvectors of H_o are degenerate, because the matrix elements of V between states with the same energy and different m_l and m_s vanish.

32 Quadratic Stark Effect

The Stark effect is the shift in energy of an atom caused by the application of a uniform static electric field. The isolated atom has the Hamiltonian H_o. The operator representing the energy of interaction of the atom with the applied electric field is given by the usual expression for the potential energy of charged particles in a uniform electric field,

$$V = -\mathbf{E} \cdot \sum q_i \hat{\mathbf{r}}_i = -\mathbf{E} \cdot \mathbf{d}. \tag{32.1}$$

The ith particle in the atom has position $\hat{\mathbf{r}}_i$ and charge q_i. The atom is assumed to be neutral, $\sum q_i = 0$, so the operator $\mathbf{d} = \sum q_i \hat{\mathbf{r}}_i$ is independent of the choice of origin of coordinates. It will be noted that $\mathbf{d}$ looks like the usual expression for the electric dipole moment of a charge distribution. Also, we are working in the abstract space of states in which position is an observable, as in section 22, so the three components of $\hat{\mathbf{r}}$ are operators.

We are assuming the unperturbed eigenvector $|n\rangle$ of H_o is not degenerate, that is, there is no other state with the same energy, so to first order in the applied electric field the perturbation to the energy of the state is

$$\delta E = \langle n|V|n\rangle = -\mathbf{E} \cdot \sum q_i \langle n|\hat{\mathbf{r}}_i|n\rangle. \tag{32.2}$$

If parity is conserved, this expectation value vanishes, by the following argument.

The parity operator, Π, is defined by the relations

$$\Pi \hat{r}_\alpha = -\hat{r}_\alpha \Pi, \qquad \Pi^2 = 1. \tag{32.3}$$

The first equation says the components of the position observable anticommute with the parity operator, as in equation (15.15) in wave mechanics. Since this anticommutation relation is unaffected if we multiply Π by a constant, we have to add the second condition, which is the same as equation (15.8).

Now let us consider a one-particle system. As discussed in section 22, an eigenvector of the position observable with eigenvalues $\mathbf{r} = x, y, z$ satisfies

$$\hat{\mathbf{r}}|\mathbf{r}\rangle = \mathbf{r}|\mathbf{r}\rangle. \tag{32.4}$$

The result of operating on both sides of this equation with Π and using the first of equations (32.3) is

$$\hat{\mathbf{r}}\Pi|\mathbf{r}\rangle = -\mathbf{r}\Pi|\mathbf{r}\rangle. \tag{32.5}$$

This means $\Pi|\mathbf{r}\rangle$ is an eigenvector of $\hat{\mathbf{r}}$ with eigenvalue $-\mathbf{r}$, that is, $\Pi|\mathbf{r}\rangle = c|-\mathbf{r}\rangle$. The second of equations (32.3) implies that $|c|^2 = 1$, that is, c is a phase factor. Since the phases of the $|\mathbf{r}\rangle$ are arbitrary, we can choose $c = 1$, so

$$\Pi|\mathbf{r}\rangle = |-\mathbf{r}\rangle. \tag{32.6}$$

The wave function $\psi(\mathbf{r})$ representing the state $|\psi\rangle$ is the set of components of $|\psi\rangle$ in the basis $|\mathbf{r}\rangle$:

$$\psi(\mathbf{r}) = \langle\mathbf{r}|\psi\rangle. \tag{32.7}$$

The wave function representing the state $\Pi|\psi\rangle$ is, by equation (32.6),

$$\begin{aligned} \Pi\psi(\mathbf{r}) &\equiv \langle\mathbf{r}|\Pi|\psi\rangle \\ &= \langle -\mathbf{r}|\psi\rangle \\ &= \psi(-\mathbf{r}). \end{aligned} \tag{32.8}$$

The first line defines the parity operator in the position representation. The last line is the definition of Π in section 15. That is, equation (32.3) agrees with the old definition of Π as an operator on wave functions.

Now if Π commutes with H_o, as is true to excellent accuracy in atomic structure (broken only by the weak interaction), parity is conserved. As discussed in section 15, this means the eigenstates of H_o can be classified as eigenstates of parity,

$$\Pi|n\rangle = \pi|n\rangle, \tag{32.9}$$

where the parity quantum number is $\pi = \pm 1$ (eq. [15.10]). Therefore, the expectation values of both sides of the first of equations (32.3) in the eigenstate $|n\rangle$ of H_o give

$$\begin{aligned} \langle n|\Pi\hat{\mathbf{r}}_i|n\rangle &= -\langle n|\hat{\mathbf{r}}_i\Pi|n\rangle, \\ \pi\langle n|\hat{\mathbf{r}}_i|n\rangle &= -\pi\langle n|\hat{\mathbf{r}}_i|n\rangle, \end{aligned} \tag{32.10}$$

where the second line follows from equation (32.9). Since $\pi \neq 0$,

$$\langle n|\hat{\mathbf{r}}_i|n\rangle = 0. \tag{32.11}$$

Therefore the expectation value of the dipole moment operator $\mathbf{d}$ (eq. [32.1]) vanishes. This means the perturbation to the energy vanishes to first order in the electric field. As will be discussed in the next section, this is a result of the assumption that the unperturbed state is not degenerate.

In second-order perturbation theory, the energy is given by equation (30.17):

$$E = E_n + \sum_{\alpha,\beta=1,2,3} \mathrm{E}_\alpha \mathrm{E}_\beta \sum_{m\neq n} \frac{\langle n|d^\alpha|m\rangle\langle m|d^\beta|n\rangle}{E_n - E_m}. \tag{32.12}$$

To avoid confusion with the energy, the components of the electric field have been written as E_α. We can find an interesting interpretation of this expression by considering the first-order perturbation to the state vector.

The perturbed state vector to first order in the electric field E is given by equation (30.15), with $V = -\mathbf{E}\cdot\mathbf{d}$. Ignoring the arbitrary phase factor, the state vector is

$$|\psi\rangle = |n\rangle - \sum_\alpha \mathrm{E}_\alpha \sum_{m\neq n} \frac{|m\rangle\langle m|d^\alpha|n\rangle}{E_n - E_m}. \tag{32.13}$$

The expectation value of the dipole moment observable $\mathbf{d}$ in the unperturbed state $|n\rangle$ vanishes, by equation (32.11). The expectation value of $\mathbf{d}$ to first order in the electric field E is, with equation (32.13),

$$\begin{aligned} \langle\psi|d^\beta|\psi\rangle = &-\sum_\alpha \mathrm{E}_\alpha \sum_{m\neq n} \frac{\langle n|d^\beta|m\rangle\langle m|d^\alpha|n\rangle}{E_n - E_m} - \\ &\sum_\alpha \mathrm{E}_\alpha \sum_{m\neq n} \frac{\langle n|d^\alpha|m\rangle\langle m|d^\beta|n\rangle}{E_n - E_m}. \end{aligned} \tag{32.14}$$

The first term on the right side of this equation is the complex conjugate of the second, for it will be recalled that

$$\langle n|d^\alpha|m\rangle^* = \langle m|d^\alpha|n\rangle, \tag{32.15}$$

because d^α is self-adjoint (eq. [20.25]). Thus, equation (32.14) may be written in the form

$$\langle\psi|d^\beta|\psi\rangle = \sum_\beta \mathrm{E}_\alpha A^{\alpha\beta}, \tag{32.16}$$

where the polarizability tensor is defined to be

$$A^{\alpha\beta} = -2\,\mathrm{Real}\sum_{m\neq n}\frac{\langle n|d^\alpha|m\rangle\langle m|d^\beta|n\rangle}{E_n - E_m}. \tag{32.17}$$

We obtain the real part of the sum because, by equation (32.15), the second term on the right-hand side of equation (32.14) is the complex conjugate of the first term.

As you can check, using equation (32.15), the polarizability tensor is symmetric, $A_{\alpha\beta} = A_{\beta\alpha}$, and is equal to the symmetric part of the last factor in equation (32.12) for the energy in second-order perturbation theory. We see therefore that the leading term in the energy perturbation is the quadratic form,

$$E = E_n - \frac{1}{2}\sum \mathrm{E}_\alpha \mathrm{E}_\beta A^{\alpha\beta}. \tag{32.18}$$

This is the expression for the energy of a neutral atom in an applied electric field, with field components E_α, in second-order perturbation theory, when the first-order part, $\propto \langle n|\mathbf{d}|n\rangle$, vanishes. The constant $A^{\alpha\beta}$ is determined by the wave function for the unperturbed atom, by equation (32.17).

A similar expression for the energy is found in classical physics. Suppose, in a classical model, the isolated atom has no dipole moment. The application of an electric field shifts the position of the ith charged particle in the atom from the unperturbed equilibrium position by a distance r_i^α. To first order, this shift is proportional to the electric field strength, E_β. Thus the applied electric field induces an electric dipole moment that we can write as

$$d^\alpha \equiv \sum q_i r_i^\alpha = \sum A^{\alpha\beta}\mathrm{E}_\beta. \tag{32.19}$$

The constant of proportionality, $A^{\alpha\beta}$, is the classical polarization tensor for the atom. The form is the same as the quantum expression in equation (32.16) for the expectation value of the dipole moment in linear perturbation theory.

To calculate the energy perturbation in the classical model, consider what happens when we increase the electric field strength by an infinitesimal amount, $\mathbf{E} \to \mathbf{E} + \delta\mathbf{E}$. The ith particle shifts position by the amount $\delta\mathbf{r}_i$, the force on the particle is $q_i\mathbf{E}$, so the work done on it is $\delta W_i = q_i\mathbf{E} \cdot \delta\mathbf{r}_i$. The net work done on the atom due to the shifts of positions of the particles is then

$$\delta W = \sum \delta W_i = \sum_i q_i \mathbf{E} \cdot \delta\mathbf{r}_i = \sum_\alpha \mathrm{E}_\alpha \delta d^\alpha. \qquad (32.20)$$

Here $\delta\mathbf{d}$ is the change in the electric dipole moment. Equation (32.19) says $\delta d^\alpha = \sum A^{\alpha\beta}\delta\mathrm{E}_\beta$, so the expression for the work done on the charges is

$$\delta W = \sum A^{\alpha\beta}\mathrm{E}_\alpha \delta\mathrm{E}_\beta. \qquad (32.21)$$

Since $A^{\alpha\beta}$ is constant, we can integrate this equation to get the net work done on increasing the electric field from zero,

$$W = \frac{1}{2}\sum A^{\alpha\beta}\mathrm{E}_\alpha\mathrm{E}_\beta. \qquad (32.22)$$

This is the work done on the springs that hold the charges in place in this classical model. To get the net change in the energy of the atom, we have to add the electrostatic potential energy, which is given by equation (32.1) with equation (32.19) for $\mathbf{d}$:

$$U = -\mathbf{E} \cdot \mathbf{d} = -\sum A^{\alpha\beta}\mathrm{E}_\alpha\mathrm{E}_\beta. \qquad (32.23)$$

The total change in energy of the atom due to the application of the electric field is the sum of the electrostatic energy and the change in internal energy:

$$E = W + U = -\frac{1}{2}\sum A^{\alpha\beta}\mathrm{E}_\alpha\mathrm{E}_\beta. \qquad (32.24)$$

This agrees with equation (32.18). We see that the quantum mechanical calculation managed to keep track of all this, to give the same form for the energy perturbation in terms of the polarization tensor defined in quantum mechanics by equation (32.16), and in classical mechanics by equation (32.19).

The effect discussed here is quadratic, that is, at small electric fields E the energy perturbation varies as the square of E. If the unperturbed energy eigenstate is degenerate there is a linear stark effect, where the energy perturbation is proportional to the first power of E at small E, as discussed next.

33 Degenerate Perturbation Theory

In some of the above examples the unperturbed energy eigenvectors are degenerate, but we could ignore that because the matrix elements $\langle n|V|m\rangle$ between different states with the same energy vanish. If that is not so, we can arrange it by using linear combinations of the degenerate states, as follows.

Suppose we start from the unperturbed energy eigenvectors $|n, c\rangle$, where n labels the eigenvalue E_n of H_o and c labels the N different states with the same value of E_n. As discussed in sections 13 and 20, the states can be arranged to be orthogonal,

$$\langle n, c|m, d\rangle = \delta_{nm}\delta_{cd}. \tag{33.1}$$

An equally good basis would include instead N different linear combinations of these states $|n, c\rangle$. The game will be to choose these combinations so the off-diagonal matrix elements of the perturbation V vanish in the new basis.

Write the N linear combinations of the $|n, c\rangle$ as

$$|n, i\rangle = \sum_c F_c^i |n, c\rangle. \tag{33.2}$$

The label i has N different values, and the constants F_c^i are the expansion coefficients for the ith of these new states, $|n, i\rangle$. The matrix elements of V among pairs of these new basis vectors are

$$\langle n, i|V|n, j\rangle = \sum_{c,d} F_c^{i*} V_{cd} F_d^j, \tag{33.3}$$

where the matrix elements of V among the original basis vectors are

$$V_{cd} = \langle n, c|V|n, d\rangle. \tag{33.4}$$

Now let us choose the expansion coefficients F_c^i in equation (33.2) to be the solutions to the matrix eigenvalue equation

$$\sum_d V_{cd} F_d^i = v^i F_c^i. \tag{33.5}$$

As discussed in sections 12 and 13, we know these N linear equations, for $c = 1, 2, \ldots, N$, have N solutions, labeled i, with eigenvalues v^i. Also, the solutions are orthogonal in the sense that, with suitably chosen normalization,

$$\sum_c F_c^{i*} F_c^j = \delta_{ij}. \tag{33.6}$$

The proof follows just as in the demonstration in section 13 that different eigenvectors can be chosen to be orthogonal. It follows from this orthogonality relation that the new set of states in equation (33.2) are orthogonal,

$$\langle n, i|n, j\rangle = \sum_c F_c^{i*} F_c^j = \delta_{ij}, \tag{33.7}$$

where the first equation follows from the orthogonality of the original $|n, c\rangle$ (eq. [33.1]). In the same way, we have from equations (33.3) and (33.5)

$$\langle n, i|V|n, j\rangle = v^i \delta_{ij}. \tag{33.8}$$

That is, we have made the matrix elements of V diagonal in the new set of N degenerate states $|n, i\rangle$, for $i = 1$ to N (and fixed n, which labels the common energy of these basis vectors).

Let us check that this solves the problem with vanishing energy denominators in equation (30.15) for the first-order perturbation to the wave function and equation (30.17) for the second-order perturbation to the energy. These were obtained from the first-order part of the energy eigenvalue equation (30.7). In terms of the $|n, i\rangle$, equation (30.8) is

$$V|n, i\rangle + \sum_{m,k} C_{m,k} E_m |m, k\rangle = \delta E(1)|n, i\rangle + E_n \sum C_{m,k}|m, k\rangle. \tag{33.9}$$

The inner product of this with $\langle n, j|$ is, with equations (33.7) and (33.8),

$$\delta E(1)\delta_{ij} = \langle n, j|V|n, i\rangle = v^i \delta_{ij}. \tag{33.10}$$

The labels i and j represent two of the N degenerate eigenvectors of H_o in the linear combinations $|n, i\rangle$ (eq. [33.2]). The delta function on the left follows from the orthogonality of the $|n, i\rangle$. This is consistent with the delta function on the right, because we have chosen the degenerate state vectors to diagonalize the matrix elements of V.

We see from equations (33.5) and (33.10) that in first-order perturbation theory the values of the energy perturbation caused by V are the eigenvalues v^i of the matrix elements V_{cd} of the operator V in the set of degenerate eigenvectors $|n, c\rangle$ with energy E_n (eq. [33.4]). If there are N degenerate eigenvectors $|n, c\rangle$ of H_o, there are N values of the energy perturbation v^i (some of which may of course be the same).

In the discussion of the quadratic Stark effect, we saw that the expectation value of the electric dipole moment $\mathbf{d}$ vanishes in a state of definite parity. That is why there is no linear Stark effect, where δE proportional to the first power of the electric field strength, in a state that is not degenerate: the unperturbed energy level has definite parity so the expectation value of V vanishes. If there are degenerate states with opposite parity, then we can form linear combinations of even and odd parity in which the expectation value of $\mathbf{d}$ is nonzero, leading to a linear Stark effect, where the degenerate states are split in energy by an amount proportional to the electric field (at small E). The classical analog of the linear Stark effect is the energy in an electric field of an object with a permanent electric dipole moment.

34 Hyperfine Structure in Atomic Hydrogen

The interaction between the magnetic dipole moments of the atomic nucleus and the electrons perturbs the energy levels of an atom. Since the effect is small it is called hyperfine structure. The hyperfine splitting of the ground state energy of atomic hydrogen is equivalent to a wavelength of 21.1 cm. This is the famous line used to drive the most stable atomic clocks, and used in radio astronomy to map out the distribution and motion (through the Doppler effect) of neutral gas in galaxies, the bulk of the gas being atomic hydrogen. The computation of the hyperfine structure in atomic hydrogen is longwinded; it is presented here as an example of a complete and fairly detailed but still manageable computation in quantum mechanics.

The Hamiltonian

The term in the Hamiltonian that represents the magnetic energy of interaction of the electron and proton dipole moments is taken from classical electromagnetism. We will need the multipole expansion of the field of a charge distribution, as follows.

The electric field at position $\mathbf{r}$ of a point charge q placed at position $\mathbf{s}$ is

$$\mathbf{E}(\mathbf{r}) = q\frac{\mathbf{r}-\mathbf{s}}{|\mathbf{r}-\mathbf{s}|^3}. \tag{34.1}$$

The Taylor series expansion of this expression in powers of $\mathbf{s}$ is

$$\mathbf{E}(\mathbf{r}) = q\frac{\mathbf{r}}{r^3} + q\left(\frac{3\mathbf{r}(\mathbf{r}\cdot\mathbf{s})}{r^5} - \frac{\mathbf{s}}{r^3}\right) + \dots, \tag{34.2}$$

where the dots represent higher powers of s/r. The electric field of a charge distribution is a sum of expressions of this form, one for each particle. We can write the sum as

$$\mathbf{E}(\mathbf{r}) = Q\frac{\mathbf{r}}{r^3} + \left(\frac{3\mathbf{r}(\mathbf{r}\cdot\mathbf{d})}{r^5} - \frac{\mathbf{d}}{r^3}\right) + \dots. \tag{34.3}$$

Here $Q = \sum q$ is the total charge, or the monopole moment, of the charge distribution. The first term on the right side of equation (34.3) is the monopole part of the field; it is what is seen at great distance from the distribution. The first-order correction in the next term is the dipole field, where the dipole moment of the charge distribution is $\mathbf{d} = \sum q\mathbf{s}$, as in equation (32.1). The next term in the expansion is the quadrupole field, which is of order s^2, and decays with distance r from the charge distribution as r^{-4}.

The static magnetic field produced by a localized steady current source has to have the same multipole expansion outside the region of the currents, because $\mathbf{E}$ and $\mathbf{B}$ satisfy the same differential equations in the region outside the sources. However, there are no (known) magnetic monopoles, so the leading term in the expansion of $\mathbf{B}$ is the dipole field in equation (34.3), with the magnetic dipole moment $\vec{\mu}$ replacing the electric dipole moment $\mathbf{d}$.

The magnetic field of a proton is a pure dipole field with moment $\vec{\mu}_p$:

$$\mathbf{B} = \frac{3\mathbf{r}(\mathbf{r}\cdot\vec{\mu}_p)}{r^5} - \frac{\vec{\mu}_p}{r^3}. \tag{34.4}$$

The energy of the electron magnetic moment in this field is (eq. [23.9])

$$V = -\vec{\mu}_e \cdot \mathbf{B} = \frac{\vec{\mu}_e \cdot \vec{\mu}_p}{r^3} - \frac{3(\vec{\mu}_e \cdot \mathbf{r})(\vec{\mu}_p \cdot \mathbf{r})}{r^5}, \tag{34.5}$$

where $\vec{\mu}_e$ is the electron magnetic dipole moment, and $\mathbf{r}$ is the separation between the electron and proton. It will be noted that this expression is symmetric in the electron and proton variables.

We will use the classical equation (34.5) for V as a guide to writing down the Hamiltonian for a hydrogen atom. The observables for the atom are the three components of the position, momentum, and spin of the electron and proton: $\mathbf{r}_e$, $\mathbf{p}_e$, $\mathbf{s}_e$, $\mathbf{r}_p$, $\mathbf{p}_p$, and $\mathbf{s}_p$. These observables satisfy the usual commutation relations,

$$\begin{aligned} [r_e^\alpha, p_e^\beta] &= i\hbar\delta_{\alpha\beta}, \\ [s_e^x, s_e^y] &= i\hbar s_e^z, \\ [r_e^\alpha, p_p^\beta] &= 0, \end{aligned} \tag{34.6}$$

and so on. As discussed in section 23 (eq. [23.11]), the proton and electron magnetic dipole moments are

$$\begin{aligned} \vec{\mu}_p &= \frac{g_p e}{2M_p c}\mathbf{s}_p, \qquad g_p = 5.59\ldots, \\ \vec{\mu}_e &= -\frac{g_e e}{2M_e c}\mathbf{s}_e, \qquad g_e = 2.002\ldots. \end{aligned} \tag{34.7}$$

The electron and proton masses have been written as M_e and M_p to avoid confusion with the spin quantum numbers. It will be noted that the charge e is positive. The magnetic moment of the negatively charged electron is antiparallel to the spin, and the magnetic moment of the positively charged proton is parallel to the spin.

The Hamiltonian for a hydrogen atom is

$$H = H_o + V, \tag{34.8}$$

where

$$H_o = \frac{p_p^2}{2M_p} + \frac{p_e^2}{2M_e} - \frac{e^2}{|\mathbf{r}_p - \mathbf{r}_e|} \tag{34.9}$$

represents the kinetic and electrostatic energies of the particles. The magnetic energy of interaction of the dipole moments is, following equation (34.5) with the magnetic moments in equation (34.7),

$$V = \frac{g_e g_p e^2}{4M_p M_e c^2}\left[\frac{3(\mathbf{s}_e\cdot\mathbf{r})(\mathbf{s}_p\cdot\mathbf{r})}{r^5} - \frac{\mathbf{s}_e\cdot\mathbf{s}_p}{r^3}\right], \tag{34.10}$$

with $\mathbf{r} = \mathbf{r}_e - \mathbf{r}_p$.

The computation of the eigenstates of H_o is discussed in sections 6 and 18. As shown in section 12, the computation is reduced to a one-body problem by changing the position variables from $\mathbf{r}_e$ and $\mathbf{r}_p$ to center of mass coordinates and the relative coordinates $\mathbf{r} = \mathbf{r}_e - \mathbf{r}_p$. This makes H_o a sum of two terms, $H_o = K + H_1$, with K representing the kinetic energy of translation of the atom. The Hamiltonian for the internal energy is

$$H_i = H_1 + V, \qquad H_1 = \frac{p^2}{2M} - \frac{e^2}{r}. \tag{34.11}$$

Here M is the reduced mass (eq. [12.17]), and $\mathbf{p}$ is the momentum conjugate to the relative position observable $\mathbf{r}$. We will start with an eigenstate of the Hamiltonian H_1, and compute the effect of the spin-spin coupling V in equation (34.10) in linear perturbation theory.

Perturbation Calculation

The first step is to introduce a basis to represent state vectors. This is done by a straightforward extension of the discussion in section 23 for a single particle with spin. Convenient basis vectors are

$$|\mathbf{r}, m_e, m_p\rangle. \tag{34.12}$$

These are simultaneous eigenstates of the three components of relative position and the z components of the electron and proton spins, with $m_e = \pm 1/2$ and $m_p = \pm 1/2$. The position observables have continuous eigenvalues, so the normalization is

$$\langle \mathbf{r}, m_e, m_p | \mathbf{r}', m_e', m_p'\rangle = \delta(\mathbf{r} - \mathbf{r}')\delta_{m_e, m_e'}\delta_{m_p, m_p'}, \tag{34.13}$$

and the corresponding completeness relation is

$$1 = \sum_{m_e m_p} \int d^3r\, |\mathbf{r}, m_e, m_p\rangle\langle \mathbf{r}, m_e, m_p|, \tag{34.14}$$

as in equations (20.53) and (20.58).

The representation of the state vector $|\psi\rangle$ in the basis (34.12) is

$$\psi(\mathbf{r})_{m_e,m_p} = \langle \mathbf{r}, m_e, m_p|\psi\rangle. \tag{34.15}$$

This is a function of the three continuous variables $\mathbf{r} = x, y, z$ and of the discrete variables m_e and m_p. The representation of the energy eigenvalue equation

$$H_1|\psi\rangle = E|\psi\rangle, \tag{34.16}$$

where H_1 is the Hamiltonian (34.11), is the inner product of both sides of this equation with the dual vector $\langle \mathbf{r}, m_e, m_p|$. As in section 22, this yields the differential equation

$$-\frac{\hbar^2}{2m}\nabla^2\psi - \frac{e^2}{r}\psi = E\psi, \tag{34.17}$$

where ψ is the wave function in equation (34.15).

Since m_e and m_p do not appear in the differential equation (34.17), we can write the solutions in the form

$$\psi = \psi(\mathbf{r})\chi_{m_e m_p}, \tag{34.18}$$

where the spin wave function $\chi_{m_e m_p}$ is an arbitrary function of the two discrete variables m_e and m_p. Because $\chi_{m_e m_p}$ has four possible sets of values of the arguments (for $m_e, m_p = \pm 1/2$), there are four linearly independent choices for this function. As discussed in section 25, these can be taken to be the triplet of states with total spin 1 and the singlet spin zero state (eq. [25.19]). We will see that the spin-spin coupling term V (eq. [34.10]) produces an energy difference between the singlet and triplet states.

We are interested in the ground state eigenfunction of H_1. The ground state wave function was obtained in section 6 (eqs. [6.21], [6.23], [18.34]); the normalized wave function is

$$\psi_0 = \frac{e^{-r/a_o}}{(\pi a_o^3)^{1/2}}, \tag{34.19}$$

where the Bohr radius is

$$a_o = \frac{\hbar^2}{Me^2}. \tag{34.20}$$

In first-order perturbation theory we must evaluate the expectation value of V in the eigenstates of H_1. The method of computation of the matrix elements of a product of operators such as appears in equation (34.10) for V was discussed in section 20. In a discrete basis $|m\rangle$, the matrix element of an operator product $V = AB$ of two operators, A and B, between states $|\psi\rangle$ and $|\phi\rangle$ is

$$\langle\psi|V|\phi\rangle = \langle\psi|AB|\phi\rangle = \sum_{mnp} \langle\psi|m\rangle\langle m|A|n\rangle\langle n|B|p\rangle\langle p|\phi\rangle, \qquad (34.21)$$

where the completeness relation $\sum |n\rangle\langle n| = 1$ has been used three times. For a continuous eigenvalue, the sum is replaced with an integral, as discussed in equation (20.57). The basis vectors here are the $|\mathbf{r}, m_e, m_p\rangle$ in equation (34.12), with the completeness relation in equation (34.14). The matrix elements of the components of the spin operators are the Pauli spin matrices that were worked out in section 24; the only difference here is that we have extra eigenvalues that have to match on either side of a matrix element such as equation (34.22) below. (It is left as an exercise to explain why.) We have for example

$$\langle\mathbf{r}, m_e, m_p|s_e^\alpha|\mathbf{r}', m_e', m_p'\rangle = \frac{\hbar}{2}\sigma^\alpha_{m_e m_e'}\delta(\mathbf{r}-\mathbf{r}')\delta_{m_p m_p'}, \qquad (34.22)$$

where the Pauli spin matrices $\sigma^\alpha_{mm'}$ are given in equation (24.20). Similarly, the matrix element of a function $f(\hat{\mathbf{r}})$ of the position observable $\hat{\mathbf{r}}$ is

$$\langle\mathbf{r}, m_e, m_p|f(\hat{\mathbf{r}})|\mathbf{r}', m_e', m_p'\rangle = f(\mathbf{r})\delta(\mathbf{r}-\mathbf{r}')\delta_{m_e m_e'}\delta_{m_p m_p'}. \qquad (34.23)$$

We need the matrix elements of the perturbation V in equation (34.10) between eigenstates of H_1. As in equation (34.21), the game is to insert the completeness relation in equation (34.14) between each factor in the matrix element. The two ends yield the wave function (eqs. [34.15] and [34.18]) and its complex conjugate. In between we have factors like equation (34.22). After eliminating the sums over Kronecker delta functions and integrals over Dirac delta functions, we get

$$\langle V\rangle = \frac{g_e g_p e^2}{4M_e M_p c^2}\sum_{\alpha\beta} I^{\alpha\beta}\frac{\hbar^2}{4}\chi^\dagger\sigma_e^\alpha\sigma_p^\beta\chi. \qquad (34.24)$$

The indices α and β are summed from 1 to 3, representing the dot products in equation (34.10). The integral over relative position $\mathbf{r}$ is

$$I^{\alpha\beta} = \int d^3r \, |\psi_0(r)|^2 \left[\frac{3r^\alpha r^\beta}{r^5} - \frac{\delta^{\alpha\beta}}{r^3} \right], \tag{34.25}$$

with $\psi_0(r)$ given by equation (34.19). The last factor $\chi^\dagger \sigma_e^\alpha \sigma_p^\beta \chi$ in equation (34.24) is the matrix product of the spin wave function χ in equation (34.18) and its Hermitian adjoint with Pauli spin matrices:

$$\chi^\dagger \sigma_e^\alpha \sigma_p^\beta \chi = \sum_{m_e m_e' m_p m_p'} \chi^*_{m_e m_p} \sigma^\alpha_{m_e m_e'} \sigma^\beta_{m_p m_p'} \chi_{m_e' m_p'}. \tag{34.26}$$

This comes from the matrix elements of the spin operators (eq. [34.22]). Now we have to work out the integral $I^{\alpha\beta}$ in equation (34.25), and the sums over spin quantum numbers in equation (34.26).

The integral has an interesting property. Since the wave function ψ_0 in the integrand is spherically symmetric, the integral cannot be affected by a rotation of the coordinate system. The only tensor with this property is the Kroneker delta function, so the integral must be of the form $I^{\alpha\beta} = I\delta^{\alpha\beta}$. (For another way to see this, note that the angle integral at fixed radius makes $I^{\alpha\beta}$ vanish when $\alpha \neq \beta$, and that the angle integrals of x^2, y^2, and z^2 are the same.) To compute the quantity I, we might be tempted to set $\alpha = \beta$ in equation (34.25) and then sum over α. But when we do that we see that the integrand vanishes everywhere except at $r = 0$, where it diverges. This does show that the only place where the value of $\psi_0(r)$ can matter is at $r = 0$, so we can write

$$\begin{aligned} I^{\alpha\beta} &= |\psi_0(0)|^2 J^{\alpha\beta}, \\ J^{\alpha\beta} &= \int d^3r \left[\frac{3r^\alpha r^\beta}{r^5} - \frac{\delta^{\alpha\beta}}{r^3} \right]. \end{aligned} \tag{34.27}$$

A safe strategy in evaluating this integral is to write the magnetic field of the proton in a way that approximates the dipole form in equation (34.4) everywhere save near $r \to 0$, where the approximation rounds off to a nonsingular value. We will evaluate the integral for this nonsingular field, and then take the limit as the field approaches a pure dipole.

The magnetic field is the curl of a vector potential, $\mathbf{A}$ (eq. [19.14]). A potential for the dipole field in equation (34.4) is

$$\mathbf{A} = \vec{\mu}_p \times \mathbf{r}/r^3. \tag{34.28}$$

The trick in evaluating the curl of this cross product, to get the magnetic field, is to use the standard vector identity for a double cross product, being careful not to change orders of differentiation, as in equations (2.7) and (2.8). That gives

$$\mathbf{B} = \nabla \times (\vec{\mu}_p \times \mathbf{r}/r^3) = \vec{\mu}_p \nabla \cdot (\mathbf{r}/r^3) - (\vec{\mu}_p \cdot \nabla)\mathbf{r}/r^3. \qquad (34.29)$$

The divergence of $\mathbf{r}/r^3$ vanishes at $\mathbf{r} \neq 0$, because this is the field of a point charge. On differentiating out the last term, one arrives at the dipole expression in equation (34.4). That is, this is a vector potential for the dipole magnetic field.

Now let us replace the vector potential with the form

$$\mathbf{A} = \vec{\mu}_p \times \mathbf{r} f(r), \qquad (34.30)$$

where $f(r) = 1/r^3$ at large r, and $f(r)$ rolls over to a finite value at $r \to 0$. Then on repeating the above calculation, one finds that the magnetic field of this potential is

$$\mathbf{B} = \vec{\mu}_p \left(2f + r\frac{df}{dr} \right) - \mathbf{r}\,\frac{\vec{\mu}_p \cdot \mathbf{r}}{r}\frac{df}{dr}. \qquad (34.31)$$

This changes the expression for the magnetic energy of interaction of the electron and proton magnetic moments in equation (34.10) to

$$V = \frac{g_e g_p e^2}{4M_p M_e c^2} \left[-\frac{(\mathbf{s}_e \cdot \mathbf{r})(\mathbf{s}_p \cdot \mathbf{r})}{r}\frac{df}{dr} + \mathbf{s}_e \cdot \mathbf{s}_p \left(2f + r\frac{df}{dr} \right) \right]. \qquad (34.32)$$

The integral in equation (34.27) becomes

$$J^{\alpha\beta} = \int d^3r \left[-\frac{r^\alpha r^\beta}{r}\frac{df}{dr} + \delta_{\alpha\beta} \left(2f + r\frac{df}{dr} \right) \right]. \qquad (34.33)$$

One readily checks that when $f = 1/r^3$ this is the same as equation (34.27).

Let us evaluate this integral in polar coordinates. The integral over angles makes the factor $r^\alpha r^\beta$ vanish when $\alpha \neq \beta$, and when $\alpha = \beta$ the value of $(r^\alpha)^2$ averaged over angles is $r^2/3$. That brings equation (34.33) to the form

$$J^{\alpha\beta} = 4\pi\delta_{\alpha\beta} \int_0^\infty r^2 dr \left[2f + \frac{2r}{3}\frac{df}{dr} \right]. \qquad (34.34)$$

The final step is to note that we can rearrange the integrand to make it a total derivative:

$$\begin{aligned} J^{\alpha\beta} &= \frac{8\pi}{3}\delta_{\alpha\beta}\int_0^\infty dr\,\left[3r^2 f + r^3\frac{df}{dr}\right] \\ &= \frac{8\pi}{3}\delta_{\alpha\beta}\int_0^\infty dr\,\frac{d}{dr}r^3 f. \end{aligned} \tag{34.35}$$

Since $f = 1/r^3$ at large r and f is finite at $r \to 0$, the integral is unity at the upper limit and vanishes at the lower limit. We have then

$$J^{\alpha\beta} = \frac{8\pi}{3}\delta_{\alpha\beta}. \tag{34.36}$$

Since this expression is independent of f (as long as $f = 1/r^3$ at large r), it is the wanted value for the integral in the limit of a pure dipole field.

With this result in equation (34.27), the expectation value of the magnetic energy in equation (34.24) is

$$\langle V\rangle = \frac{2\pi}{3}\frac{g_e g_p e^2}{M_e M_p c^2}|\psi_0(0)|^2\frac{\hbar^2}{4}\chi^\dagger\vec{\sigma}_e\cdot\vec{\sigma}_p\chi. \tag{34.37}$$

The trick in evaluating the spin matrix product in equations (34.37) and (34.26) is to recall that, to apply first-order perturbation theory, we must choose the eigenstates of H_1 so that the matrix elements of V between degenerate eigenstates are diagonal (eq. [33.10]). We saw how to do this in the present case in section 25 (eqs. [25.23] and [25.27]). Introduce the total spin operator,

$$\hat{\mathbf{s}} = \hat{\mathbf{s}}_e + \hat{\mathbf{s}}_p. \tag{34.38}$$

On squaring this sum and rearranging, we have

$$\hat{\mathbf{s}}_e\cdot\hat{\mathbf{s}}_p = \frac{1}{2}(\hat{s}^2 - \hat{s}_e^2 - \hat{s}_p^2). \tag{34.39}$$

Since $\hat{s}^2$ commutes with H_1 (because H_1 does not contain $\hat{\mathbf{s}}_e$ or $\hat{\mathbf{s}}_p$), we can take the eigenstates of H_1 to be simultaneous eigenstates of the total spin $\hat{s}^2$ and of the z component $\hat{s}_z$. It will be recalled from section 25 that the eigenvalues of $\hat{s}^2$ are $\hbar^2 s(s+1)$, with $s = 0$ and 1, representing the singlet and triplet states (25.19). We see from equation (34.39) that the spin wave function $\chi(s,m)_{m_e,m_p}$ for the ground state wave functions in the hydrogen atom (eq. [34.18]) belonging to a state with total spin s and z component m satisfies

$$\frac{\hbar^2}{4}\vec{\sigma}_e \cdot \vec{\sigma}_p \; \chi(s,m) = \frac{1}{2}\hbar^2(s^2 + s - 3/2) \; \chi(s,m). \tag{34.40}$$

(Recall that we are representing the spin observable $\hat{\mathbf{s}}_e$ as $\hbar\vec{\sigma}_e/2$, where the σ_α are the Pauli spin matrices, so the matrix on the left side of this equation is the representation of $\hat{\mathbf{s}}_e \cdot \hat{\mathbf{s}}_p$. Also, since the electron and proton have spin 1/2, $\hat{s}_e^2 = \hat{s}_p^2 = 3\hbar^2/4$.)

Equation (34.40) says $\chi(s,m)$ is an eigenvector of $\vec{\sigma}_e \cdot \vec{\sigma}_p$. That means the matrix product $\chi^\dagger\vec{\sigma}_e \cdot \vec{\sigma}_p\chi$ in equation (34.37) vanishes unless the quantum numbers s and m in χ are the same on the left and the right, that is, the matrix elements of V between states with the same unperturbed energy are diagonal. The diagonal matrix elements (those with the same s and m) are the values of the first-order perturbation to the energy.

Collecting equations (34.19) and (34.20) for $\psi_0(0)$, and equations (34.37) and (34.40) for $\langle V \rangle$, we find that the first-order perturbation to the ground state energy in atomic hydrogen due to the interaction between the electron and proton magnetic dipole moments is

$$\delta E_s = \langle V \rangle = \frac{g_e g_p e^8 M_e^2}{3 M_p c^2 \hbar^4}(s^2 + s - 3/2). \tag{34.41}$$

The reduced mass has been approximated as $M = M_e$.

For the three terms in the triplet state, with $s = 1$, the last factor in equation (34.41) is 1/2, and for the singlet state with $s = 0$ the factor is $-3/2$. This means the value of the energy perturbation averaged over the four initially degenerate states vanishes: the magnetic spin-spin interaction pushes the three triplet states up, and the one singlet state down three times as far.

The difference between triplet and singlet state energies can be written as

$$\delta E_1 - \delta E_0 = 2\pi\hbar\nu. \tag{34.42}$$

This gives the frequency ν of the radiation emitted in the decay from the excited triplet state to the ground singlet state, as discussed in section

3 and in section 37 below. From equation (34.41), the frequency is

$$\nu = \frac{g_e g_p e^8 M_e^2}{3\pi M_p c^2 \hbar^5}. \tag{34.43}$$

As in section 9, we can simplify this expression by introducing convenient physical constants. The fine-structure constant is

$$\alpha = \frac{e^2}{\hbar c} = \frac{1}{137.04}, \tag{34.44}$$

and the Compton wavelength of the electron is

$$\lambda_e = \frac{\hbar}{M_e c} = 3.86 \times 10^{-11}\,\text{cm}. \tag{34.45}$$

With these expressions equation (34.43) becomes

$$\nu = \frac{g_e g_p \alpha^4}{3\pi} \frac{M_e}{M_p} \frac{c}{\lambda_e}. \tag{34.46}$$

With equation (34.7) for the gyromagnetic ratios, we get $\nu = 1420\,\text{MHz}$, corresponding to wavelength $\lambda = c/\nu = 21\,\text{cm}$.

The ratio of the energy difference between the singlet and triplet states in equation (34.41) to the binding energy B of the electron in the hydrogen atom (eq. [6.24]) is

$$\frac{\delta E_1 - \delta E_0}{B} = \frac{4}{3} \frac{M_e}{M_p} g_e g_p \alpha^2. \tag{34.47}$$

This is a product of the small numbers $\alpha^2 = 1/137^2$ and $M_e/M_p = 1/2000$. That is, the energy shift due to the interaction of the electron and proton magnetic moments is a very small perturbation to the net energy of the atom.

35 Time-Dependent Perturbation Theory

The formalism that has been developed so far successfully predicts the energy levels of an isolated atom, but it says an isolated atom in an excited state stays there indefinitely, which certainly is not right: the atom decays with the emission of one or more photons. The problem is that we have not used a quantum mechanical description of the electromagnetic field. The way this is done is described in this and the following two sections. The first step is to consider the behavior of an atom placed in a given classical electromagnetic field that is oscillating with time at a definite frequency. We will see that the atom is induced to make transitions between what would be the stationary energy levels if it were isolated. Because this electromagnetic field acts like a simple harmonic oscillator, we can use the quantum mechanics of an oscillator to find the spontaneous transition rate in the absence of the applied field.

Perturbation Expansion

Consider a Hamiltonian of the form

$$H = H_o + V(t). \tag{35.1}$$

The time-independent operator H_o represents an isolated system, and $V(t)$ can represent the effect of an applied time varying field, such as electromagnetic radiation. Since this latter part of the Hamiltonian can be a function of time, we must consider here solutions to Schrödinger's time-dependent equation.

Suppose the eigenvectors of H_o are discrete and normalized,

$$H_o|n\rangle = E_n|n\rangle, \qquad \langle n|m\rangle = \delta_{nm}. \tag{35.2}$$

Because H_o is independent of time, the vectors $|n\rangle$ and energies E_n are independent of time. The solution to Schrödinger's time-dependent equation for the isolated system,

$$i\hbar\frac{\partial|\psi\rangle}{\partial t} = H_o|\psi, \tag{35.3}$$

can be written as an expansion in this basis,

$$|\psi\rangle = \sum c_n|n\rangle e^{-iE_nt/\hbar}, \tag{35.4}$$

where the expansion coefficients c_n are independent of time.

We are interested in solutions to Schrödinger's equation for the full Hamiltonian in equation (35.1),

$$i\hbar\frac{\partial|\phi\rangle}{\partial t} = (H_o + V(t))|\phi\rangle. \tag{35.5}$$

Since the eigenstates $|n\rangle$ of H_o are a complete set, we can express a solution to this equation as a linear combination of the $|n\rangle$. Following equation (35.4), let us write this expansion in the form

$$|\phi\rangle = \sum c_n(t)|n\rangle e^{-iE_nt/\hbar}. \tag{35.6}$$

In the limit $V \to 0$ the c_n are constants, as in equation (35.4), but in general the c_n are functions of time. Equation (35.6) in Schrödinger's equation (35.5) is

$$\sum_n (i\hbar\frac{dc_n}{dt} + E_n c_n)e^{-iE_nt/\hbar}|n\rangle = \sum_n (E_n + V(t))c_n e^{-iE_nt/\hbar}|n\rangle. \tag{35.7}$$

On taking the inner product of this expression with the dual basis vector $\langle m|$, using the orthogonality condition $\langle m|n\rangle = \delta_{mn}$, and rearranging, we get

$$i\hbar\frac{dc_m}{dt} = \sum_n \langle m|V|n\rangle c_n(t)e^{i(E_m-E_n)t/\hbar}. \tag{35.8}$$

This set of coupled differential equations determines the expansion coefficients $c_n(t)$ in the expression (35.6) for the state vector.

Suppose that, at time $t = 0$, the system is in the eigenstate $|i\rangle$ of H_o. This means the initial conditions at $t = 0$ are that $c_i = 1$ and all the other coefficients vanish. Then if t is positive but small enough we can set the c_n in the right-hand side of equation (35.8) equal to their initial values, to get the first approximation to the expansion coefficients c_f,

$$i\hbar\frac{dc_f}{dt} = \langle f|V|i\rangle e^{i(E_f-E_i)t/\hbar}. \tag{35.9}$$

If $f \neq i$, $c_f(0) = 0$, so the solution to equation (35.9) is

$$c_f(t) = -\frac{i}{\hbar}\int_0^t dt\langle f|V|i\rangle e^{i(E_f-E_i)t/\hbar}, \tag{35.10}$$

for $f \neq i$. This is the wanted expression for the expansion coefficients in equation (35.6) in first-order time-dependent perturbation theory. These coefficients are determined by the transition matrix elements $\langle f|V|i\rangle$.

It will be noted that the operator V need not be a function of time. If V is constant, the time-independent perturbation theory in the preceding sections gives approximations to the eigenstates and eigenvectors of H_o+

V, while equation (35.10) in equation (35.6) gives the time evolution of the state vector from the initial condition $|\phi\rangle = |n\rangle$, which is an eigenstate of H_o, not $H_o + V$.

In second order time-dependent perturbation theory, one substitutes the expression (35.10) for the coefficients $c_f(t)$ back into the right-hand side of equation (35.8), and then one integrates the equation to get an expression for the amplitudes $c_f(t)$ that contains terms with products of two matrix elements $\langle n|V|m\rangle$. On iterating this procedure, one develops an expression for the amplitudes as a power series in the matrix elements of V.

As we will now discuss, the probability amplitude for finding that the system at time t is in the eigenstate $|f\rangle$ of H_o is $\langle f|\phi\rangle = c_f(t)$, and the square $|c_f(t)|^2$ is the probability of finding that V has induced a transition from the initial eigenstate $|i\rangle$ of H_o to the final eigenstate $|f\rangle$ at time t.

Transition Probabilities

Suppose the time-dependent perturbing term in equation (35.1) is an oscillating function of time,

$$V(t) = V_o \cos(\omega t). \tag{35.11}$$

The amplitude V_o is a constant operator, and ω is the real constant frequency of oscillation of the perturbation $V(t)$. On using $\cos \omega t = (e^{i\omega t} + e^{-i\omega t})/2$, and writing the energy difference between the eigenstates $|i\rangle$ and $|f\rangle$ of H_o as

$$E_f - E_i \equiv \hbar\omega_o, \tag{35.12}$$

we see that the first-order perturbation solution in equation (35.10) is

$$c_f(t) = -\frac{1}{2\hbar}\langle f|V_o|i\rangle \left[\frac{e^{i(\omega_o+\omega)t} - 1}{\omega_o + \omega} + \frac{e^{i(\omega_o-\omega)t} - 1}{\omega_o - \omega}\right]. \tag{35.13}$$

When the applied frequency ω in $V(t)$ is close to the resonant frequency ω_o defined in equation (35.12), the term with the small denominator in equation (35.13) dominates. Thus if $\omega_o > 0$, which means $E_f > E_i$ so the atom gains energy, and if ω is close to resonance, $\omega \sim \omega_o$, we can approximate equation (35.13) by keeping only the term with the small denominator $(\omega_o - \omega)$:

$$c_f(t) = -\frac{1}{2\hbar}\langle f|V_o|i\rangle \frac{e^{i(\omega_o-\omega)t} - 1}{\omega_o - \omega}. \tag{35.14}$$

If instead $E_f < E_i$, so $\omega_o < 0$ and the atom loses energy to the field, equation (35.13) is dominated by the other term, with $\omega \sim -\omega_o$ (if we adopt the rule that ω in eq. [35.11] is positive).

By the rules of quantum mechanics, the probability that the atom is found to be in the state $|f\rangle$ at time t is $P_f = |\langle f|\phi\rangle|^2 = |c_f(t)|^2$. On squaring equation (35.14), and using the identity

$$\left|e^{i\theta} - 1\right|^2 = 2(1 - \cos\theta) = 4\sin^2\theta/2, \tag{35.15}$$

we get

$$P_f = \frac{|\langle f|V_o|i\rangle|^2}{\hbar^2}\frac{\sin^2[(\omega - \omega_o)t/2]}{(\omega - \omega_o)^2}. \tag{35.16}$$

This is the first-order perturbation theory approximation to the probability for finding that a near resonance perturbation has induced a transition. The initial state vector at time $t = 0$ is the eigenstate $|i\rangle$ of H_o, and P_f is the probability that a measurement of H_o at time t reveals that the system is in the eigenstate $|f\rangle$.

The function of $\omega_o - \omega$ in equation (35.16) has width $\delta\omega \sim 1/t$ at time t. This says the probability of inducing a transition by applying the oscillating field for a time interval t is about independent of the choice of applied frequency ω if ω differs from the resonance frequency ω_o by no more than about $1/t$. In the picture developed below, one thinks of the oscillating electromagnetic field as a sea of photons with energy $\hbar\omega$. The absorption of one of these photons promotes the energy of the system from E_i to $E_f = E_i + \hbar\omega_o$. Equation (35.16) says the absorption probability in the time interval t is relatively high if the photon energy differs from the energy difference between initial and final states of the system by no more than $\delta E \sim \hbar\delta\omega \sim \hbar/t$. This relation between the time interval and energy uncertainty, $t\delta E \sim \hbar$, is similar in appearance to the uncertainty principle relating position and momentum (eq. [10.50]).

36 Induced Transitions between the Hyperfine Levels in Atomic Hydrogen

In the example to be considered here, a hydrogen atom is placed in electromagnetic radiation with frequency ω that is close to the resonance for the hyperfine transition. This means the wavelength $\lambda = 2\pi c/\omega$ of the radiation is about 21 cm. Since this wavelength is very much larger than the size of the atom, we can take it that the atom has been placed in a homogeneous magnetic field that is oscillating with amplitude B_o and frequency ω:

$$\mathbf{B} = B_o \vec{\epsilon} \cos \omega t. \tag{36.1}$$

The constant unit vector $\vec{\epsilon}$ (with $|\vec{\epsilon}| = 1$) defines the polarization of the radiation field. It will be recalled that $\vec{\epsilon}$ is perpendicular to the propagation vector of the electromagnetic wave, as discussed in section 2 (eq. [2.13]).

The important part of the perturbation to the atom is the coupling of the magnetic field to the magnetic dipole moment of the electron. (We can ignore the smaller coupling to the magnetic dipole moment of the proton, because the much larger mass makes the proton magnetic moment much smaller.) As discussed in section 23, this magnetic coupling to the electron magnetic moment is (eq. [23.11])

$$V(t) = -\vec{\mu} \cdot \mathbf{B} = \frac{ge}{2M_e c} \mathbf{s}_e \cdot \mathbf{B}, \tag{36.2}$$

where $\mathbf{s}_e$ is the electron spin vector operator and M_e is the electron mass. With equation (36.1) this is

$$V(t) = \frac{ge}{2M_e c} B_o \, \mathbf{s}_e \cdot \vec{\epsilon} \cos \omega t. \tag{36.3}$$

Then the transition probability in equation (35.16) becomes

$$P_f = |\langle f | \mathbf{s}_e \cdot \vec{\epsilon} | i \rangle|^2 \left(\frac{geB_o}{2M_e c\hbar} \right)^2 \frac{\sin^2(\omega - \omega_o)t/2}{(\omega - \omega_o)^2}. \tag{36.4}$$

Now we have to evaluate the matrix element in this equation. Let the polarization vector $\vec{\epsilon}$ be placed in the $x - z$ plane and tilted at angle θ to the z axis, so

$$\mathbf{s}_e \cdot \vec{\epsilon} = s_z \cos \theta + s_x \sin \theta. \tag{36.5}$$

Suppose the atom initially is in the ground singlet state, and consider the probability of excitation to the triplet state with $m = 1$. The initial and final states are

$$|i\rangle = |0,0\rangle = \frac{1}{2^{1/2}}(|+-\rangle - |-+\rangle), \qquad (36.6)$$
$$|f\rangle = |1,1\rangle = |++\rangle.$$

As in equation (25.19), the arguments in the ket vectors $|0,0\rangle$ and $|1,1\rangle$ are the total spin quantum number $s = 0, 1$ and the quantum number m for the z component of the total spin, and the vectors $|m_e, m_p\rangle$ are eigenvectors of the z components of the electron and proton spins.

The matrix element $\langle f|s_z(e)|i\rangle$ of the z component of the electron spin between the initial and final states in equation (36.6) vanishes because $|i\rangle$ and $|f\rangle$ are orthogonal and $|f\rangle$ is an eigenvector of $s_z(e)$. The matrix element of the operator $s_x(e)$ is readily evaluated by using the Pauli spin matrices (eq. [24.20]):

$$\langle f|s_x(e)|i\rangle = -\frac{1}{2^{1/2}}\langle ++|s_x(e)|-+\rangle = -\frac{\hbar}{2^{3/2}}. \qquad (36.7)$$

Then with equation (36.5), the transition probability (36.4) is

$$P_f = \frac{(geB_o)^2}{32(M_e c)^2}\frac{\sin^2(\omega-\omega_o)t/2}{(\omega-\omega_o)^2}\sin^2\theta. \qquad (36.8)$$

This is the probability in first-order perturbation theory for the excitation from the ground singlet state of the atom to the triplet state with $m = 1$ by an applied magnetic field that oscillates at frequency ω with amplitude B_o. The magnetic field is tilted at angle θ to the z axis.

The angular dependence of the transition probability in equation (36.8) is worth noting. If $\theta = 0$ the perturbing magnetic field $\mathbf{B}(t)$ is parallel to the z axis, so the system is invariant under rotations around this axis, and the z component of angular momentum is conserved. Since $|i\rangle$ and $|f\rangle$ have different z components of total spin, P_f in equation (36.8) has to vanish when $\theta = 0$. The probability for a transition in which m changes by unity is largest when $\mathbf{B}$ is perpendicular to the z axis, $\theta = \pi/2$.

It is also interesting to consider a circularly polarized electromagnetic wave. This means the magnitude B_o of the magnetic field at the position of the atom is constant, while the direction $\vec{\epsilon}$ of the field rotates in a circle (normal to the propagation vector) at constant frequency ω. In a right-hand circular polarized wave propagating along the $+z$ axis, the magnetic field at a fixed position is

$$\mathbf{B}(t) = B_o(\mathbf{i}\cos\omega t + \mathbf{j}\sin\omega t), \tag{36.9}$$

where $\mathbf{i}$ and $\mathbf{j}$ are unit vectors along the x and y axes. With the usual Pauli spin matrices, a calculation like the one that led from equation (36.4) to (36.8) shows that the matrix elements between the singlet ground state and the triplet states are

$$\begin{aligned} \langle 1,1|\mathbf{s}\cdot\mathbf{B}\,|0,0\rangle &= -\frac{B_o\hbar}{2^{3/2}}e^{-i\omega t}, \\ \langle 1,0|\mathbf{s}\cdot\mathbf{B}\,|0,0\rangle &= 0, \\ \langle 1,-1|\mathbf{s}\cdot\mathbf{B}\,|0,0\rangle &= \frac{B_o\hbar}{2^{3/2}}e^{i\omega t}. \end{aligned} \tag{36.10}$$

The matrix element $\langle f|V|i\rangle \propto \langle 1,1|\mathbf{s}\cdot\mathbf{B}\,|0,0\rangle$ in the first of equations (36.10) has only a positive frequency term, $\propto e^{-i\omega t}$, while in the previous examples where the potential varies with time as

$$V \propto \cos\omega t \propto e^{i\omega t} + e^{-i\omega t}, \tag{36.11}$$

the matrix elements have both positive and negative frequencies. (From the convention that a state with positive energy E varies with time as $e^{-iEt/\hbar}$, one says that $e^{-i\omega t}$ with $\omega > 0$ has positive frequency, while $e^{i\omega t}$ has negative frequency.) This means the two terms in equation (35.13) are replaced here with only one term, which in the first of equations (36.10) is proportional to

$$\frac{e^{i(\omega_o-\omega)t} - 1}{\omega_o - \omega}, \tag{36.12}$$

with $\hbar\omega_o = E_f - E_i$. In the transition from singlet to triplet states the atom absorbs energy from the field, $E_f > E_i$, so $\omega_0 > 0$. Thus, if ω is close to the resonance frequency, $\omega \sim \omega_o$, an appreciable transition probability develops with increasing time.

In the third of the matrix elements in equations (36.10), the transition probability amplitude is proportional to

$$\frac{e^{(\omega_o+\omega)t}-1}{\omega_o+\omega}. \tag{36.13}$$

Since $\omega_o > 0$ for the transition from singlet to triplet states, there is no resonance: the amplitude $c_f(t)$ just oscillates with increasing time. That means the right-hand circular polarized wave in equation (36.9) has negligible probability for excitation from the singlet state to the triplet state with $m = -1$ (or to the triplet state with $m = 0$, as we see from the second of the matrix elements in eq. [36.10]). In the picture to be discussed next this is because right-hand circular polarized electromagnetic radiation consists of photons with spin 1 and z component of spin equal to $+\hbar$, parallel to the direction of propagation of the radiation. By conservation of the z component of angular momentum, the absorption of a photon with spin $+\hbar$ along the z axis can only promote the singlet state of the atom to a triplet state with $m = +1$, corresponding to a gain of angular momentum $+\hbar$ along the z axis. This is the allowed transition in the first line of equation (36.10).

37 Spontaneous Transitions between the Hyperfine Levels in Atomic Hydrogen

Electromagnetic Radiation as a Collection of Simple Harmonic Oscillators

Now we are ready to discuss the quantum treatment of the electromagnetic radiation field. As in section 2, it is convenient here to suppose the universe is periodic in a volume $V_u = L^3$, so we can represent the electromagnetic field as a Fourier sum rather than integral. A mode of oscillation, or term in this Fourier sum, is specified by its propagation vector,

$$\mathbf{k} = 2\pi\mathbf{n}/L, \tag{37.1}$$

where $\mathbf{n} = n_x, n_y, n_z$ is a triplet of integers, and by the constant polarization vector, $\vec{\epsilon}$, for the magnetic field. Since $\vec{\epsilon}$ is perpendicular to $\mathbf{k}$, there are two linearly independent modes for a given $\mathbf{k}$, corresponding to the two constant vectors $\vec{\epsilon}_1$ and $\vec{\epsilon}_2$ we can choose to be orthogonal to $\mathbf{k}$ and to each other.

In the classical picture of electromagnetic radiation, the magnetic field of a given mode of oscillation evaluated at a fixed position in space varies with time as

$$\mathbf{B} = \vec{\epsilon} B_o \cos(\omega t + \phi), \tag{37.2}$$

where the angular frequency is $\omega = kc$, and B_o and ϕ are the real constant amplitude and phase. Since the mode acts like a simple harmonic oscillator, let us quantize it according to the rules for a simple harmonic oscillator. This will lead us to replace the magnetic field in equation (37.2) with the field operator in equation (37.18) below.

As discussed in section 6, the position observable corresponding to the displacement from equilibrium in a one-dimensional simple harmonic oscillator is (eq. [6.34])

$$\hat{x} \propto a + a^\dagger, \tag{37.3}$$

where a is the lowering operator and the adjoint, $a^\dagger$, is the raising operator for the energy eigenvectors of the oscillator. The Hamiltonian (eq. [6.37]) is

$$H = a^\dagger a + \hbar\omega/2. \tag{37.4}$$

The raising and lowering operators satisfy the commutation relation (eq. [6.39])

$$[a, a^\dagger] = \hbar\omega. \tag{37.5}$$

In quantum field theory, it is conventional to eliminate the factor $\hbar\omega$ from the commutation relation by setting $a = (\hbar\omega)^{1/2} b$. Then the operator b satisfies

$$[b, b^\dagger] = 1, \qquad H = \hbar\omega(b^\dagger b + 1/2). \tag{37.6}$$

It also is conventional to go to a Heisenberg representation for the time evolution of the operator we are going to introduce to represent the magnetic field. It will be recalled that in the Schrödinger representation we have been using, the evolution of a system is expressed by the time dependence of the state vector, through Schrödinger's equation, while observables like angular momentum and the potential energy V are constant (in the absence of a time-variable applied field of the sort discussed in the last two sections). As discussed in section 14 (eq. [14.30]), the Heisenberg representation is obtained by a unitary transformation from the Schrödinger representation that makes the state vectors independent of time, and the observables time-dependent, in such a way that the matrix elements $\langle\psi|Q|\phi\rangle$ are the same in the two representations. In the Heisenberg representation, the time evolution of the operator b is (eq. [14.31])

$$i\hbar\frac{db}{dt} = [b, H]. \tag{37.7}$$

With equations (37.6), this is

$$i\hbar\frac{db}{dt} = \hbar\omega[b, b^\dagger b] = \hbar\omega b. \tag{37.8}$$

The solution to this differential equation is

$$b(t) = c\,e^{-i\omega t}, \qquad b^\dagger(t) = c^\dagger e^{i\omega t}. \tag{37.9}$$

The constant of integration is the constant operator, c. The second equation follows from the usual rules for the adjoint of an operator.

Finally, it is the standard convention to write the constant operator as a instead of c. This brings equations (37.3) and (37.6) to

$$\begin{aligned} \hat{x} &= K(a\,e^{-i\omega t} + a^\dagger e^{i\omega t}), \\ [a, a^\dagger] &= 1, \qquad H = \hbar\omega(a^\dagger a + 1/2). \end{aligned} \tag{37.10}$$

This is the wanted form for the operator $\hat{x}$ that represents the magnetic field in a chosen mode of oscillation of the electromagnetic field. The time dependence of the operator can be compared to the time dependence of the classical expression for the magnetic field in equation (37.2). We will get the real normalizing constant K by demanding consistency with the classical limit, where the fields satisfy Maxwell's equations.

To compute in time-dependent perturbation theory, we will need the matrix elements of the operators a and $a^\dagger$ between eigenvectors of the oscillator Hamiltonian H. Since the operator a lowers the energy, we know the ground state $|0\rangle$ for the simple harmonic oscillator satisfies

$$a|0\rangle = 0, \tag{37.11}$$

as in equation (6.48), for otherwise $|0\rangle$ would not be the state of lowest energy. The first excited state is found by applying the raising operator to the ground state:

$$|1\rangle = a^\dagger|0\rangle. \tag{37.12}$$

The normalized n^{th} excited state is

$$|n\rangle = \frac{1}{(n!)^{1/2}}(a^\dagger)^n|0\rangle. \tag{37.13}$$

To understand the normalizing constant, you are invited to check that the commutation relation in equation (37.10) implies

$$[a, (a^\dagger)^n] = n(a^\dagger)^{n-1}, \tag{37.14}$$

and to show from this and equation (37.11) that

$$\langle 0|a^n(a^\dagger)^n|0\rangle = n!. \tag{37.15}$$

This means the state in equation (37.13) is normalized, $\langle n|n\rangle = 1$.

You can use the same method to see that the only nonzero matrix elements of a and $a^\dagger$ are of the form

$$\langle n-1|a|n\rangle = n^{1/2}, \qquad \langle n+1|a^\dagger|n\rangle = (n+1)^{1/2}. \tag{37.16}$$

This means the nonzero matrix elements of the observable $\hat{x}$ in equation (37.10) between energy eigenstates are

$$\begin{aligned} \langle n-1|\hat{x}|n\rangle &= n^{1/2}Ke^{-i\omega t}, \\ \langle n+1|\hat{x}|n\rangle &= (n+1)^{1/2}Ke^{i\omega t}, \end{aligned} \tag{37.17}$$

all others vanishing. These matrix elements will be used to get the relation between the excitation and decay rates of an atom in an electromagnetic field (eqs. [37.28] and [37.29] below).

Returning to the electromagnetic radiation field, we noted above that each mode of oscillation of the field can be labeled by its propagation vector, $\mathbf{k}$, along with an index $\alpha = 1, 2$ to represent the two orthogonal choices of the polarization vector $\vec{\epsilon}$ normal to $\mathbf{k}$. Because each mode is a simple harmonic oscillator, quantum theory replaces the field strength in the mode with an operator constructed out of the raising and lowering operators $a_{\mathbf{k},\alpha}$ and $a^\dagger_{\mathbf{k},\alpha}$, as in equation (37.10). That is, quantum field theory replaces the classical magnetic field in the mode (eq. [37.2]) with a field operator,

$$B_o\cos(\omega t + \phi) \rightarrow K_{\mathbf{k},\alpha}(a_{\mathbf{k},\alpha}e^{-i\omega t} + a^\dagger_{\mathbf{k},\alpha}e^{i\omega t}). \tag{37.18}$$

The frequency is related to the propagation vector by the usual expression,

$$\omega = kc. \tag{37.19}$$

As noted above, the constant of proportionality, $K_{\mathbf{k},\alpha}$, will be fixed by demanding that this quantum expression have the right classical limit. The index $(\mathbf{k},\alpha)$ in the constant of proportionality and the raising and lowering operators in equation (37.18) indicates that these objects refer to a particular mode, with propagation vector $\mathbf{k}$ and polarization vector $\vec{\epsilon}_\alpha$. To get the full magnetic field operator, we would introduce the space dependence of the magnetic field, $\propto e^{i\mathbf{k}\cdot\mathbf{r}}$, and then sum over modes. The space dependence is not needed because in the example to be worked here the wavelength is very long compared to the size of the atom.

The vector $\propto (a^\dagger_{\mathbf{k},\alpha})^n|0\rangle$ in equation (37.13) is an eigenvector of the energy operator (37.11) for the mode:

$$\begin{aligned} H_{\mathbf{k},\alpha}(a^\dagger_{\mathbf{k},\alpha})^n|0\rangle &= \hbar\omega_{\mathbf{k}}(a^\dagger_{\mathbf{k},\alpha}a_{\mathbf{k},\alpha} + 1/2)(a^\dagger_{\mathbf{k},\alpha})^n|0\rangle \\ &= (n+1/2)\hbar\omega_{\mathbf{k}}(a^\dagger_{\mathbf{k},\alpha})^n|0\rangle. \end{aligned} \tag{37.20}$$

The last line follows from the algebra in equation (37.14), or more simply by recalling that $a^\dagger_{\mathbf{k},\alpha}$ raises the energy by the amount $\hbar\omega_{\mathbf{k}}$. The fact that the energy in the mode can only change by integral multiples of $\hbar\omega_{\mathbf{k}}$ agrees with the phenomena discussed in sections 2 and 3, that indicate the energy in electromagnetic radiation is quantized in units $\hbar\omega_{\mathbf{k}}$ that act as particles, photons. Thus $a^\dagger_{\mathbf{k},\alpha}$ is called the creation operator for photons in the mode labeled by $\mathbf{k}$ and α, the photon annihilation operator is $a_{\mathbf{k},\alpha}$, and $(a^\dagger_{\mathbf{k},\alpha})^n|0\rangle/(n!)^{1/2}$ is a state containing n photons in the mode $\mathbf{k},\alpha$. Equation (37.18) is the expression for the magnetic field observable for the mode in terms of these creation and annihilation operators.

Induced and Spontaneous Decay

Equation (36.3) gives the magnetic coupling to the hyperfine spin states of a hydrogen atom in terms of the classical expression for the magnetic field. Quantum theory replaces the classical magnetic field of a mode of oscillation of the radiation with the field operator in equation (37.18). The net radiation field is the sum over all modes of oscillation, so the coupling must be summed over all mode labels $\mathbf{k}, \alpha$. This gives

$$V = \frac{ge}{2M_e c} \sum_{\mathbf{k},\alpha} \mathbf{s}_e \cdot \vec{\epsilon}_{\mathbf{k},\alpha} K_{\mathbf{k},\alpha} (a_{\mathbf{k},\alpha} e^{-i\omega t} + a^{\dagger}_{\mathbf{k},\alpha} e^{i\omega t}). \tag{37.21}$$

The frequency is $\omega = kc$ in the mode with propagation vector $\mathbf{k}$.

Now let us compute transition probabilities using this interaction term. Suppose a hydrogen atom initially is in the singlet ground state $s = 0$, and consider the probability that at a later time t it is found to be in a specific one of the excited triplet $s = 1$ states. In first-order perturbation theory, this probability is proportional to the square of the matrix element between initial and final states $|i\rangle$ and $|f\rangle$. The state $|i\rangle$ is labeled by the quantum number $s = 0$ for the initial state of the atom, and by the photon numbers for each mode of the electromagnetic field. Since each term in the interaction potential V in equation (37.21) contains a raising or lowering operator, we see that in first-order perturbation theory the probability amplitude is nonzero only if there is unit difference between the photon numbers of just one of the modes in the initial and final states. Let us consider the transition probability amplitude in the case that the number of photons in the specific mode $\mathbf{k}$, α has changed from n to $n-1$, the numbers of photons in all other modes being the same in initial and final states. The change $n \to n-1$ means the energy in the electromagnetic field has decreased by the amount $\hbar\omega_{\mathbf{k},\alpha}$.

Leaving out the photon numbers for all the unaffected modes, the initial and final states can be written as

$$\begin{aligned} |i\rangle &= |s, n\rangle, \\ |f\rangle &= |t, n-1\rangle. \end{aligned} \tag{37.22}$$

Here s and t stand for specific singlet and triplet states of the atom, and n is the number of photons in the particular mode that loses a photon. The matrix element of the interaction term (37.21) between these initial and final states is

$$\langle f|V|i\rangle = \frac{ge}{2M_e c} \langle \mathbf{s}_e \rangle_{ts} \cdot \vec{\epsilon}_{\mathbf{k},\alpha} K_{\mathbf{k},\alpha} n^{1/2} e^{-i\omega t}. \tag{37.23}$$

In the sum over modes in equation (37.21), the only term that survives with nonzero matrix element between the initial and final states in equation (37.22) is the one containing the annihilation operator $a_{\mathbf{k},\alpha}$ for the mode $(\mathbf{k}, \alpha)$, for it is this operator that connects the photon numbers in the initial and final states. The factor $n^{1/2}$ is the matrix element of the annihilation operator, as in equation (37.16). The annihilation operator in equation (37.21) is multiplied by the positive frequency factor $e^{-i\omega t}$. If the photon number had increased from n in the initial state to $n+1$ in the final state, the nonzero matrix element would have had negative frequency from the creation operator in equation (37.21). Finally, the factor $\langle \mathbf{s}_e \rangle_{ts}$ in equation (37.23) represents the matrix elements of the three components of the electron spin operator between the initial singlet and final triplet state of the atom.

In linear perturbation theory, the probability amplitude $c_f(t)$ for the final state $|f\rangle$ is given by equation (35.10). In the present example, we have

$$\begin{aligned} i\hbar \frac{dc_f}{dt} &= \langle f|V|i\rangle e^{i(E_f - E_i)t/\hbar} \\ &= \frac{ge}{2M_e c} \langle \mathbf{s} \rangle_{ts} \cdot \vec{\epsilon}_{\mathbf{k},\alpha} K_{\mathbf{k},\alpha} n^{1/2} e^{i(E_t - E_s - \hbar\omega)t/\hbar}. \end{aligned} \tag{37.24}$$

In the second line, E_i has been written as the initial singlet state energy E_s, and E_f as the final triplet energy E_t.

As discussed in section 35, the probability amplitude $c_f(t)$ can only grow to an appreciable value if the argument of the exponential is close to zero, which means here that

$$E_t - E_s \approx \hbar\omega. \tag{37.25}$$

This just means the annihilation of a photon with frequency ω is accompanied by an energy increase $\hbar\omega$ in the system that absorbed the photon.

It is easy to see that the increase in energy of the system, from singlet to triplet, has to be accompanied by annihilation rather than creation of a photon. If instead of the transition in equation (37.22) we had considered the transition from singlet to triplet with the production of a photon, $|i\rangle = |s, n\rangle \rightarrow |f\rangle = |t, n+1\rangle$, the creation operator in equation (37.21) would have been needed for a nonzero matrix element. That would have given the matrix element (37.23) a negative frequency, so equation (37.24) for dc_f/dt would be proportional to $e^{i(E_t - E_s + \hbar\omega)t/\hbar}$. Since $E_t > E_s$, this factor oscillates with time whatever the frequency, so dc_f/dt cannot integrate up to an appreciable probability amplitude c_f.

As one would expect, the excitation of the atom has to be accompanied by the loss of a photon.

In the decay from an excited triplet state to the ground singlet state of the atom, a photon is created in a mode that initially contained n photons. Here the initial and final states are

$$\begin{aligned} |i\rangle &= |t, n\rangle, \\ |f\rangle &= |s, n+1\rangle. \end{aligned} \tag{37.26}$$

Only the negative frequency term from equation (37.21) survives here, because the creation operator $a^{\dagger}_{\mathbf{k},\alpha}$ connects the photon number $n \to n+1$. The matrix element of the creation operator between initial and final states yields the factor $(n+1)^{1/2}$, in place of the factor $n^{1/2}$ in equation (37.24) (eq. [37.16]). Thus, in place of equation (37.24) we get

$$i\hbar \frac{dc_f}{dt} = \frac{ge}{2M_e c} \langle \mathbf{s}_e \rangle_{st} \cdot \vec{\epsilon}_{\mathbf{k},\alpha} K_{\mathbf{k},\alpha} (n+1)^{1/2} e^{i(E_s - E_t + \hbar\omega)t/\hbar}, \tag{37.27}$$

for the decay from the excited state in equation (37.26). Here also the probability amplitude $c_f(t)$ can only grow to an appreciable value if $E_t - E_s \approx \hbar\omega$. That is, in decaying from the excited state to the ground state the atom creates a photon with energy $\hbar\omega$ nearly equal to the loss of energy by the atom.

The transition probabilities for these two processes, absorption and emission of a photon, are found by integrating equations (37.24) and (37.27) from time 0 to t and then squaring. The results differ only in the factor n in the probability for absorption of a photon from a specified mode and $n+1$ in the probability of emission of a photon into the mode. This does not depend on the specific process of emission and absorption; it comes from the matrix elements of the creation and annihilation operators, that contain respectively the factors $(n+1)^{1/2}$ and $n^{1/2}$ (eq. [37.16]). Thus we have the following handy rule.

Let $P(a \to b)$ be the probability that in some fixed time interval a system is found to be excited from state a to b by the absorption of a photon from a specific mode of oscillation of the electromagnetic field. As we see in equation (37.24), this probability is proportional to the number n of photons initially in the mode,

$$P(a \to b) = nA, \tag{37.28}$$

where A is the constant of proportionality. As in equation (37.27), the probability for decay from the excited state b to a during the same time interval, with n photons initially present in the mode, is

$$P(b \to a) = (n + 1)A. \tag{37.29}$$

This is proportional to $n + 1$ because that is the square of the matrix element of the creation operator. The constant of proportionality is the same as in the excitation probability in equation (37.28), because the rest of the computation goes the same way in either case.

The result for absorption, $P(a \to b)$ proportional to the photon number, n, is familiar: the excitation probability is proportional to the number of photons available to be absorbed. If $n = 0$, equation (37.29) says the decay probability is A. This is the probability for spontaneous decay of the system accompanied by production of a photon in a specified mode. Equation (37.29) also says that the decay probability with production of a photon in a specific mode is enhanced if photons already are present in the mode. This induced decay is the basis for a maser amplifier: radiation tends to induce decays to amplify the radiation already present. The relations (37.28) and (37.29) among absorption, induced decay, and spontaneous decay probabilities were discovered by Einstein before the discovery of quantum mechanics, from the statistical mechanics of Planck's blackbody radiation theory.

Decay Rate for the Hydrogen Hyperfine State

To compute the half-life of a hydrogen atom in the excited hyperfine state, we must sum the decay probabilities $P(t \to s)$ over all modes for the decay photon, and we need the constants $K_{\mathbf{k},\alpha}$ in equations (37.18) and (37.27). We could get the $K_{\mathbf{k},\alpha}$ by going a little deeper into quantum field theory, but an easier way is to demand that the theory agree with classical electromagnetism in the limit of large photon number.

In the classical limit, the transition probability associated with a specific mode of the electromagnetic field is given by equation (36.8), where B_o is the classical amplitude of the magnetic field in the mode. To connect that to the photon number n in the mode, note that in the classical limit the net electromagnetic energy in the mode is

$$U = \frac{B_o^2}{8\pi} \times \frac{1}{2} \times 2 \times V_u = n\hbar\omega. \tag{37.30}$$

The first factor is the usual magnetic energy per unit volume, $B^2/8\pi$. The factor $1/2$ corrects the square of the amplitude of $\mathbf{B}$ to the mean square value of the oscillating field. (The time average of $\sin^2 \omega t$ is $1/2$.) The factor of two takes account of the fact that there is as much energy in the electric field as in the magnetic field. The last factor is the volume of the universe with our periodic boundary conditions. The second equation says this total energy in the mode is equivalent to n photons of energy $\hbar\omega$.

We see from equation (37.30) that in the classical limit of large photon number n the electromagnetic field in a mode with frequency ω and magnetic field amplitude B_o is equivalent to photon number

$$n = B_o^2 V_u/(8\pi\hbar\omega). \tag{37.31}$$

(As discussed in problem [II.23], in a state with a well-defined magnetic field the photon number is a random variable; eq. [37.31] is the mean photon number, which is what we need to compute the decay probability.)

The transition probability in equation (36.8), which treats the radiation as a classical electromagnetic field, is proportional to B_o^2. Thus, equation (37.31) says that in the classical limit the transition probabilities from t to s or s to t are proportional to the photon number, n. Equations (37.28) and (37.29) say the Einstein transition probabilities are proportional to n or $n+1$, depending on whether a photon is annihilated or created. This consistent, because the relation between n and B_o^2 applies only in the limit $n \gg 1$, where there are enough photons to define the electric and magnetic fields. And we see that the constant of proportionality A in the transition probabilities in equations (37.28) and (37.29) is just the factor multiplying n in the classical probability (36.8), with B_o^2 expressed in terms of n. That is, on substituting equation (37.31) for B_o^2 into equation (36.8) for the transition probability, and then setting $n = 1$, we get

$$A_{\mathbf{k},\alpha} = \frac{8\pi\hbar\omega}{V_u} \frac{(ge)^2}{32(M_e c)^2} \frac{\sin^2(\omega - \omega_o)t/2}{(\omega - \omega_o)^2} \sin^2\theta. \tag{37.32}$$

This is the wanted expression for the probability for spontaneous decay of the atom from the excited triplet state with $m = 1$ (eq. [36.6]) to the ground singlet state, with the creation of a photon in the mode $\mathbf{k}, \alpha$.

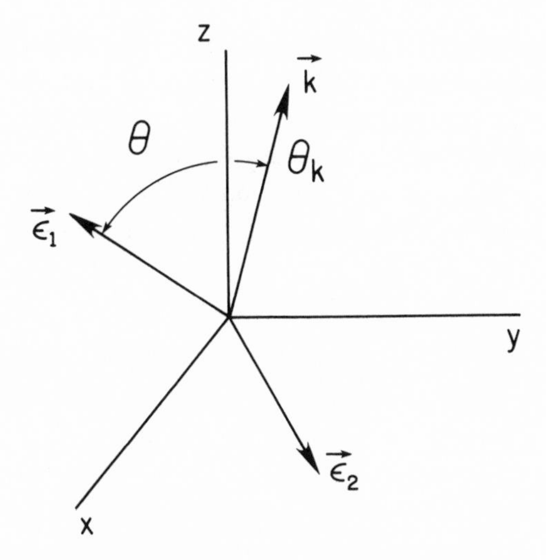

Fig. 37.1 Magnetic field polarization vectors. The propagation vector, $\mathbf{k}$, for the mode of the electromagnetic field is tilted at angle $\theta_{\mathbf{k}}$ to the z axis. The polarization vector $\vec{\epsilon}_2$ is in the xy plane and normal to $\mathbf{k}$. The polarization vector $\vec{\epsilon}_1$ is perpendicular to $\mathbf{k}$ and $\vec{\epsilon}_2$, tilted at angle $\theta = \pi/2 - \theta_{\mathbf{k}}$ to the z axis.

Since this calculation might at first seem unconvincing, it may be well to review the logic. The classical equation (36.8) that applies in the limit of large photon number n says the transition probability is proportional to n. The Einstein probabilities in equations (37.28) and (37.29) say the transition probabilities are proportional to n and $n + 1$. This is consistent, because the classical result assumes $n \gg 1$. Since the Einstein probabilities apply whether n is large or small, we can use the classical probability to determine the constant of proportionality, A. But this constant is just the spontaneous part of the decay probability (37.29).

We have now the probability $A_{\mathbf{k},\alpha}$ for spontaneous decay from the triplet to the singlet state, with production of a photon in a definite mode. To get the total decay probability, we need to sum the $A_{\mathbf{k},\alpha}$ over all modes, as was done in section 2 for blackbody radiation. The sum is over the values of the propagation vector, $\mathbf{k}$, and the polarization, $\alpha = 1, 2$.

The transition probability $A_{\mathbf{k},\alpha}$ in equations (36.8) and (37.32) is proportional to $\sin^2 \theta$, where θ is the angle between the z axis and the polarization vector $\vec{\epsilon}$ along the direction of the magnetic field in the mode. Let the propagation vector $\mathbf{k}$ be tilted at angle $\theta_{\mathbf{k}}$ to the z axis, as indicated in figure 37.1. Since $\vec{\epsilon}$ has to be perpendicular to $\mathbf{k}$, we can take one polarization vector to be perpendicular to $\mathbf{k}$ and to the z axis, so $\vec{\epsilon}$ is in the xy plane and $\sin\theta = 1$. Then the orthogonal polarization vector is in the plane of $\mathbf{k}$ and the z axis, and tilted from the z axis at angle $\theta = 90^\circ - \theta_\mathbf{k}$, so this polarization vector also is perpendicular to $\mathbf{k}$. Here $\sin\theta = \cos\theta_\mathbf{k}$. Thus the sum over polarizations for fixed $\mathbf{k}$ gives

$$\sum_\alpha \sin^2\theta = 1 + \cos^2\theta_\mathbf{k}. \tag{37.33}$$

In the limit where the size of the periodic universe goes to infinity, the sum over $\mathbf{k}$ becomes an integral, as in sections 1 and 2 (eq. [1.58]):

$$\sum_\mathbf{k} \rightarrow \int \frac{V_u d^3k}{(2\pi)^3}. \tag{37.34}$$

The sum of the decay probabilities $A_{\mathbf{k},\alpha}$ in equation (37.32) over polarizations and values of the propagation vector $\mathbf{k}$ thus is

$$P = \int \frac{V_u k^2 dk \sin\theta_\mathbf{k} d\theta_\mathbf{k} d\phi_\mathbf{k}}{8\pi^3} \frac{\pi\hbar\omega g^2 e^2}{4V_u M_e^2 c^2} \frac{\sin^2(\omega - \omega_o)t/2}{(\omega - \omega_o)^2}(1 + \cos^2\theta_\mathbf{k}). \tag{37.35}$$

It will be recalled that $\hbar\omega_o$ is the energy difference between ground and excited states (eqs. [35.12], [37.24]). The integral over $\mathbf{k}$ has been written in polar coordinates, with $\theta_\mathbf{k}$ the polar angle, as in equation (37.33), and volume element $d^3k = k^2 dk \sin\theta_\mathbf{k} d\theta_\mathbf{k} d\phi_\mathbf{k}$. The integrals over the polar angles $\theta_\mathbf{k}$, $\phi_\mathbf{k}$ are easy. The variable of integration k is the magnitude of the propagation vector, and the frequency of the mode is $\omega = kc$. We can simplify the integral over k by noting that at large t the integral is dominated by the small denominator at $\omega \approx \omega_o$.[1]

To evaluate the integral over the resonance at $\omega \sim \omega_o$, note that everywhere except in the last factor in equation (37.35) we can set $\omega = kc$

[1] The term $\sin^2(\omega - \omega_o)t/2$ in the numerator of eq. [37.35] cuts off the diverging integral over $(\omega - \omega_o)^{-2}$ at $|\omega - \omega_o| \sim 1/t$, so the integral over the resonance grows in proportion to the time interval t. Therefore, at large enough t the integral is dominated by the integral over the small denominator at $\omega \sim \omega_o$. It might be noted that the integral over k in equation (37.35) diverges at $k \rightarrow \infty$. However, that is removed by the fact that at large k the wavelength is smaller than the atom, so the matrix element is suppressed by the factor $e^{i\mathbf{k}\cdot\mathbf{r}}$ in the photon wave function. The coupling between the field and the electron spin must be integrated over the width of the electron wave function. At large k, the oscillation of the space part of the field makes the space part of the integral negligibly small.

equal to the value ω_o at the resonance. Then all we have to evaluate is the integral

$$I = \int_{-\infty}^{\infty} d\omega \frac{\sin^2(\omega - \omega_o)t/2}{(\omega - \omega_o)^2}. \tag{37.36}$$

With the change of variables $\omega = \omega_0 + 2x/t$ we get

$$I = \frac{t}{2} \int_{-\infty}^{\infty} \frac{\sin^2 x}{x^2} dx = \frac{\pi t}{2}. \tag{37.37}$$

Then equation (37.35) is

$$\lambda = \frac{P}{t} = \frac{4\pi V_u}{8\pi^3} \frac{\omega_o^2}{c^3} \frac{\pi \hbar \omega_o g^2 e^2}{4V_u M_e^2 c^2} \frac{\pi}{2} \frac{4}{3} = \frac{g^2 \omega_o^3 \hbar e^2}{12 M_e^2 c^5}. \tag{37.38}$$

Two pleasant features of this result will be noted. First, the volume of the universe, V_u, cancels, as we must require because the atom cannot know we have chosen to compute in some finite but enormous periodic universe. Second, the decay probability P is proportional to the time interval, t, so we can define the decay rate, λ, as the decay probability per unit time. This constant decay rate is the result of two competing factors. Exactly at resonance, where $\omega = \omega_o$, the time derivative dc_f/dt of the probability amplitude per mode is independent of time, so $c_f(t) \propto t$. The probability per mode exactly at resonance therefore varies as t^2. However, the width of the resonance is proportional to $1/t$, as discussed in section 35. The net decay probability varies as the product of these factors, or as $P \propto t$.

Before evaluating equation (37.38), let us simplify it as in section 34 by using the fine-structure constant, $\alpha = e^2/(\hbar c)$, the electron Compton wavelength $\lambda_c = \hbar/(M_e c)$, and the wavelength of the transition,

$$\lambda_o = 2\pi c/\omega_o = 21.1\,\text{cm}. \tag{37.39}$$

This gives

$$\lambda = \frac{2\pi^3}{3} g^2 \alpha c \lambda_c^2 / \lambda_o^3 = 2.9 \times 10^{-15}\,\text{sec}^{-1}. \tag{37.40}$$

The half-life is $t_{1/2} = \log(2)/\lambda \sim 10^7$ years. This is a slow decay, but there are so many hydrogen atoms in a galaxy that the 21 cm radiation from decay of atoms in the triplet state (that are placed there by collisions with other atoms) is detected from galaxies 100 million light years away.

Problems

V.1) The Hamiltonian H_o has eigenstates $|n\rangle$ with energies E_n that are not degenerate. The $|n\rangle$ are discrete and normalized,

$$\langle n|m\rangle = \delta_{mn}. \tag{V.1}$$

As discussed in section 30, to terms of first order in the perturbation ϵV the eigenstates of the Hamiltonian $H = H_o + \epsilon V$ are

$$|\psi_n\rangle = |n\rangle + \epsilon \sum_{m \neq n} \frac{|m\rangle\langle m|V|n\rangle}{E_n - E_m} + \text{order}\, \epsilon^2. \tag{V.2}$$

a) Show that, to terms of first order in ϵ, the state vectors in equation (V.2) are orthogonal and normalized,

$$\langle \psi_n|\psi_m\rangle = \delta_{nm} + \text{order}\, \epsilon^2. \tag{V.3}$$

b) By substituting equation (V.2) into the equation

$$H|\psi\rangle = E|\psi, \tag{V.4}$$

and rearranging the expression, verify that, correct to terms of first order in ϵ, equation (V.2) in fact solves the eigenvalue equation (V.4).

c) Suppose there is an operator G such that the perturbation V satisfies the operator relation

$$V = [G, H_o]. \tag{V.5}$$

Show how this expression can be used to simplify equation (V.2) by eliminating the sum over m.

d) If V satisfies equation (V.5), find an expression for the nth energy level of H in second-order perturbation theory in terms of expectation values of V, G, and their product.

V.2) A particle that moves in one dimension is confined to the interval $-L/2 \leq x \leq L/2$ by the potential well

$$\begin{aligned} &U(x) \to \infty, \quad \text{at} \quad |x| > L/2, \\ &U(x) = 0, \quad \text{at} \quad |x| \leq L/2. \end{aligned} \tag{V.6}$$

a) Find the ground state energy for the Hamiltonian

$$H_o = \frac{p^2}{2m} + U(x). \tag{V.7}$$

b) Suppose the potential is changed by the addition of the term

$$V(x) = cx^2, \tag{V.8}$$

where c is a real constant. Find in first-order perturbation theory the ground state energy for the Hamiltonian

$$H = H_o + V. \tag{V.9}$$

Express the answer in terms of the relevant physical constants multiplied by a dimensionless integral you need not bother to evaluate.

V.3) Consider the Hamiltonian

$$H = H_o + V, \tag{V.10}$$

where H_o represents a one-dimensional simple harmonic oscillator,

$$H_o = \frac{p^2}{2m} + \frac{1}{2}\kappa x^2, \tag{V.11}$$

and the perturbation is

$$V = cx, \tag{V.12}$$

with c a real constant.

a) Find the energy levels of the Hamiltonian H by writing the full potential energy term $\kappa x^2/2 + cx$ as a completed square plus a constant. This makes H the sum of a simple harmonic oscillator Hamiltonian and a constant.

b) Use perturbation theory starting from the eigenstates $|n\rangle$ of H_o to find the ground state energy of H to terms of second order in V. You will need matrix elements of the form $\langle n|x|0\rangle$, which you can compute by operator techniques using the ladder operators a and $a^\dagger$ in section 6, as was done in problem (II.22). If all goes well, your answer should be consistent with the result from part (a).

V.4) A particle of mass m and spin $1/2$ moves in three dimensions with the Hamiltonian

$$H = H_o + V, \tag{V.13}$$

where H_o represents a three-dimensional simple harmonic oscillator,

$$H_o = \frac{p^2}{2m} + \frac{1}{2}m\omega^2 r^2, \tag{V.14}$$

and

$$V = \lambda \mathbf{s} \cdot \mathbf{r}, \tag{V.15}$$

where $\mathbf{s}$ is the spin vector operator and λ is a real constant. Compute the ground state energy through terms of second order in the perturbation V.

V.5) In section 6, the binding energy and ground state wave function of the electron in a hydrogen atom were obtained under the assumption that the proton acts as a point charge, so the potential energy is $-e^2/r$ all the way to $r = 0$. In this approximation, the wave function is

$$\psi(r) \propto e^{-r/a_o}, \tag{V.16}$$

where a_o is the Bohr radius. In a better model, the charge of the proton is uniformly distributed within a sphere with radius characteristic of the size of a proton, so the charge density is

$$\begin{aligned} \rho &= \frac{3e}{4\pi r_o^3}, \qquad r < r_o = 1 \times 10^{-13}\,\text{cm}, \\ &= 0, \qquad r > r_o. \end{aligned} \tag{V.17}$$

a) To see how this distributed charge affects the predicted binding energy of the electron, use classical physics to find an expression for the potential energy $V_1(r)$ of a point charge $-e$ as a function of distance r from the origin of the charge distribution in equation (V.17). Let $V_0 = -e^2/r$ be the potential used to derive the wave function in equation (V.16), treat the difference $V(r) = V_1(r) - V_0(r)$ as a perturbation, and use first-order perturbation theory to find the correction δB to the binding energy B in the ground state due to the finite size of the proton. Obtain a numerical value for the fractional correction $\delta B/B$. You can simplify the calculation by noting that r_o is much smaller than the Bohr radius, a_o.

b) Explain why the corrections to the energy levels of a hydrogen atom for the finite width of the charge distribution in a proton are a good deal larger in the s-wave states (orbital angular momentum quantum number $l = 0$) than for the states with $l > 0$.

V.6) A particle of mass M and zero spin moves in one dimension with periodic boundary conditions, period L. The particle is almost free, so the energy eigenstates are well approximated as

$$\psi_n(x) \propto e^{2\pi i n x/L}, \tag{V.18}$$

where n is an integer (that can be positive or negative). The particle is perturbed by a periodic potential

$$V(x) = V_o \sin(2\pi m x/L). \tag{V.19}$$

Here V_o is a real constant and m is a positive even integer.

Find in first-order perturbation theory the effect of V on the energy of each of the unperturbed states in equation (V.18). Note that you will have to be a little careful because each of the unperturbed energy levels except for $n = 0$ is doubly degenerate, corresponding to a positive and negative n.

V.7) The first excited energy levels in atomic hydrogen have principal quantum number $n = 2$. The four degenerate states (neglecting relativistic corrections) with this energy have angular momentum quantum numbers $l = 0$ and $l = 1$ with $m = 0, \pm 1$. The electron wave functions for these states are (eq. [18.34])

$$\begin{aligned}
&\phi_{2s} \propto [1 - r/(2a_o)]e^{-r/(2a_o)}, \\
&\phi_{2p,1} \propto r \sin\theta e^{i\phi} e^{-r/(2a_o)} = (x + iy)e^{-r/(2a_o)}, \\
&\phi_{2p,0} \propto r \cos\theta e^{-r/(2a_o)} = z e^{-r/(2a_o)}, \\
&\phi_{2p,-1} \propto r \sin\theta e^{-i\phi} e^{-r/(2a_o)} = (x - iy)e^{-r/(2a_o)}.
\end{aligned} \tag{V.20}$$

The polar coordinates r, θ, and ϕ are defined in equation (17.42).

Suppose a uniform electric field E parallel to the z axis is applied to the atom, adding to the Hamiltonian the constant term

$$V = e\mathrm{E}z. \tag{V.21}$$

(The electron charge is $-e$; spins can be ignored, and the proton can be taken to be fixed in space, so this is a single-particle problem.) To find the shifts in energies of the originally degenerate $n = 2$ states, we need the matrix elements

$$\langle l, m|V|l', m'\rangle, \tag{V.22}$$

for the four unperturbed wave functions.

a) Show that the only nonzero one of these matrix elements is

$$\langle 0, 0|V|1, 0\rangle, \tag{V.23}$$

along with its complex conjugate.

b) Find the perturbation to the energies of the $n = 2$ levels to terms of first order in the electric field, E.

V.8) The degenerate perturbation theory in section 33 indicates how to deal with a situation in which two or more eigenstates of the unperturbed Hamiltonian have the same energy. The question naturally arises, what does one do if two or more unperturbed eigenstates have energies that are close but not exactly equal?

Suppose the Hamiltonian H_o has the complete set of discrete eigenstates $|n\rangle$, with eigenvalues E_n. The energy E_1 is close to but not necessarily equal to E_2, and the other energy levels are well removed from E_1 and E_2:

$$|E_n - E_1| \gg |E_2 - E_1|, \tag{V.24}$$

for all other n. In computing the perturbations of the energy eigenvalues away from E_1 and E_2 in the Hamiltonian

$$H = H_o + \epsilon V, \tag{V.25}$$

a convenient expansion for the eigenstates $|\psi\rangle$ of H is

$$|\psi\rangle = \alpha|1\rangle + \beta|2\rangle + \epsilon \sum_{n\neq 1,2} C_n|n\rangle. \tag{V.26}$$

Here α, β, and the C_n are constants, and as indicated the *ansatz* is that the dominant parts of $|\psi\rangle$ are $|1\rangle$ and $|2\rangle$, with only small admixtures from the states $|n\rangle$ with $n \neq 1, 2$.

By substituting equation (V.26) into

$$H|\psi\rangle = E|\psi\rangle, \tag{V.27}$$

discarding terms of order ϵ^2 on the left-hand side, and then taking the inner product of the equation with $\langle 1|$ and with $\langle 2|$, find the energy E in first-order perturbation theory, in terms of E_1, E_2, and the matrix elements

$$V_{11} = \langle 1|V|1\rangle, \quad V_{12} = \langle 1|V|2\rangle, \quad V_{22} = \langle 2|V|2\rangle. \tag{V.28}$$

Check that your result for E is consistent with degenerate first-order perturbation theory when $E_1 = E_2$, and with nondegenerate first-order perturbation theory in the limit of sufficiently large $|E_2 - E_1|$.

V.9) In a magnetic resonance experiment, a sample is placed in a constant magnetic field and a much weaker perpendicular oscillating magnetic field. Ignoring the particle motion and the interactions with other particles, we can write the Hamiltonian as

$$H = -\frac{gq}{2mc}(B_0 s_z + B_1 s_x \cos\omega t). \tag{V.29}$$

Here q and m are the charge and mass of the particle, and g is its dimensionless gyromagnetic ratio, as in equation (23.11). The constant magnetic field, along the z axis, is B_0, and B_1 is the amplitude of the perpendicular oscillating magnetic field directed along the x axis.

Suppose the particle has spin 1/2. A convenient basis is the set of two eigenstates of s_z,

$$\chi_+ = \begin{bmatrix} 1 \\ 0 \end{bmatrix}, \qquad \chi_- = \begin{bmatrix} 0 \\ 1 \end{bmatrix}. \tag{V.30}$$

These are the energy eigenstates when $B_1 = 0$.

Find in first-order perturbation theory the probability that the oscillating field applied for a time t causes a spin flip, that is, find the probability $P(t)$ that, if the system is in the state χ_- at time $= 0$, it is found to be in the state χ_+ at a chosen time $t > 0$.

V.10) A particle moves in one dimension with the Hamiltonian

$$H = H_o + V(t), \tag{V.31}$$

where H_o represents a simple harmonic oscillator,

$$H_o = \frac{p^2}{2m} + \frac{1}{2}m\omega^2 x^2, \tag{V.32}$$

and

$$V(t) = \alpha x\, e^{-(t/\tau)^2}, \tag{V.33}$$

with α and τ real constants.

At time $t \to -\infty$, when $V(t)$ is negligibly small, the system is in the ground state of H_o. Use first-order perturbation theory to find the probability P_n that at time $t \to +\infty$ the system is in the nth eigenstate of H_o.

V.11) A hydrogen atom in the $2s$ state (principal quantum number $n = 2$, angular momentum quantum number $l = 0$) has a long half-life against decay to the ground state compared to the half-life of the $2p$ states (with $n = 2$ and $l = 1$), because decay from the $2s$ state violates the selection rules for electric dipole radiation discussed in problem (V.21). In a laboratory gas, a hydrogen atom in the $2s$ state typically decays through a collision with another atom that induces a transition to a $2p$ state from which it quickly decays to the ground state.

To approximate the effect of a collision on a hydrogen atom in the $2s$ state, we can imagine that a uniform electric field pointing along the $+z$ axis is turned on to the constant value E_o for the time interval t_o. That is, the applied electric field at the position of the hydrogen atom vanishes at times $t < 0$ and $t > t_o$, and the electric field is $\mathrm{E} = \mathrm{E}_o$ at $0 < t < t_o$.

Given that at $t < 0$ the electron is in the $2s$ state, compute in first-order time-dependent perturbation theory the probabilities that the electron ends up in each of the $2p$ states. The perturbation to the Hamiltonian is given by equation (V.21), where here the electric field E is a function of time, and the electron wave functions for the four $n = 2$ levels of a hydrogen atom are given in equation (V.20). It will be recalled that these four states have very nearly the same energy.

V.12) As discussed in section 34, an isolated hydrogen atom has a singlet ground state (where the spins of the electron and proton are combined to total spin $s = 0$), and three degenerate triplet (total spin $s = 1$) hyperfine states at energy E above the singlet state.

When a magnetic field B is applied along the x axis, it adds to the Hamiltonian for the atom the term

$$V_1 = \frac{ge}{2M_e c} B s_x^e, \tag{V.34}$$

where s_x^e is the x component of the spin operator for the electron. Suppose the magnetic field is applied for time t_o, so the magnetic field vanishes at times $t < 0$ and $t > t_o$, and the field is constant at B_o at $0 \le t \le t_o$.

Before the field is applied, the atom is in the excited hyperfine state $s = 1$ and $m = 1$.

a) Use first-order time-dependent perturbation theory to find the probability that the atom ends up in the excited hyperfine state with $m = -1$, that is, find the transition probability for

$$s = 1,\ m = 1 \rightarrow s = 1,\ m = -1. \tag{V.35}$$

b) Use first-order perturbation theory to find the probability that the atom ends up in the excited hyperfine state with $m = 0$, that is, find the transition probability for

$$s = 1,\ m = 1 \rightarrow s = 1,\ m = 0. \tag{V.36}$$

V.13) Consider a single spin zero particle in a spherically symmetric potential well. Transitions between energy eigenstates $|i\rangle \rightarrow |f\rangle$ of this system are induced by the addition of the sinusoidally varying potential energy term

$$V_1 = A\vec{\epsilon} \cdot \mathbf{L} \cos \omega t. \tag{V.37}$$

Here A and ω are real constants, $\vec{\epsilon}$ is a constant unit vector, and $\mathbf{L}$ is the orbital angular momentum operator for the particle. (This models the magnetic dipole interaction of the particle with the radiation field, as in problem [II.20].)

a) For states $|i\rangle$ and $|f\rangle$ that have definite parities, π_i and π_f, what are the selection rules for this transition, that is, what are the values of π_i and π_f such that the transition probability in first-order perturbation theory is not forced to vanish?

b) Suppose the initial and final states have definite orbital angular momentum quantum numbers, $|i\rangle = |l_i, m_i\rangle$, and $|f\rangle = |l_f, m_f\rangle$. By considering what happens when each component of $\mathbf{L}$ operates on $|l_i, m_i\rangle$, find the the selection rules for l_f and m_f for given l_i and m_i, such that the transition probability is not forced to vanish in first-order perturbation theory.

V.14) The method of coupling of angular momenta is discussed in sections 17 and 25, and applied in problems (II.19) and (III.9). This problem is meant to give you some more practice.

It will be recalled that, if J_x, J_y, and J_z obey the angular momentum commutation relations (17.7), there is a complete set of simultaneous eigenstates of J^2 and J_z:

$$J^2|j,m\rangle = \hbar^2 j(j+1)|j,m\rangle, \qquad J_z|j,m\rangle = m\hbar|j,m\rangle, \tag{V.38}$$

with $-j \leq m \leq j$. The ladder operators,

$$J_\pm = J_x \pm iJ_y, \tag{V.39}$$

change m by ± 1 (eq. [17.26]), so we have

$$J_+|j,m\rangle = C|j,m+1\rangle. \tag{V.40}$$

The normalizing constant, C, is computed in wave mechanics notation in problem (II.18). In the abstract space notation, one takes the inner product of the expression (V.40) with its dual. As you are invited to check, the identity

$$J_-J_+ = J^2 - J_z^2 - \hbar J_z. \tag{V.41}$$

gives

$$|C|^2 = \hbar^2(j^2 + j - m^2 - m). \tag{V.42}$$

The square root has an arbitrary phase factor; the standard convention is to take C to be real and positive. This gives

$$J_+|j,m\rangle = \hbar(j^2 + j - m^2 - m)^{1/2}|j,m+1\rangle. \tag{V.43}$$

By operating on both sides of equation (V.43) by J_-, and using equation (V.41) again, you can check that the normalizing factor in the lowering operator equation is

$$J_-|j,m\rangle = \hbar(j^2 + j - m^2 + m)^{1/2}|j,m-1\rangle. \tag{V.44}$$

Note that once the phase factor in equation (V.43) has been fixed, there is no freedom in the phase of equation (V.44).

For an example, consider a system of two spins, one spin 3/2, the other spin 1. A basis for this system is the set of simultaneous eigenstates of the z components of the two spin operators,

$$|m_1, m_2\rangle, \tag{V.45}$$

with

$$-3/2 \leq m_1 \leq 3/2, \qquad -1 \leq m_2 \leq 1. \tag{V.46}$$

Let the total spin operator for the system be $\mathbf{s} = \mathbf{s}_1 + \mathbf{s}_2$, where $\mathbf{s}_1$ and $\mathbf{s}_2$ are the two vector spin operators.

a) By considering the results of operating on the vector $|m_1 = 3/2, m_2 = 1\rangle$ in equation (V.45) with $s_z = s_z(1) + s_z(2)$, and the results of operating with $s_- s_+$, show that $|3/2, 1\rangle$ is an eigenstate of total spin with $s = 5/2$ and $m = 5/2$:

$$|s = 5/2, m = 5/2\rangle = |m_1 = 3/2, m_2 = 1\rangle. \tag{V.47}$$

b) By operating on equation (V.47) with $s_- = s_-(1) + s_-(2)$, find an expression for the state $|5/2, 3/2\rangle$ with total spin $s = 5/2$ and $m = 3/2$ as a linear combination of the basis vectors in equation (V.45).

c) Show that the state $|3/2, 3/2\rangle$ with total spin $s = 3/2$ and $m = 3/2$ has to be a linear combination of the vectors in equation (V.45) with $m_1 + m_2 = 3/2$.

d) By using the fact that the state $|3/2, 3/2\rangle$ has to satisfy

$$s_+ |3/2, 3/2\rangle = 0, \tag{V.48}$$

find the expression for $|3/2, 3/2\rangle$ as a linear combination of the $|m_1, m_2\rangle$. As a check, you ought to find that the result is orthogonal to the expression for $|5/2, 3/2\rangle$ from part (b).

e) Suppose the Hamiltonian for the system is the dot product of the two spin vector operators,

$$H = A\, \mathbf{s}_1 \cdot \mathbf{s}_2, \tag{V.49}$$

where A is a positive real constant. By using the trick in equation (34.39), find the energy eigenvalues, and find the degeneracy of each, that is, the number of different states all with the same energy. The total number of different energy eigenstates ought to be equal to the number of vectors in the basis in equation (V.45), $4 \times 3 = 12$.

f) Suppose the system is perturbed by adding to the Hamiltonian in equation (V.49) the term

$$V_a = b\, s_z(2). \tag{V.50}$$

Here b is a real constant, and $s_z(2)$ is the z component of the spin operator for particle 2 (with spin 1). This removes the degeneracy in the largest eigenvalue of H in equation (V.49). Find the new energies of these states in first-order perturbation theory.

g) Suppose V_a is removed and instead the system is perturbed by applying a weak oscillating magnetic field, adding to the Hamiltonian the term

$$V_b = c s_x(2) \cos \omega t, \tag{V.51}$$

where c and ω are real constants. Here $s_x(2)$ is the x component of the spin of particle 2 (not the z component from the last part). The frequency is tuned to induce transitions from the state with total spin $s = 5/2$ and $m = 1/2$ to the states with $s = 3/2$. Find, in first-order time-dependent perturbation theory, the final values of m for which there is a nonzero transition probability amplitude, and find the ratio of transition probabilities to each of these final $s = 3/2$ states.

V.15) After relativistic corrections, the $n = 2$ states in atomic hydrogen with orbital angular momentum quantum number $l = 1$ and total angular momentum $j = 1/2$ are degenerate with the $n = 2$ states with orbital angular momentum quantum number $l = 0$ and total angular momentum $j = 1/2$ (from the spin of the electron). The relativistic correction gives the states with $l = 1$ and $j = 3/2$ slightly higher energy.

Use first-order perturbation theory to find the shifts in energy of these degenerate $j = 1/2$ states when an electric field E is applied along the z axis, adding to the Hamiltonian the potential energy term

$$V_1 = e\mathrm{E}z, \tag{V.52}$$

where e is the magnitude of the charge of the electron. The $n = 2$ electron wave functions are given in equation (V.20).

V.16) A deuterium atom consists of an electron bound to a deuteron (which is the bound state of a proton and neutron). The difference in reduced mass in a hydrogen and deuterium atom is negligible; the only important differences for the hyperfine structures of the two atoms are that a proton has spin 1/2 and a deuteron has spin 1, and the magnetic moments of the two nuclei are

$$\vec{\mu}_p = \frac{g_p e}{2M_p c}\mathbf{s}_p, \qquad \vec{\mu}_d = \frac{g_d e}{2M_p c}\mathbf{s}_d, \tag{V.53}$$

where

$$g_p = 5.59, \qquad g_d = 0.86. \tag{V.54}$$

The proton and deuteron spin vectors are $\mathbf{s}_p$ and $\mathbf{s}_d$. The mass in both expressions is the proton mass.

a) Find the wavelength in centimeters of the hyperfine line in atomic deuterium by scaling from the known line in atomic hydrogen, $\lambda = 21.1$ cm, and suitably adapting the calculation in section 34.

b) Suppose atomic deuterium is placed in a uniform static magnetic field. Find the Zeeman shifts in the energies of the six spin states in the ground electronic state of the atom in the limit that the energy shifts due to the applied magnetic field are small compared to the hyperfine splitting. As for atomic hydrogen, it is a good approximation to ignore the magnetic moment of the deuteron compared to that of the electron.

V.17) A system of three spin 1/2 particles with spin vector operators $\mathbf{s}(1)$, $\mathbf{s}(2)$, and $\mathbf{s}(3)$ was considered in problem (III.9). Suppose the Hamiltonian for this three-spin system is

$$H = as^2 + bW(t)s_x(1). \tag{V.55}$$

Here a and b are real constants, s^2 is the square of the total spin operator, $s_x(1)$ is the x component of the spin of particle 1, and the function $W(t)$ signifies that this second term is present only from time $t = 0$ to $t = \tau$:

$$\begin{aligned} W(t) &= 0 \quad \text{for} \quad t < 0, \\ &= 1 \quad \text{for} \quad 0 \le t \le \tau, \\ &= 0 \quad \text{for} \quad t > \tau. \end{aligned} \tag{V.56}$$

If initially the system is in the state $|i\rangle = |+++\rangle$, find in first-order perturbation theory the probability that the system ends up in the state $|-++\rangle$.

V.18) Consider a configuration of the two electrons in a helium atom that has total orbital angular momentum quantum number $l = 2$ and total spin angular momentum quantum numbers $s = 1$. The spin-orbit coupling to be discussed in section 42 breaks this configuration into three energy levels, with total angular momentum quantum numbers $j = 1$, 2, and 3. If a weak magnetic field B is applied to the atom, the energy level with total angular momentum quantum number j is further split into $2j + 1$ states with equally separated energies, at energy differences

$$\delta E_j = \frac{g_j e\hbar}{2m_e c} B, \tag{V.57}$$

where $-e$ and m_e are the electron charge and mass. This equation defines the Landé g factor, g_j.

To compute the g factor in linear perturbation theory, we need the matrix elements (eq. [31.6])

$$\langle j, m | L_z + 2s_z | j, m' \rangle, \tag{V.58}$$

when the z axis is placed along the magnetic field. This matrix element vanishes unless $m' = m$, so we are doing nondegenerate perturbation theory. It is easy to evaluate the matrix element

$$\langle j, m | L_z + s_z | j, m \rangle, \tag{V.59}$$

so we need

$$\langle j, m | s_z | j, m \rangle. \tag{V.60}$$

This can be computed by writing $|j, m\rangle$ as a linear combination of simultaneous eigenstates of L_z and s_z, by the methods used in problem (V.14), but it is also worth knowing about a less laborious method.

The Wigner-Eckart theorem says the matrix elements of the components of the total spin operator $\mathbf{s}$ between states with total angular momentum quantum number j and z components m and

m' are related to the matrix elements of the components of the total angular momentum operator $\mathbf{J}$ by the equation

$$\langle j, m'|s_\alpha|j, m\rangle = f(j)\langle j, m'|J_\alpha|j, m\rangle. \tag{V.61}$$

The factor $f(j)$ is a function of the quantum number j alone. (That is, the dependence on m and m' appears in the matrix elements only.) The derivation of this equation, and its generalizations, are to be found in more complete books on quantum mechanics. The content of the equation can be understood by considering the matrix elements of s_α and J_α computed in two coordinate systems that are rotated at some given angle relative to each other. In both coordinate systems the total angular momentum quantum number has the same value, j, but the rotation mixes the m values. Equation (V.61) says that since the vectors $\mathbf{J}$ and $\mathbf{s}$ transform the same way under a rotation of the coordinates, the matrix elements of $\mathbf{J}$ and $\mathbf{s}$ transform the same way under the rotation.

a) To find the function $f(j)$ in equation (V.61), consider the matrix element

$$\langle j, m'|\mathbf{s}\cdot\mathbf{J}|j, m\rangle = \sum_{j'', m'', \alpha} \langle j, m'|s_\alpha|j'', m''\rangle\langle j'', m''|J_\alpha|j, m\rangle. \tag{V.62}$$

The right-hand side is simplified by noting that $\langle j'', m''|J_\alpha|j, m\rangle$ vanishes unless $j'' = j$, and by using equation (V.61). The left-hand side is simplified by using the relation

$$\mathbf{L}^2 = (\mathbf{J} - \mathbf{s})^2 = J^2 + s^2 - 2\mathbf{J}\cdot\mathbf{s}. \tag{V.63}$$

Show that the result is the relation

$$\langle j, m|s_z|j, m\rangle = \frac{m\hbar}{2j(j+1)}[j(j+1) + s(s+1) - l(l+1)]. \tag{V.64}$$

b) Use the result from part (a) to find Landé's g-factor for the states in helium with $l = 2$, $s = 1$, and $j = 2$.

V.19) The electric dipole approximation to a radiative transition between states of a system of charged particles assumes the particle velocities are small, so the size of the radiating system is negligibly small compared to the wavelength of the radiation, and it neglects the interaction of the system with the magnetic field of the radiation. The energy of a particle with charge q at position $\mathbf{r}$ in an electric field $\mathbf{E}$ is

$$V = -q\mathbf{r} \cdot \mathbf{E}. \tag{V.65}$$

In the electric dipole approximation the variation of the electromagnetic radiation field across the system is taken to be negligibly small, so the electric field at the system can be written as

$$\mathbf{E}(t) = E_o \vec{\epsilon} \cos \omega t, \tag{V.66}$$

where E_o is a real constant, ω is the real constant frequency of the radiation, and $\vec{\epsilon}$ is a unit polarization vector. The term in the Hamiltonian describing the interaction of a single particle with the radiation is then

$$V = -qE_o \vec{\epsilon} \cdot \mathbf{r} \cos \omega t. \tag{V.67}$$

In the first-order approximation to the photoionization of atomic hydrogen, the initial electron wave function is $\psi_i \propto e^{-r/a_o}$, representing an electron in the ground state in a hydrogen atom, and the final state is approximated as a plane wave,

$$\psi_f \propto e^{i\mathbf{k}_f \cdot \mathbf{r}}, \tag{V.68}$$

representing an unbound electron with final momentum $\hbar \mathbf{k}_f$.

Use first-order time-dependent perturbation theory, with the perturbation in equation (V.67), and the above initial and final wave functions, to find the angular distribution of the ejected electron. That is, find how the probability for observing that the electron ends up in the final state with momentum $\hbar \mathbf{k}_f$ varies with the direction of $\mathbf{k}_f$ relative to the fixed direction $\vec{\epsilon}$ of the electric field of the electromagnetic radiation. Do not bother to compute factors that are independent of direction.

The computation is simplified by using polar coordinates with polar axis along $\mathbf{k}_f$. A useful result from spherical trigonometry is that if $\vec{\epsilon}$ has polar angles θ_f and ϕ_f in this coordinate system, and $\mathbf{r}$ has polar angles θ and ϕ, then the angle $\theta_{\epsilon r}$ between $\vec{\epsilon}$ and $\mathbf{r}$ satisfies

$$\cos \theta_{\epsilon r} = \cos \theta_f \cos \theta + \sin \theta_f \sin \theta \cos(\phi_f - \phi). \tag{V.69}$$

If you find yourself working lengthy integrals you have missed the simple approach.

V.20) It is interesting to compare the electric dipole radiation rate in quantum mechanics to the expression for the electric dipole radiation rate in classical electromagnetic theory.

Consider a particle with charge q that is moving as a simple harmonic oscillator with frequency ω_o along the direction of the z axis. If the electric field polarization vector makes an angle θ with the z axis, then the electric dipole interaction in equation (V.67) for an applied electromagnetic field with frequency ω and electric field amplitude E_o is

$$V = -qE_o z \cos\omega t \cos\theta. \tag{V.70}$$

a) Show that, in first-order perturbation theory, the probability per unit time for spontaneous decay from the initial state n of the simple harmonic oscillator to the final state $n-1$, with energy change $E_i - E_f = \hbar\omega_o$, is

$$\lambda = \frac{4}{3}\frac{q^2\omega_o^3}{\hbar c^3}|\langle n-1|z|n\rangle|^2. \tag{V.71}$$

The calculation proceeds as in section 37. One first computes the transition probability λt for a given applied radiation field (given E_o, ω, and θ in eq. [V.70]). The prescription for converting to a spontaneous transition rate is the same as in section 37, except that E_o replaces B_o in equation (37.30). Here the transition probability is proportional to $\cos\theta$, instead of $\sin\theta$, so you have to rework the sum over polarizations in equation (37.33).

b) By evaluating the matrix element in equation (V.71), by the methods in section 6 and problem (II.22), show that equation (V.71) becomes

$$\lambda = \frac{2}{3}\frac{q^2\omega_o}{mc^3\hbar}n, \tag{V.72}$$

for the transition from the energy level $E_n = (n+1/2)\hbar\omega$ to $(n-1/2)\hbar\omega$.

To compare equation (V.72) to the classical electric dipole radiation rate, let us write the initial energy of the oscillator as

$$E_n = (n+1/2)\hbar\omega_o = \frac{1}{2}m\omega_o^2 z_o^2. \tag{V.73}$$

The last expression is the energy of a classical oscillator with classical amplitude (maximum displacement from equilibrium) z_o. This expression only makes sense when $n \gg 1$, so we can ignore the difference between n and $n+1$. Then the result of eliminating n from equations (V.72) and (V.73) is

$$\lambda \hbar \omega_o = \frac{q^2 \omega_o^4 z_o^2}{3c^3}. \tag{V.74}$$

The left side is the probability per unit time for emission of a photon, multiplied by the photon energy, $\hbar\omega$. That is, it is the mean energy radiation rate in quantum mechanics. The right-hand side is the classical expression for the mean energy radiation rate in electric dipole approximation.

V.21) The selection rules that govern the allowed changes of parity and angular momentum quantum numbers in an electric dipole radiation transition may be studied as follows.

If the system has charges q_i at positions $\mathbf{r}_i$, the electric dipole moment of the system is

$$\mathbf{d} = \sum q_i \mathbf{r}_i, \tag{V.75}$$

and the coupling to a classical electromagnetic radiation field with electric field amplitude E_o and electric polarization vector $\vec{\epsilon}$ is

$$V = -E_o \vec{\epsilon} \cdot \mathbf{d} \cos \omega t. \tag{V.76}$$

Neglecting spins, the stationary states of the unperturbed atom can be classified as eigenstates of the Hamiltonian H_o for the isolated system, the total orbital angular momentum operators L^2, L_z, and parity:

$$|i\rangle = |n, l, m, \pi\rangle. \tag{V.77}$$

a) The definition (32.3) of the parity operator generalizes to

$$\Pi \mathbf{d} = -\mathbf{d}\Pi. \tag{V.78}$$

By considering the matrix elements of this relation, in the basis of equation (V.77), show that the matrix elements for transitions among the eigenstates of H_o induced by the perturbation V in equation (V.76) vanish unless the initial and final states satisfy the selection rule for parity in an electric dipole transition,

$$\pi' = -\pi. \tag{V.79}$$

b) Let

$$d_{\pm} = d_x \pm i d_y, \tag{V.80}$$

where the d_α are the components of the electric dipole moment (V.75). Evaluate (reduce to simplest possible forms) the commutators

$$\begin{aligned} &[L_z, d_z], \\ &[L_z, d_{\pm}], \\ &[L_{\pm}, d_z], \\ &[L_{\pm}, d_{\pm}]. \end{aligned} \tag{V.81}$$

Here L_z and $L_{\pm} = L_x \pm i L_y$ are the usual total orbital angular momentum operators.

c) Use the commutators to show that $(d_x + i d_y)|l, m\rangle$ is an admixture of (linear combination of) states $|l', m'\rangle$ with $m' = m + 1$.

d) Find the electric dipole selection rule for the quantum numbers m and m' in the transition $|l, m\rangle \to |l', m'\rangle$ when the electric polarization vector $\vec{\epsilon}$ is directed along the x axis.

e) A right-hand circular polarized wave propagating toward the positive z axis has electric polarization vector at the position of the atom that is given by the equation

$$\vec{\epsilon}(t) = \mathbf{i}\cos(\omega t) + \mathbf{j}\sin(\omega t), \tag{V.82}$$

where $\mathbf{i}$ and $\mathbf{j}$ are unit vectors along the x and y axes, and ω is the constant frequency of the radiation.

Find the selection rule for m and m' for this case. (Here the selection rule follows not from the vanishing of the matrix element but from the resonance condition, that allows the transition amplitude $c_f(t)$ to grow with time. As discussed in section 36, you will have to consider separately the cases of absorption and emission of radiation.)

f) Suppose the radiation is linearly polarized, so the electric polarization vector $\vec{\epsilon}$ in equation (V.76) has a fixed direction, at angle θ to the z axis in the xz plane. Consider a transition for which $m' = m$. Find how the transition probability in electric dipole approximation varies with the angle θ.

g) Consider a state $|i\rangle = |0, 0\rangle$ with zero orbital angular momentum: $l = m = 0$. Use the commutation relations from part (b) to show that $d_z|0, 0\rangle$ and $d_\pm|0, 0\rangle$ are eigenvectors of the total orbital angular momentum operators L^2 and L_z, find the quantum numbers, and use the results to find the selection rules for an electric dipole transition between a state with $l = 0$ and a state with angular momentum quantum numbers l' and m'.

h) We are labeling state vectors by their orbital angular momentum quantum numbers: $|l, m\rangle$. Use the commutation relations (V.81) to show

$$d_+|l, l\rangle \propto |l + 1, l + 1\rangle. \tag{V.83}$$

i) Use the commutation relations (V.81) with equation (V.83) to show that the vector $|l + 1, l\rangle$ can be expressed as a linear combination of $d_z|l, l\rangle$ and $d_+|l, l - 1\rangle$.

j) Use the result from the last part to show that the vector $|l, l\rangle$ can be expressed as a linear combination of $d_z|l, l\rangle$ and $d_+|l, l - 1\rangle$.

k) Use the commutation relations with the result from the last two parts to show that the vectors $|l+1, l-1\rangle$, $|l, l-1\rangle$ and $|l-1, l-1\rangle$ all can be expressed as linear combinations of $d_-|l, l\rangle$, $d_z|l, l-1\rangle$, and $d_+|l, l - 2\rangle$.

l) You will note that the last four parts are identical to the procedure in section 17 for the addition of angular momenta l and l'. Use this to find the electric dipole selection rules for l and l'.

CHAPTER 6

ATOMIC AND MOLECULAR STRUCTURE

This chapter deals with the structures of the lighter atoms and the simplest molecule, molecular hydrogen. The main approximation method used here is the energy variational principle in section 38. This chapter also introduces the Pauli exclusion principle, that governs the symmetry of the state vector for a system of identical particles such as electrons.

38 Energy Variational Principle

A powerful technique for computing the low-lying energies of a system such as an atom or molecule is based on the following simple theorem. Let the Hamiltonian for the system be H, and let E_n and $|n\rangle$ be the complete set of energy eigenfunctions and eigenvectors,

$$H|n\rangle = E_n|n\rangle. \tag{38.1}$$

It will be supposed that the eigenvectors are discrete, so we can normalize them to

$$\langle n|m\rangle = \delta_{nm}. \tag{38.2}$$

Let $|\phi\rangle$ be a trial approximation to the ground state solution $|0\rangle$ to equation (38.1). Completeness allows us to express this trial vector as a sum over the eigenvectors of H:

$$|\phi\rangle = \sum c_n|n\rangle. \tag{38.3}$$

In terms of this expansion, the expectation value of the energy in the vector $|\phi\rangle$ is

$$\langle H\rangle_\phi = \frac{\langle\phi|H|\phi\rangle}{\langle\phi|\phi\rangle} = \frac{\sum |c_n|^2 E_n}{\sum |c_n|^2}. \tag{38.4}$$

The denominator, $\langle\phi|\phi\rangle$, has been inserted because $|\phi\rangle$ need not have been properly normalized. Since the ground state energy is less than any other, $E_0 \le E_n$, we can only reduce this sum by replacing E_n with E_0, so

$$\langle H\rangle_\phi = \frac{\langle\phi|H|\phi\rangle}{\langle\phi|\phi\rangle} \ge E_0. \tag{38.5}$$

That is, the expectation value of the energy in the trial vector $|\phi\rangle$ never is less than the ground state energy. This gives a computational technique for finding an approximation to the ground state energy: the best approximation to E_0 among a set of trial functions $|\phi\rangle$ is the one that gives the minimum value of $\langle H\rangle_\phi$.

It will be noted that a mediocre approximation to $|0\rangle$ can give a good approximation to E_0, because the difference between $|\phi\rangle$ and $|0\rangle$ is measured by the size of the ratios c_n/c_0, while we see from equation (38.4) that the difference between $\langle H\rangle_\phi$ and E_0 is measured by the squares, $(c_n/c_0)^2$. Thus if the c_n/c_0 are fairly small the energy error may be quite small. Note also that if the ground state has some definite quantum number, q, in addition to energy, then we can estimate the energy of the lowest exited state with quantum number $q' \ne q$ by using trial vectors with quantum number q'.

As an example, consider a three-dimensional simple harmonic oscillator, with Hamiltonian

$$\begin{aligned} H &= \frac{p^2}{2m} + \frac{1}{2}m\omega^2 r^2, \\ &= -\frac{\hbar^2}{2m}\nabla^2 + \frac{1}{2}m\omega^2 r^2. \end{aligned} \tag{38.6}$$

This is the sum of three one-dimensional Hamiltonians for each of the three position components x, y, and z, so we know the ground state wave function is a Gaussian and the ground state energy is $3\hbar\omega/2$. Suppose, for the purpose of the example, the energy and wave function for the ground state of the oscillator are not known. The wave function might be expected to be roughly like that of a hydrogen atom,

$$\phi = e^{-\alpha r}. \tag{38.7}$$

We will seek the value of the parameter α that gives the best approximation to the oscillator ground state wave function, by varying α to minimize the expectation value of the energy, $\langle H\rangle_\phi$, in the wave function (38.7).

The normalizing denominator is

$$\begin{aligned}\langle\phi|\phi\rangle &= \int d^3r|\phi(r)|^2 \\ &= 4\pi\int_0^\infty r^2 dr e^{-2\alpha r} \\ &= \pi/\alpha^3.\end{aligned} \tag{38.8}$$

A handy integral here and below is

$$\int_0^\infty dx\, x^n e^{-x} = n!. \tag{38.9}$$

For the kinetic energy we need $(\phi, \nabla^2\phi) = -(\nabla\phi, \nabla\phi)$.[1] The gradient of the trial wave function is

$$\nabla\phi = -\alpha e^{-\alpha r}\mathbf{r}/r. \tag{38.10}$$

This gives $(\phi, \nabla^2\phi) = -\int|\nabla\phi|^2 d^3r = -\pi/\alpha$. Thus the kinetic energy part of the numerator of the expectation value of H is

$$(\phi, \frac{p^2}{2m}\phi) = \frac{\pi\hbar^2}{2m\alpha}. \tag{38.11}$$

The potential energy part is

$$(\phi, \frac{1}{2}m\omega^2r^2\phi) = \frac{3\pi m\omega^2}{2\alpha^5}. \tag{38.12}$$

Equations (38.8) to (38.12) give

$$\langle H\rangle_\alpha = \frac{\hbar^2\alpha^2}{2m} + \frac{3m\omega^2}{2\alpha^2}. \tag{38.13}$$

As usual, one finds the minimum of this function by setting the derivative with respect to α^2 equal to zero. At the minimum, the parameter is

$$\alpha^2 = 3^{1/2}m\omega/\hbar, \tag{38.14}$$

[1] The first derivative of $\phi = e^{-\alpha r}$ is discontinuous at $r \to 0$, so the second derivative is singular. One sees that this does not affect the relation $(\phi, \nabla^2\phi) = -(\nabla\phi, \nabla\phi)$ by imagining ϕ to be the limit of a sequence of smooth functions. By computing $(\nabla\phi, \nabla\phi)$, we avoid having to think about the singular second derivative of ϕ.

and the expectation value of H at the minimum is

$$\langle H \rangle_{\min} = 3^{1/2}\hbar\omega = 1.73\hbar\omega. \tag{38.15}$$

The result is greater than the correct value, $1.5\hbar\omega$, as it must be, but it is reasonably close considering the crude approximation of the trial wave function. It is left as an exercise to check that the characteristic width of the trial function, $\sim \alpha^{-1}$ with α given by equation (38.14), approximates the width of the ground state simple harmonic oscillator Gaussian wave function.

39 The Ground State of Helium

Here there are two electrons bound to a nucleus with charge $2e$. We will treat the nucleus as a point charge fixed at the center of mass. The Hamiltonian for the electrons is[2]

$$H = \frac{p_1^2}{2m} + \frac{p_2^2}{2m} - \frac{2e^2}{r_1} - \frac{2e^2}{r_2} + \frac{e^2}{r_{12}}. \tag{39.1}$$

The first two terms are the kinetic energy, with m the electron mass. The next two terms are the potential energies of the electrons of charge $-e$ at distances r_1 and r_2 from the nucleus with charge $2e$. The last term is the potential energy of interaction of the two electrons, with separation $r_{12} = |\mathbf{r}_1 - \mathbf{r}_2|$. If the Hamiltonian contained only the first four terms (of the five shown), we could find exact solutions for the energy eigenvalues and eigenfunctions. We will take the forms of these exact eigenfunctions to be the trial wave functions for the full Hamiltonian.

Neglecting the last term in equation (39.1), that represents the energy of interaction of the two electrons, the Hamiltonian is the sum of two single-particle hydrogen atom Hamiltonians, one for each electron. These operators commute, so we know their simultaneous eigenfunctions are products of hydrogen atom wave functions. The ground state eigenfunction is

$$\psi(\mathbf{r}_1, \mathbf{r}_2) = \phi(r_1)\phi(r_2) = e^{-\alpha(r_1+r_2)}. \tag{39.2}$$

[2] In section 41 below we take account of the exclusion principle that determines the symmetry of the electron wave function. For atoms with larger numbers of electrons this principle restricts the possible eigenfunctions of the Hamiltonian (39.1), but in helium it restricts only the spin part of the wave function; the calculation given here is unaffected. In section 42 we will discuss a relativistic correction that adds a small spin interaction term to the Hamiltonian in equation (39.1).

We will use this functional form, with α an adjustable parameter, as a trial wave function to estimate the ground state energy in the full Hamiltonian.

Some of the integrals needed to compute $\langle H\rangle_\psi$ (eq. [38.4]) in the trial wave function ψ have already been done. We have from equation (38.8) the normalizing integral,

$$(\psi, \psi) = \pi^2/\alpha^6, \tag{39.3}$$

and from equation (38.11) the kinetic energy for one of the electrons is

$$(\psi, \frac{p_1^2}{2m}\psi) = \frac{\pi}{\alpha^3}\frac{\pi\hbar^2}{2m\alpha}. \tag{39.4}$$

The potential energy of each electron in the field of the nucleus is determined by the integral

$$(\psi, r_2^{-1}\psi) = \int d^3r_1 e^{-2\alpha r_1}\int d^3r_2 e^{-2\alpha r_2}/r_2 = \pi^2/\alpha^5. \tag{39.5}$$

For the interaction between the electrons, we need the integral

$$\begin{aligned}(\psi, r_{12}^{-1}\psi) &= \int d^3r_1 d^3r_2 e^{-2\alpha(r_1+r_2)}/r_{12} \\ &= \frac{1}{32\alpha^5}\int d^3x_1 d^3x_2 e^{-(x_1+x_2)}/x_{12},\end{aligned} \tag{39.6}$$

with $\mathbf{x} = 2\alpha\mathbf{r}$. As a first step, note that the integral over $\mathbf{x}_2$ at fixed $\mathbf{x}_1$ is

$$I(\mathbf{x}_1) = \int d^3x_2 e^{-x_2}/x_{12}. \tag{39.7}$$

This is the form of the potential energy of a unit point charge at $\mathbf{x}_1$ due to a charge distribution e^{-x_2}. We know from electrostatics that a particle at radius x in this distribution sees the charge within x as a point charge at the center, so the potential at x_1 is

$$I(x_1) = \frac{1}{x_1}\int_0^{x_1} d^3x_2 e^{-x_2} + \int_{x_1}^{\infty}\frac{1}{x_2}d^3x_2 e^{-x_2}. \tag{39.8}$$

This works out to

$$I(x_1) = 4\pi\left[\frac{2}{x_1} - \frac{2e^{-x_1}}{x_1} - e^{-x_1}\right]. \tag{39.9}$$

Now it is easy to work the integral over x_1, to get

$$(\psi, r_{12}^{-1}\psi) = \frac{5\pi^2}{8\alpha^5}. \tag{39.10}$$

Collecting equations (39.3) to (39.5) and (39.10), we get

$$\langle H \rangle_\alpha = \frac{\hbar^2\alpha^2}{m} - \frac{27e^2\alpha}{8}. \tag{39.11}$$

On setting the derivative of this expression with respect to α equal to zero, we find that at the minimum of $\langle H \rangle_\alpha$ the parameter is

$$\alpha = \frac{27e^2 m}{16\hbar^2}. \tag{39.12}$$

This value for α in equation (39.11) gives the estimate of the ground state energy,

$$E_0 = -\left[\frac{27}{16}\right]^2 \frac{me^4}{\hbar^2}. \tag{39.13}$$

To get a numerical result, recall from section 6 that the binding energy of atomic hydrogen is

$$B = \frac{me^4}{2\hbar^2} = 13.6\,\text{eV}, \tag{39.14}$$

so our approximation to the ground state energy of helium is

$$E_0 = -\left[\frac{27}{16}\right]^2 \times 2 \times 13.6\,\text{eV} = -77.5\,\text{eV}. \tag{39.15}$$

The measured value is $-78.97\,\text{eV}$. This is a little lower, as it must be.

Our approximation to the ground state wave function in helium can be compared to the wave function for a hydrogen-like atom consisting of a single electron bound to a nucleus of charge ze. The calculation in section 6 is easily generalized to show that the latter is

$$\phi \propto e^{-zr/a_o}, \tag{39.16}$$

where the Bohr radius is $a_o = \hbar^2/(me^2)$. Thus in the trial function for helium in equation (39.2) each electron has the wave function of a hydrogen-like atom with effective charge z_{eff} defined by the equation

$$\alpha = z_{\text{eff}}/a_o. \tag{39.17}$$

By equation (39.12), the effective charge is

$$z_{\text{eff}} = 27/16. \tag{39.18}$$

That is, each electron behaves as if it were bound to a positive charge ez_{eff} that is less than the full charge of the nucleus, $2e$, but greater than e. In effect, the negative charge of each electron partially screens the other electron from the full charge of the nucleus.

More accurate calculations allow for adjustment of the shape of the trial wave function, and also take account of the fact that the wave function is not a product of single-particle functions, as in equation (39.2), but rather a single function of the six position variables for the two electrons. For high precision one also has to take account of relativistic corrections as well as the fact that the nucleus is not quite at the center of mass. The measured binding energy of the two electrons in a helium atom is the sum of the work required to remove one electron, derived from measured wavelengths of spectral lines connecting the ground state and a fully removed electron, and the work to remove the second electron, computed from the theory of hydrogen-like atoms. It is important as a demanding test of quantum mechanics that theory and measurement agree within the joint uncertainties, about one part in 10^7.

40 The Lowest Excited States of Helium

The ground state of helium has zero angular momentum. As we noted in section 38, a trial wave function with $l = 1$ is guaranteed to make the expectation value of the Hamiltonian not less than the energy of the lowest state in helium with $l = 1$.

We can classify energy eigenfunctions according to yet another quantum number. Since the Hamiltonian in equation (39.1) is symmetric under exchange of the positions of the two electrons, the energy eigenfunctions must be either even or odd under exchange of the particle position arguments in the wave function. Thus we can approximate the lowest energy even and odd $l = 1$ states by using even and odd trial wave functions.

We will again use trial functions constructed out of single-particle hydrogen-like functions. We have already used the ground state single particle function,

$$\phi_{1s} = e^{-\alpha r}. \tag{40.1}$$

The first index means this is the wave function belonging to principal quantum number $n = 1$ in hydrogen. The second index, s, says this is an s-wave function, with angular momentum quantum number $l = 0$. The degenerate first excited states in hydrogen, with principal quantum number $n = 2$, have angular momentum quantum numbers $l = 0$ and $l = 1$. The former is designated ϕ_{2s}. The $n = 2$ p-wave functions, with $l = 1$, are of the form (eq. [18.34])

$$\phi_{2p} = re^{-\beta r}\cos\theta, \tag{40.2}$$

for $m = 0$, and

$$\phi_{2p} = re^{-\beta r}\sin\theta e^{\pm i\phi}, \tag{40.3}$$

for $m = \pm 1$.

As we have noted, the wave function is either even or odd under exchange of the two electrons. We will approximate the lowest energy even and odd states with angular momentum $l = 1$ as

$$\psi_{\pm} = \phi_{1s}(r_1)\phi_{2p}(\mathbf{r}_2) \pm \phi_{1s}(r_2)\phi_{2p}(\mathbf{r}_1). \tag{40.4}$$

This trial function has two parameters, α and β, in ϕ_{1s} and ϕ_{2p}.

Working out the expectation value of the Hamiltonian (39.1) in this trial function is a good exercise, but not one that is appropriate here, so the computation will only be outlined. With equation (40.2) for ϕ_{2p}, the normalizing integral for $\psi_{\pm}$ works out to

$$(\psi_{\pm}, \psi_{\pm}) = \frac{2\pi^2}{\alpha^3\beta^5}. \tag{40.5}$$

The kinetic energy integral is

$$(\psi_{\pm}, \frac{p_1^2}{2m}\psi_{\pm}) = \frac{\pi^2\hbar^2}{2m}\left[\frac{1}{\alpha\beta^5} + \frac{1}{\alpha^3\beta^3}\right]. \tag{40.6}$$

The term representing the potential energy of an electron in the field of the nucleus is fixed by the integral

$$(\psi_{\pm}, \frac{1}{r_1}\psi_{\pm}) = \pi^2\left[\frac{1}{\alpha^2\beta^5} + \frac{1}{2\alpha^3\beta^4}\right]. \tag{40.7}$$

Using tricks like those in equations (39.7) and (39.8), one finds that the term representing the interaction between the electrons is

$$(\psi_\pm, \frac{1}{r_{12}}\psi_\pm) = \pi^2 \left[\frac{1}{\alpha^3\beta^4} - \frac{3\alpha+\beta}{\alpha^3(\alpha+\beta)^5} \pm \frac{56}{3(\alpha+\beta)^7} \right] . \tag{40.8}$$

With the energy unit

$$\frac{e^2}{2a_o} = 13.605\,\text{eV} \equiv 1\,\text{Rydberg}, \tag{40.9}$$

where a_o is the Bohr radius, and the definitions

$$a = \alpha a_o, \qquad b = \beta a_o, \tag{40.10}$$

the expectation value of the energy in the trial wave function in equation (40.4) finally works out to

$$E_\pm = \frac{(\psi_\pm, H\psi_\pm)}{(\psi_\pm, \psi_\pm)} = \left[a^2 + b^2 - 4a - b - \frac{(3a+b)b^5}{(a+b)^5} \pm \frac{56a^3b^5}{3(a+b)^7} \right] \text{Ry}. \tag{40.11}$$

Now one has to do a numerical search for the minimum value of this expression as a function of the parameters a and b. The result for the even state is

$$a_+ = 2.003, \qquad b_+ = 0.483, \tag{40.12}$$

and the estimate of the energy in the even state is

$$E_+ = -4.245\text{ Ry} = -57.75\,\text{eV}. \tag{40.13}$$

The measured value is

$$E_+ = -57.76\,\text{eV}, \tag{40.14}$$

which is lower, as expected, but the approximation is quite close.

As in equations (39.17) and (39.18), we can interpret α and β in terms of the effective charge seen by each electron. Equations (39.17), (40.10), and (40.12) say the effective charge for the $1s$ electron wave function is $z_{\text{eff}} = a_+ = 2.003$. This is quite close to the charge of the nucleus. The definition of the $2p$ wave function in equations (40.2) and (40.3) is such that in a hydrogen-like atom with charge ze the parameter is $\beta = z/(2a_o)$ (eq. 18.34]). Therefore, equation (40.12) says the effective charge seen by the $2p$ electron is $z_{\rm eff} = 2b_+ = 0.97$. That is, the inner $1s$ electron sees almost exactly the bare nucleus with charge $2e$, and this inner electron almost exactly covers the nucleus so the negative charge of the inner electron reduces the charge seen by the outer electron to very nearly e.

The minimum of equation (40.11) for the odd function ψ_- is at

$$a_- = 1.991, \qquad b_- = 0.544, \tag{40.15}$$

and the approximation to the energy is

$$E_- = -4.261 \text{ Ry}. \tag{40.16}$$

This is lower than the even state energy by 0.22 eV; the measured energy difference is 0.25 eV. The odd ψ_- function actually is split into three closely spaced energy levels. This is caused by the spin-orbit coupling to be discussed in section 42.

The lower energy of the odd function ψ_- results from the fact that the odd wave function vanishes when $\mathbf{r}_1 = \mathbf{r}_2$ (that is the only way to satisfy the condition that the wave function changes sign when $\mathbf{r}_1$ and $\mathbf{r}_2$ are exchanged). The wave function ψ_- therefore is small when $\mathbf{r}_1$ is close to $\mathbf{r}_2$, and this reduces the mean value of the positive energy term e^2/r_{12} representing the interaction of the two electrons. The even wave function ψ_+ is large when $\mathbf{r}_1$ is close to $\mathbf{r}_2$, increasing the mean value of e^2/r_{12}, and making the energy higher than in the odd state.

The lowest excited states of helium with zero orbital angular momentum can be approximated by the the trial wave functions

$$\chi_\pm = \phi_{1s}(r_1)\phi_{2s}(r_2) \pm \phi_{1s}(r_2)\phi_{2s}(r_1). \tag{40.17}$$

As before, ϕ_{1s} has the form of the single-particle ground state wave function in atomic hydrogen, and the single-particle function ϕ_{2s} has the form of the hydrogen $2s$ wave function, $\phi_{2s} \sim (1 - \beta r)e^{-\beta r}$ (eq. [18.34]). We know the even function in equation (40.17) represents a new state in helium, rather than just a poor approximation to the ground state, because it has a new feature, the zero from the factor $\sim (1 - \beta r)$ in ϕ_{2s}. As for the $l = 1$ states, the coulomb repulsion between the two electrons makes the energy of the odd state χ_- lower than the even state χ_+.

The states approximated by equation (40.17) are said to have electron configuration $1s2s$. In the same way, the ground state electron configuration (eq. [39.2]) is $1s^2$, and the configuration of the lowest energy state with $l = 1$ in equation (40.4) is $1s2p$.

Since in atomic hydrogen the $2s$ and $2p$ energy levels are degenerate (apart from small relativistic corrections), it is not surprising that the states approximated by the $1s2s$ configuration have energies fairly close to the $1s2p$ configuration. The latter have the higher energy (less tight binding energy), because the angular momentum contribution to the effective potential suppresses the $2p$ wave function near the nucleus ($\phi_{2p} \propto r$ at small r in eqs. [40.2] and [40.3], while ϕ_{2s} is nonzero at $r = 0$). In a classical picture, the $2s$ orbit with no angular momentum plunges toward the nucleus, while the orbits with nonzero angular momentum avoid the nucleus. Either way, we see that the $1s$ electron is more effective at screening a $2p$ electron from the attractive charge of the nucleus than it is at screening a $2s$ electron, because the $2p$ electron is less likely to be found near the nucleus. Thus the $1s2p$ configuration is less tightly bound.

41 Pauli Exclusion Principle

To complete the physics we must take account of the symmetry imposed on the state vector belonging to a system containing more than one electron. A complete set of compatible observables for n electrons includes the position observables and one component of the spin of each particle. The simultaneous eigenstates of these observables are the basis vectors

$$|\mathbf{r}_1, m_1, \mathbf{r}_2, \ldots, m_n\rangle. \tag{41.1}$$

The operator P_{12} exchanges the quantum numbers belonging to particles 1 and 2:

$$P_{12}|\mathbf{r}_1, m_1, \mathbf{r}_2, m_2, \mathbf{r}_3, \ldots\rangle = |\mathbf{r}_2, m_2, \mathbf{r}_1, m_1, \mathbf{r}_3, \ldots\rangle. \tag{41.2}$$

The operator P_{ij} is defined the same way. To find the effect of P_{ij} on any state vector $|\psi\rangle$, expand $|\psi\rangle$ in the basis in equation (41.1) and then apply P_{ij} to each basis vector, as in equation (41.2).

Since the exchange operator applied twice does nothing, we know $P_{ij}^2 = 1$. This means the eigenvalues of P are $p = \pm 1$, corresponding to eigenstates even or odd under P_{ij}. Since electrons are identical, exchanging the electrons labeled i and j cannot affect the Hamiltonian, that is, P_{ij} has to commute with the Hamiltonian. That means energy eigenstates can be classified as even or odd under P_{ij}, and the symmetry does not change with time.

The Pauli exclusion principle is that for electrons (and all other fermions, including neutrons and protons) Nature allows only the odd state. That is, the allowed state vectors for a system of electrons must satisfy

$$P_{ij}|\psi\rangle = -|\psi\rangle, \tag{41.3}$$

for all i and j and all states $|\psi\rangle$ of the electrons. Within quantum mechanics this principle has to be taken as a postulate, justified because it helps account for the observed properties of atoms, as will be seen in some examples in this and the following sections. In quantum field theory, that develops along the lines of section 37, one finds under fairly general conditions that spin 1/2 particles moving in three dimensions have to obey the exclusion principle (41.3).

Let us apply this condition to the states of helium discussed in the last two sections. It is a good first approximation to leave the spin operators out of the Hamiltonian, as in equation (39.1). In this approximation, the solutions to the energy eigenvalue problem in the basis of equation (41.1) are of the form

$$\langle \mathbf{r}_1, m_1, \mathbf{r}_2, m_2|\psi\rangle = \psi(\mathbf{r}_1, \mathbf{r}_2)\chi_{m_1,m_2}. \tag{41.4}$$

The spin variables m_1 and m_2 appear in an arbitrary multiplicative function χ_{m_1,m_2}, because the electron spins do not appear in Schrödinger's equation. However, this spin wave function is constrained by the Pauli principle, which says

$$P_{12}|\psi\rangle = -|\psi\rangle. \tag{41.5}$$

Using equation (41.2), we see that the inner product of this equation with a basis vector (eq. [41.1]) is

$$\langle \mathbf{r}_1, m_1, \mathbf{r}_2, m_2|\psi\rangle = -\langle \mathbf{r}_2, m_2, \mathbf{r}_1, m_1|\psi\rangle. \tag{41.6}$$

That is, the wave function (eq. [41.4]) is antisymmetric:

$$\psi(\mathbf{r}_1, \mathbf{r}_2)\chi_{m_1,m_2} = -\psi(\mathbf{r}_2, \mathbf{r}_1)\chi_{m_2,m_1}. \tag{41.7}$$

We saw in the last two sections that the spatial part $\psi(\mathbf{r}_1, \mathbf{r}_2)$ is either even or odd under exchange of the position quantum numbers $\mathbf{r}_1$ and $\mathbf{r}_2$, because the Hamiltonian is symmetric under exchange of the electron positions. The exclusion principle in equation (41.7) says the spin wave function $\chi_{m_1 m_2}$ has to be odd when the spatial part is even, and χ_{m_1,m_2} has to be even when the spatial part is odd.

In the ground state of helium, the space part $\psi(\mathbf{r}_1, \mathbf{r}_2)$ is even (as in eq. [39.2]), so χ has to be odd:

$$\chi_{m_1,m_2} = -\chi_{m_2,m_1}. \tag{41.8}$$

The solution to this functional equation (up to a phase factor) is

$$\chi_{m_1,m_2} = \frac{1}{2^{1/2}}(\delta_{m_1,1/2}\delta_{m_2,-1/2} - \delta_{m_2,1/2}\delta_{m_1,-1/2}). \tag{41.9}$$

This is equivalent to the expression

$$\frac{|+-\rangle - |-+\rangle}{2^{1/2}} \tag{41.10}$$

for the singlet spin zero state (eq. [25.19]). That is, the ground state of helium has zero net spin.

The term symbol used to describe the ground state of a helium atom is 1^1S. The letter S (for $l = 0$) indicates that, in the approximation of the wave function as a product of a spin wave function and a function of position, the electrons have zero net orbital angular momentum. The superscript means the state has a singlet spin wave function, that is, zero net spin. The prefactor means that, in the approximation of the space part of the wave function as a product of two single particle wave functions, both the radial wave functions have no nodes (no zeros apart from the one at $r = 0$.) The term symbol for the even $1s2s$ configuration in equation (40.17) is 2^1S, the prefactor meaning that in the approximation of equation (40.17) the radial wave function for the excited electron has one node.

For the odd $1s2s$ configuration in equation (40.17), the exclusion principle in equation (41.7) says the spin wave function χ_{m_1,m_2} has to be even. The three even functions of m_1 and m_2 are the triplet functions in equation (25.19), with total spin $s = 1$ and z components $m_s = 1$, 0, and -1. The term symbol for this triplet of states is 2^3S. The term symbols for the $1s2p$ configurations in equation (40.4) similarly are 2^1P and 2^3P for the even and odd spatial functions, where the spin function has to be respectively odd (singlet) and even (triplet). The letter P indicates the electrons have unit orbital angular momentum.

The term symbol assigns to a state of the atom orbital and spin angular momentum quantum numbers. This is correct in the approximation that the Hamiltonian H contains no spin observables, as in equation (39.1), so the eigenstates of H can be classified by definite spin and orbital angular momentum quantum numbers. In the full Hamiltonian for an atom spins do appear in the relativistic spin-orbit coupling to be discussed in the next section, so spin and orbital angular momentum are not separately conserved. That means an energy eigenstate does not really have definite spin and orbital angular momentum, but rather is a linear combination of wave functions with different spin and orbital angular momenta that add up to the conserved total angular momentum of the state. However, in low atomic weight atoms the spin-orbit interaction is weak, so states are well approximated by wave functions with definite spin and orbital angular momentum. In this case, the states are conveniently labeled by the term symbol

$$ {}^{2s+1}T_j. \tag{41.11} $$

The multiplicity in the superscript is the number of spin components belonging to the total electron spin, s. The letter represents the total orbital angular momentum, S for $l = 0$, P for $l = 1$, D for $l = 2$. The subscript j is the quantum number for total angular momentum, $\mathbf{J} = \mathbf{L} + \mathbf{s}$.

Because spin is so close to being conserved in helium, atoms with spin 0 and spin 1 act as different kinds of helium, called respectively parahelium and orthohelium. In the spectrum of helium it is almost impossible to see lines corresponding to transitions between the ortho and para states. Such radiative transitions require a matrix element $\langle f|V|i\rangle$ that connects spin 0 and 1. The spins are coupled to the magnetic part of the radiation field, as we saw in the radiative transitions among hyperfine states in hydrogen, but the coupling is weak and further suppressed by the fact that the two electrons see almost the same magnetic field, and a field difference is needed to change the total spin (by making one spin precess at a different rate from the other). A result is that the lowest energy ${}^3S$ state, at $\sim 20\,\mathrm{eV}$ above the ground state, is metastable, having nowhere it can readily decay by emission of radiation. Instead, helium in the ^3S state tends to be deexcited by collisions with other atoms.

Two general features of the exclusion principle are worth noting. First, although the spins make only a very weak contribution to the Hamiltonian for helium, the lowest energy state with spin one is $\sim 20\,\mathrm{eV}$ above the spin zero ground state, which is a considerable difference. This is a result of the symmetry imposed on the spatial part of the wave function by the exclusion principle. Second, an electron arriving as a cosmic ray particle from a distant galaxy has to have a wave function antisymmetric with respect to the local electrons, even though the new electron has been away from us for a long time (and in the inflation paradigm for the early universe never before was in causal contact with local electrons). As in the EPR effect discussed in section 27, this curious result cannot be used for acausal communication: the only way to see the antisymmetry is to bring the electrons together.

42 Lithium

Wave Functions

The dominant terms in the Hamiltonian for atomic lithium are the kinetic and electric potential energies, as in equation (39.1) for helium. In the approximation that the weak spin-dependent terms to be discussed below are ignored, the energy eigenstates are products of a spatial wave function for the three electrons and a spin wave function, as in equation (41.4). The spatial part is usefully approximated as a product of single-particle functions. In these approximations, a solution to Schrödinger's equation for the ground state energy is of the form

$$\psi = \phi_{1s}(\mathbf{r}_1)\phi_{1s}(\mathbf{r}_2)\phi_{2s}(\mathbf{r}_3)\chi_{m_1,m_2,m_3}. \tag{42.1}$$

The electron configuration in this expression is written $1s^2 2s$, meaning two of the electrons have wave functions ϕ_{1s} like the hydrogen atom ground state function, and one has a wave function ϕ_{2s}, like the excited $2s$ wave function in hydrogen with zero orbital angular momentum and principal quantum number $n = 2$.

As in equation (41.7), the exclusion principle says the wave function has to be antisymmetric under exchange of the quantum numbers belonging to any two of the electrons,

$$\psi(\mathbf{r}_1, m_1, \mathbf{r}_2, m_2, \mathbf{r}_3, m_3) = -\psi(\mathbf{r}_2, m_2, \mathbf{r}_1, m_1, \mathbf{r}_3, m_3), \tag{42.2}$$

and so on. To make the trial wave function (42.1) satisfy the exclusion principle, we have to add or subtract the results of exchanging the quantum numbers of the electrons, so as to make the function completely antisymmetric. This yields the wanted approximation to the ground state function,

$$\begin{aligned}\psi = \frac{1}{3^{1/2}}\Big[&\phi_{1s}(\mathbf{r}_1)\phi_{1s}(\mathbf{r}_2)\chi^{\text{singlet}}_{m_1,m_2}\phi_{2s}(\mathbf{r}_3)\chi_{m_3}+ \\ &\phi_{1s}(\mathbf{r}_2)\phi_{1s}(\mathbf{r}_3)\chi^{\text{singlet}}_{m_2,m_3}\phi_{2s}(\mathbf{r}_1)\chi_{m_1}+ \\ &\phi_{1s}(\mathbf{r}_3)\phi_{1s}(\mathbf{r}_1)\chi^{\text{singlet}}_{m_3,m_1}\phi_{2s}(\mathbf{r}_2)\chi_{m_2}\Big].\end{aligned} \tag{42.3}$$

The last factor χ_m in each term in this expression is an arbitrary function of one spin component. The other spin wave function, $\chi^{\text{singlet}}_{m1,m2}$, is the antisymmetric singlet two-spin function with total spin zero, as in equation (41.9). The antisymmetry of this latter function makes each term in equation (42.3) antisymmetric under the exchange of the first pair of arguments. As you can check, it also makes the wave function ψ completely antisymmetric.

The normalization in equation (42.3) assumes the single particle functions are orthogonal and normalized, so we have

$$\begin{aligned}&\int d^3r|\phi_{1s}|^2 = 1, \\ &\int d^3r|\phi_{2s}|^2 = 1, \\ &\int d^3r\phi^*_{2s}\phi_{1s} = 0.\end{aligned} \tag{42.4}$$

Then in the expression for the inner product of ψ with itself the cross terms vanish and the sum of the square terms cancels the factor 1/3, to make $(\psi, \psi) = 1$.

The total spin in the wave function (42.3) is 1/2, because in each term two of the electrons couple to zero spin. Since the orbital angular momentum is $l = 0$, the total angular momentum is $j = 1/2$, and the term symbol (41.11) is $1s^2 2s\ ^2S_{1/2}$.

The configuration $1s^3$, with all three electrons placed in ϕ_{1s} functions, is not allowed because the antisymmetrization in equation (42.3) would make the function vanish. That is why one of the electrons has to be promoted to the $2s$ single particle function.

The first excited states in lithium have electron configuration $1s^2 2p\ ^2P_{1/2,3/2}$, with ϕ_{2s} in equation (42.3) replaced by ϕ_{2p}. The total spin is $s = 1/2$, the orbital angular momentum is $l = 1$, and by the triangle rule (17.74) these can couple to total angular momentum $j = 1/2$ and $3/2$. In hydrogen, the $2s$ and $2p$ wave functions have the same energy. As was discussed in section 40, the $1s^2 2s$ configuration in lithium is the ground state, with lower energy than the $1s^2 2p$ configuration, because the ϕ_{2s} wave function is less well shielded from the charge of the nucleus by the two inner $1s$ electrons.

In the $1s^2 2p$ configuration, and the higher energy configurations $1s^2 3s$, $1s^2 3p$, $1s^2 3d$, and so on, the outer electron sees the nucleus of charge $3e$ closely surrounded by the two $1s$ electrons. This acts as a nearly pointlike charge e, so the wave function for the outer electron is close to that of a hydrogen atom with charge e. Thus, the energy difference between the $1s^2 3d$ and $1s^2 2p$ configurations in lithium corresponds to a wavelength of 6104 Å, just slightly shorter than the wavelength of the $3d$ to $2p$ transition in atomic hydrogen, 6563 Å.

The configurations with orbital angular momentum greater than unity are split into pairs of states with slightly different energies. This fine-structure is caused by the spin-orbit interaction to be discussed now.

Spin-Orbit Interaction

The expression for the Hamiltonian as a sum of kinetic and potential energy terms (as in eq. [39.1]) has to be corrected for relativistic effects, one of which introduces an interaction term between the electron spin and its orbital motion. This term is derived in chapter 8; a crude approach (that is off by a factor of two), goes as follows.

Recall that a particle moving through an electric field sees a magnetic field. This is because the charges that are the source of the electric field are moving in the rest frame of the particle. For example, an observer moving parallel to a line of charge sees a current, which produces a magnetic field. To order v/c in the velocity, the magnetic field seen by a particle moving with velocity $\mathbf{v}$ through an electric field $\mathbf{E}$ is

$$\mathbf{B} = \mathbf{E} \times \mathbf{v}/c. \tag{42.5}$$

As discussed in section 23, an electron has magnetic dipole moment

$$\vec{\mu} = -\frac{ge}{2mc}\mathbf{s}, \tag{42.6}$$

and the energy of this dipole moment in a magnetic field $\mathbf{B}$ is (eqs. [23.9] and [23.11])

$$V = -\vec{\mu} \cdot \mathbf{B}. \tag{42.7}$$

In the approximation that the electron sees a spherically symmetric charge distribution, we can write the electrostatic potential as a function $\phi(r)$ of distance r from the nucleus, so the electric field is

$$\mathbf{E} = -\nabla\phi = -\frac{\phi'\mathbf{r}}{r}, \tag{42.8}$$

with $\phi' = d\phi/dr$. Collecting, and setting $g = 2$, we get

$$V = -\frac{e}{mc^2}\frac{\phi'}{r}\mathbf{r} \times \mathbf{v} \cdot \mathbf{s}. \tag{42.9}$$

Since the angular momentum of the particle in classical mechanics is $\mathbf{L} = m\mathbf{r} \times \mathbf{v}$, we can rewrite this as

$$V = -\frac{e}{m^2c^2}\frac{\phi'}{r}\mathbf{L} \cdot \mathbf{s}. \tag{42.10}$$

The only problem with this expression is that it is off by a factor of two. The full relativistic calculation in chapter 8 gives

$$V = -\frac{e}{2m^2c^2}\frac{\phi'}{r}\mathbf{L} \cdot \mathbf{s}. \tag{42.11}$$

This form, with $\mathbf{L}$ and $\mathbf{s}$ operators, is the quantum expression for the spin-orbit interaction.

In the approximation that the electron sees a nearly pointlike charge e, as for the $2p$ electron in lithium, the electrostatic potential is $\phi = e/r$, and equation (42.11) becomes

$$V = \frac{1}{2m^2c^2}\frac{e^2}{r^3}\mathbf{L} \cdot \mathbf{s}. \tag{42.12}$$

Now to find the fine-structure in the $1s^2 2p\ ^2P_{1/2,3/2}$ configuration in lithium in linear perturbation theory, we must compute the expectation value of this expression in a suitable approximation to the electron wave function.

Fine-Structure

The useful first approximation to the wave function in the $1s^2 2p\ ^2P$ configuration is similar to equation (42.3):

$$\psi = \frac{1}{3^{1/2}}\Big[\phi_{1s}(\mathbf{r}_1)\phi_{1s}(\mathbf{r}_2)\chi^{\text{singlet}}_{m_1,m_2}\phi_{2p}(\mathbf{r}_3)\chi_{m_3} + \phi_{1s}(\mathbf{r}_2)\phi_{1s}(\mathbf{r}_3)\chi^{\text{singlet}}_{m_2,m_3}\phi_{2p}(\mathbf{r}_1)\chi_{m_1} + \phi_{1s}(\mathbf{r}_3)\phi_{1s}(\mathbf{r}_1)\chi^{\text{singlet}}_{m_3,m_1}\phi_{2p}(\mathbf{r}_2)\chi_{m_2}\Big]. \tag{42.13}$$

The spin-orbit interaction (42.12) for the three electrons is of the form

$$V = \sum f(r_i)\mathbf{L}_i \cdot \mathbf{s}_i, \tag{42.14}$$

where the function f for the $2p$ electron is

$$f = \frac{1}{2m^2c^2}\frac{e^2}{r^3}. \tag{42.15}$$

The computation of the expectation value of V in the wave function ψ in equation (42.13) is greatly simplified by the fact that $\mathbf{L}_1$ operating on $\phi_{1s}(r_1)$ vanishes, because ϕ_{1s} is spherically symmetric. That means $f(r_1)\mathbf{L}_1 \cdot \mathbf{s}_1$ operating on ψ in equation (42.13) leaves only one term, where the argument of ϕ_{2p} is $\mathbf{r}_1$. As you can check, the inner product of $f(r_1)\mathbf{L}_1 \cdot \mathbf{s}_1\psi$ with ψ has vanishing cross terms. The sum over the three operators in the potential eliminates the factor $1/3$ from the prefactor in ψ, giving finally

$$\langle V\rangle = \int d^3r_1 f(r_1)\phi^*_{2p}(r_1)\mathbf{L}_1\phi_{2p}(r_1) \cdot \sum \chi^*_{m_1}\frac{\hbar}{2}\vec{\sigma}_{m_1,m_1'}\chi_{m_1'}. \tag{42.16}$$

In the last factor, $\hbar\vec{\sigma}/2$ is the Pauli spin matrix representation of the spin operator. The two electrons in the inner helium-like shell make no contribution to the fine-structure.

The trick in simplifying equation (42.16) was used in computing the hyperfine structure of hydrogen (eq. [34.39]). The total angular momentum operator for the $2p$ electron is the sum of the orbital and spin parts,

$$\mathbf{J}_1 = \mathbf{L}_1 + \mathbf{s}_1. \tag{42.17}$$

The square of this vector operator is

$$J_1^2 = L_1^2 + s_1^2 + 2\mathbf{L}_1 \cdot \mathbf{s}_1. \tag{42.18}$$

Thus we can rewrite the spin-orbit interaction term for the $2p$ electron as

$$f(r_1)\mathbf{L}_1 \cdot \mathbf{s}_1 = f(r_1)(J_1^2 - L_1^2 - s_1^2)/2. \tag{42.19}$$

This form makes it apparent that V commutes with each component of the total angular momentum $\mathbf{J} = \sum \mathbf{J}_i$ (for we recall that, for example, L_x commutes with L^2 and of course it commutes with s^2). This is a good thing, because the total angular momentum ought to be conserved.

The $2p$ electron has spin $s = 1/2$ and orbital angular momentum quantum number $l = 1$. Because the square of the total angular momentum J_1^2 commutes with L_1^2 and s_1^2 we can arrange that the $2p$ electron also is an eigenstate of J_1^2 with eigenvalue $\hbar^2 j(j+1)$. The triangle rule in section 17 says the allowed values of the total angular momentum quantum number j are $|l - s|$ to $l + s$ (eq. [17.74]). For a $2p$ electron this is $j = 1/2$ and $j = 3/2$. The two $1s^2 2p$ energy levels thus are $^2P_{1/2}$ and $^2P_{3/2}$, (in the notation of eq. [41.11]).

Equation (42.18) says a state with quantum numbers j, $l = 1$, and $s = 1/2$ is an eigenstate of $\mathbf{L}_1 \cdot \mathbf{s}_1$ with eigenvalue

$$\frac{\hbar^2}{2}(j^2 + j - l^2 - l - s^2 - s) = \frac{\hbar^2}{2}(j^2 + j - 11/4). \tag{42.20}$$

This means the matrix elements of $\mathbf{L} \cdot \mathbf{s}$ in the degenerate energy eigenstates with different total angular momentum are diagonal, so can use ordinary first-order perturbation theory, where the energy perturbation is the expectation value of V. On replacing the operator $\mathbf{L}_1 \cdot \mathbf{s}_1$ in equation (42.16) with the eigenvalue in equation (42.20), we get

$$\delta E_j = \frac{\hbar^2}{2}(j^2 + j - 11/4) \int d^3r |\phi_{2p}(r)|^2 f(r). \tag{42.21}$$

This is the wanted expression for the perturbation to the energy of the state from the $2s^2 2p$ configuration with total angular momentum quantum number j.

To evaluate this expression for δE_j, we need the $2p$ wave function $\phi_{2p}(r)$. Since we have observed that the $2p$ electron behaves very much as it does in a hydrogen atom, it is a reasonable approximation to use the hydrogen atom $2p$ wave function with charge e (eq. [18.34]),

$$\phi_{2p} = \frac{r e^{-r/(2a_o)} Y_1^m}{(24)^{1/2} a_o^{5/2}}. \tag{42.22}$$

This function is normalized. (It will be recalled that $\int d\Omega \, |Y_l^m|^2 = 1$.) The Bohr radius is $a_o = \hbar^2/(me^2)$. With equation (42.22), the integral in equation (42.21) is easy to work. On collecting all the pieces, one finds that the energy difference between the $j = 3/2$ and $j = 1/2$ states $1s^2 2p \; {}^2P_j$ is (in the approximation of the wave functions in eqs. [42.13] and [42.22])

$$E_{3/2} - E_{1/2} = \frac{\alpha^4 mc^2}{32}. \tag{42.23}$$

This is a relatively small energy because the fine-structure constant α (eq. [34.44]) enters to a high power.

It is customary to express the fine-structure energy difference in units of wavenumbers. (The wavenumber λ^{-1} is the reciprocal of the wavelength of a photon that would be produced or absorbed in a radiative transition between the states. The frequency is c/λ, so the energy difference is hc/λ.) The numerical result for the fine-structure splitting in the $1s^2 2p$ configuration in lithium, in our approximation, is

$$E_{3/2} - E_{1/2} = 0.36 \, \mathrm{cm}^{-1}, \tag{42.24}$$

close to the measured value. This can be compared to the energy difference between the $1s^2 2p \; {}^2P$ and ground state $1s^2 2s \; {}^2S$ configurations in lithium, $15000 \, \mathrm{cm}^{-1}$.

43 Beryllium to Carbon*

The rich structures of the heavier atoms are well beyond the scope of this book, but is worth pausing to look at an interesting calculation that appears at carbon.

The atom after lithium in the periodic table is beryllium, with charge in the nucleus $z = 4$. The four electrons in the ground state have the configuration $1s^2 2s^2$. In the approximation of the wave function as products of single-particle functions, the spatial part is symmetric under exchange of the position variables of the two $2s$ electrons, so the exclusion principle says the spin wave function for these two electrons has to be the antisymmetric singlet function with zero total spin. The term symbol (eq. [41.11]) for the ground state thus is

$$1s^2 2s^2\,{}^1S_0. \qquad (43.1)$$

The first excited states of beryllium, at $\sim 2.7\,\text{eV}$ above the ground state, are

$$1s^2 2s 2p\,{}^3P_{0,1,2}. \qquad (43.2)$$

As for helium and lithium, this configuration has higher energy than the $1s^2 2s^2$ ground state in equation (43.1) because the other three electrons in beryllium more strongly shield the $2p$ electron from the nucleus than the $2s$ electron with its larger wave function at small radius. In the ${}^3P$ states in equation (43.2) the spin part of the wave function is the even triplet function with $s = 1$, so the orbital part is antisymmetric under exchange of the positions of the $2s$ and $2p$ electrons.

The subscripts in equation (43.2) are the values of the total angular momentum quantum number, j. Since the total spin of the atom is $s = 1$, and the orbital angular momentum is $l = 1$, the triangle rule (eq. [17.74]) says the total angular momentum can be $j = 0$, 1, or 2. The spin-orbit interaction energy increases with increasing j in this atom, as in lithium (eq. [42.21]), so $3P_0$ has the lowest energy.

The next states in beryllium, at $\sim 5.3\,\text{eV}$ above the ground state, are $1s^2 2s 2p\,{}^1P_1$. These have the same electron configuration as in equation (43.2), but higher energy because the coulomb energy of repulsion of the two electrons is larger in the spatially symmetric ${}^1P_1$ states.

In boron, with $z = 5$, the ground state configuration of the five electrons is $1s^2 2s^2 2p$. The orbital angular momentum is $l = 1$, and the spin is $s = 1/2$, so the total angular momentum of the atom can be $j = 1/2$ and $3/2$. Because of the spin-orbit coupling energy, which varies as $j(j+1)$, the ground state is ${}^2P_{1/2}$; the state ${}^2P_{3/2}$ is $\sim 0.002\,\text{eV}$ above the ground state.

In carbon, with $z = 6$, the lowest energy configuration is $1s^2 2s^2 2p^2$. Each $2p$ electron has $l_i = 1$ and $s_i = 1/2$, so the triangle rule says the total orbital angular momentum can be $l = 0$, 1, and 2, the total spin can be $s = 0$ and 1, and the total angular momentum can be $j = 0$, 1, 2, or 3. We can understand the actual values of s, l, and j allowed by the exclusion principle by constructing properly antisymmetric angular momentum eigenstates out of the single-particle functions, as follows.

We can assign each $2p$ electron to any of three spatial functions, for the three possible values of m for $l = 1$. Let us write these functions as

$$u^m(r), \qquad m = -1,\ 0,\ 1. \tag{43.3}$$

Consider the state with approximate wave function

$$\psi_1 \propto u^1(r_1)u^0(r_2) - u^0(r_1)u^1(r_2). \tag{43.4}$$

This is supposed to be multiplied by functions ϕ_{1s} and ϕ_{2s} for the inner four electrons, and by a suitable spin wave function. Let $\mathbf{L}(1)$ and $\mathbf{L}(2)$ be the single-particle orbital angular momentum operators for the two $2p$ electrons, and let $\mathbf{L} = \mathbf{L}(1)+\mathbf{L}(2)$ be the total orbital angular momentum operator for these two electrons. In the usual way, the z component is

$$L_z = L_z(1) + L_z(2), \tag{43.5}$$

and the raising and lowering operators are

$$L_\pm = L_\pm(1) + L_\pm(2). \tag{43.6}$$

Since ψ_1 in equation (43.4) is odd under exchange of the two $2p$ electrons, it satisfies

$$L_+\psi_1 = 0, \qquad L_z\psi_1 = \hbar\psi_1. \tag{43.7}$$

This means the total orbital angular momentum in the state ψ_1 is $l = 1$. Since ψ_1 is odd it has to be multiplied by an even spin wave function, which is the triplet function with $s = 1$. This means ψ_1 approximates states with $l = s = 1$, giving

$$^3P_{0,1,2}. \tag{43.8}$$

The subscripts are the values of the total angular momentum, j, obtained by adding spin $s = 1$ and orbital angular momentum $l = 1$. The spatial part ψ_1 of the wave function in equation (43.4) has $m = 1$. The operator L_- applied ψ_1 gives the spatial wave functions for $m = 0$ and -1.

Next, let us seek states with total spin $s = 0$. Here the spin wave function is odd, so the space part has to be even. One possibility is

$$\psi_2 = u^1(r_1)u^1(r_2). \tag{43.9}$$

This satisfies

$$L_+\psi_2 = 0, \qquad L_z\psi_2 = 2\hbar\psi_2, \tag{43.10}$$

so the total angular momentum is $l = 2$. With $s = 0$ and $l = 2$ we get the states

$$^1D_2, \tag{43.11}$$

with total angular momentum $j = 2$.

Finally, consider the symmetric linear combination of single-particle functions

$$\psi_0 = a\,u^1(r_1)u^{-1}(r_2) + b\,u^0(r_1)u^0(r_2) + a\,u^{-1}(r_1)u^1(r_2). \tag{43.12}$$

If the ratio of the constants, a/b, is chosen so L_+ operating on this combination gives zero, then we know we have chosen the linear combination with zero orbital angular momentum. Since this symmetric combination requires a singlet spin wave function, we get the state

$$^1S_0, \tag{43.13}$$

with zero spin, zero orbital angular momentum, and $j = 0$.

A counting argument tells us we have exhausted the possibilities. Each of the $2p$ electrons can be assigned one of three possible values for the z component of orbital angular momentum, $m_l = 1$, 0, and -1, and one of two possible values for the spin component, $m_s = 1/2$ and $-1/2$. That is, there are $3 \times 2 = 6$ different possible assignments of single-particle functions. The exclusion principle says each electron must have a different assignment, so the number of different possible assignments is $6 \times 5/2 = 15$. (The product is divided by 2 because it can't matter which of the identical electron has which assignment.) We can find linear combinations of these assignments that have definite values of l, s, and j. Since the states with definite l, s, and j are orthogonal, the linear combinations that produce them must be linearly independent. Therefore, there are just 15 possible states to be produced out of independent combinations of the 15 assignments. One is the $^1S_0$ state in equation (43.13). The $^1D_2$ state in equation (43.11) has $j = 2$, so there are 5 possible values for m_j. The $^3P_{0,1,2}$ states in equation (43.8) include one with $j = 0$, three with $j = 1$, and five with $j = 2$. These add to 15, so we know we are done.

The $^3P_0$ state has the lowest energy, because the antisymmetry of the spatial part of the wave function for the two $2p$ electrons lowers their coulomb repulsion energy. Just above $^3P_0$ are the $^3P_1$ and $^3P_2$ states at 0.002 eV and 0.005 eV above the ground state, their energies being increased by the spin-orbit interaction. The $^1D_2$ states are 1.26 eV above the ground state, the $^1S_0$ another 1.42 eV higher.

44 Molecular Hydrogen

Born-Oppenheimer Approximation

In a hydrogen molecule we have to deal with four particles: two electrons and two protons. The standard approximation that greatly simplifies the analysis commences by writing the Hamiltonian as

$$H = K_p + H_e. \tag{44.1}$$

The first term is the kinetic energy of the protons,

$$K_p = \frac{p_a^2}{2M} + \frac{p_b^2}{2M} = -\frac{\hbar^2}{2M}(\nabla_a^2 + \nabla_b^2). \tag{44.2}$$

The proton mass is M, and the two proton momenta are $\mathbf{p}_a$ and $\mathbf{p}_b$. The part H_e contains the kinetic energies of the electrons and all the potential energy terms:

$$\begin{aligned} H_e &= \frac{p_1^2}{2m} + \frac{p_2^2}{2m} + V, \\ V &= -\frac{e^2}{|\mathbf{r}_1 - \mathbf{r}_a|} - \frac{e^2}{|\mathbf{r}_1 - \mathbf{r}_b|} - \frac{e^2}{|\mathbf{r}_2 - \mathbf{r}_a|} - \frac{e^2}{|\mathbf{r}_2 - \mathbf{r}_b|} \\ &\quad + \frac{e^2}{|\mathbf{r}_1 - \mathbf{r}_2|} + \frac{e^2}{|\mathbf{r}_a - \mathbf{r}_b|}. \end{aligned} \tag{44.3}$$

The electron mass is m and the electron positions are $\mathbf{r}_1$ and $\mathbf{r}_2$. The proton positions are $\mathbf{r}_a$ and $\mathbf{r}_b$. The first four terms in the potential are the energies of each electron in the field of each proton, and the last two are the energies of interaction of the electrons and of the protons.

The first step in the Born-Oppenheimer approximation is to imagine that the proton positions $\mathbf{r}_a$ and $\mathbf{r}_b$ are fixed and consider a solution to the wave equation

$$H_e\phi = E_e(r)\phi. \tag{44.4}$$

As indicated, the eigenvalue E_e of the electron energy operator H_e (eq. [44.3]) is a function of the separation $r = |\mathbf{r}_a - \mathbf{r}_b|$ of the protons. The wave function ϕ depends on the positions of the electrons relative to the protons:

$$\phi = \phi(\mathbf{r}, \mathbf{r}_1 - \mathbf{r}_a, \mathbf{r}_2 - \mathbf{r}_a), \tag{44.5}$$

where $\mathbf{r} = \mathbf{r}_a - \mathbf{r}_b$ is the vector between the positions of the two protons.

Now we seek an approximate solution to the energy eigenvalue equation for the full system,

$$H\Psi = E\Psi, \tag{44.6}$$

of the form

$$\Psi = \psi(\mathbf{r}_a, \mathbf{r}_b)\phi. \tag{44.7}$$

In the Born-Oppenheimer approximation, the function ψ depends only on the positions $\mathbf{r}_a$ and $\mathbf{r}_b$ of the protons. The proton kinetic energy part K_p of H (eq. [44.2]) operating on this function gives

$$K_p\Psi = \phi K_p\psi - \frac{\hbar^2}{M}\nabla_a\phi \cdot \nabla_a\psi - \frac{\hbar^2}{M}\nabla_b\phi \cdot \nabla_b\psi + \psi K_p\phi. \tag{44.8}$$

The second derivatives with respect to position in K_p differentiate either ψ or ϕ or both. Because the protons are considerably more massive, their momenta are larger than the electron momenta, and since the momentum operator is proportional to a space derivative this means the space derivatives of the proton wave function ψ are larger than the derivatives of the electron function ϕ. Therefore, the dominant term in the right-hand side of equation (44.8) is the first, $\phi K_p\psi$, containing the second derivatives of the proton function. In the Born-Oppenheimer approximation one drops the sub-dominant terms, leaving

$$K_p\Psi = K_p\psi\phi = \phi K_p\psi. \tag{44.9}$$

Then Schrödinger's equation with $H = K_p + H_e$ becomes

$$H\Psi = (K_p + H_e)\psi\phi = \phi K_p\psi + \psi H_e\phi = E\psi\phi. \tag{44.10}$$

Since H_e (eq. [44.3]) has derivatives only with respect to the electron positions $\mathbf{r}_1$ and $\mathbf{r}_2$, we can move $\psi(\mathbf{r}_a, \mathbf{r}_b)$ past H_e in the second expression in equation (44.10). We have to keep the second space derivatives of ϕ in this second term, because they have a small denominator, the electron mass.

The final step is to use equation (44.4) to replace $H_e\phi$ with $E_e(r)\phi$. This makes the electron wave function ϕ a common factor that can be divided out of Schrödinger's equation (44.10), leaving

$$K_p\psi + E_e(r)\psi = E\psi. \tag{44.11}$$

This says the proton part of the wave function, $\psi(\mathbf{r}_a, \mathbf{r}_b)$, acts like a two-body system with the potential energy $E_e(r)$ found by solving equation (44.4).

To summarize, the idea behind the Born-Oppenheimer approximation is that the electrons see the more massive protons as slowly moving, so the electron wave function can smoothly adjust to the changing positions of the protons. This is an adiabatic approximation, like the WKB computation in section 7. The prescription that follows from the Born-Oppenheimer approximation has two steps. First, one solves equation (44.4) for the electron wave function ϕ for fixed and given proton positions $\mathbf{r}_a$ and $\mathbf{r}_b$. The eigenvalue $E_e(r)$ is the "electronic energy" of the electrons, kinetic and potential, plus the potential energy $e^2/|\mathbf{r}_a - \mathbf{r}_b|$ of the protons. This energy, which is a function of the separation $r = |\mathbf{r}_a - \mathbf{r}_b|$ of the protons, is the total energy of the molecule when the proton positions are given. In the second step, the energy $E_e(r)$ serves as the potential energy term in the Schrödinger equation (44.11) for the motion of the protons. The eigenvalue E in this last equation is the total energy of the molecule.

Low Energy Electronic States

We can make considerable progress to understanding the solutions to Schrödinger's equation (44.4) for the electron wave function for fixed proton positions $\mathbf{r}_a$ and $\mathbf{r}_b$ by considering the symmetries of the Hamiltonian. Since H_e is symmetric under reflection through the midpoint of $\mathbf{r} = \mathbf{r}_a - \mathbf{r}_b$, the electron eigenfunctions ϕ can be classified as g for even (the German word is gerade) or u for odd (ungerade) under this spatial reflection. Since H_e is symmetric under rotation about the axis $\mathbf{r}$ defined by the protons, the component of the electron angular momentum along

$\mathbf{r}$ is conserved. The states with zero angular momentum along the axis are labeled Σ; there also are low-lying excited states Π with one unit of angular momentum along the axis.

We will consider only the lowest energy states, Σ_g and Σ_u. First approximations to their electron spatial wave functions are

$$\phi_\pm = \phi_{1s}(\mathbf{r}_1 - \mathbf{r}_a)\phi_{1s}(\mathbf{r}_2 - \mathbf{r}_b) \pm \phi_{1s}(\mathbf{r}_1 - \mathbf{r}_b)\phi_{1s}(\mathbf{r}_2 - \mathbf{r}_a), \tag{44.12}$$

where ϕ_{1s} has the form of the ground state hydrogen atom wave function. The combination ϕ_+ is even under exchange of $\mathbf{r}_1$ and $\mathbf{r}_2$, so to satisfy the exclusion principle the spin wave function has to be odd under the exchange of m_1 and m_2. That means this is the singlet $s = 0$ state. The state thus is more completely labeled $^1\Sigma_g$. The odd ϕ_- state is $^3\Sigma_u$.

Application of the variational principle to estimate $E_e(r)$ shows that $^1\Sigma_g$ has lower energy than $^3\Sigma_u$; a qualitative explanation goes as follows. The $^3\Sigma_u$ state, having an odd spatial part, is favored by the lower value of the energy of interaction of the two electrons, $\langle e^2/r_{12}\rangle$. However, that is overbalanced by the fact that in the even $^1\Sigma_g$ state the charge density of the electrons is relatively large between the two protons, as a result of the positive cross term in $|\phi_+|^2$, while in the odd state there is a hole in the mean charge distribution halfway between the protons because the antisymmetry makes ϕ_- vanish when r_1 and r_2 both are near the center of the molecule. Because of this hole, the coulomb energy of repulsion of the two protons makes $E_e(r)$ for $^3\Sigma_u$ always exceed that of two well-separated hydrogen atoms. That is, there is no triplet bound state. In the spatially even $^1\Sigma_g$ state, the concentration of the electron charge between the protons attracts them, lowering the energy. This is the source of the binding energy in a hydrogen molecule.

The electronic energy in the $^1\Sigma_g$ state at proton separations r near the minimum of $E_e(r)$ can be approximated as a parabola,

$$E_e(r) = E_e(r_o) + \frac{1}{2}K(r - r_o)^2, \tag{44.13}$$

where the separation at the minimum is

$$r_o = 0.74\,\text{Å} = 7.4 \times 10^{-9}\,\text{cm}, \tag{44.14}$$

the curvature is

$$K = 35\,\text{eV}\,\text{Å}^{-2}, \tag{44.15}$$

and the minimum value of E_e is

$$E_e(r_o) = -4.746\,\text{eV}. \tag{44.16}$$

This energy is referred to zero energy for two hydrogen atoms well separated and at rest. It is left as an exercise to use these quantities to work out the low-lying energy levels of the molecule.

Problems

VI.1) Suppose a particle of mass m is confined to a rectangular box. The potential energy of the particle is $V = 0$ in the region $0 < x < a$ and $0 < y < b$ and $0 < z < c$, and V is positive and very large elsewhere. The sides are arranged so

$$a < b < c. \tag{VI.1}$$

The potential ensures that the wave function $\psi(\mathbf{r})$ of the particle is negligibly small outside the box.

a) Using the boundary condition on ψ, write down the normalized ground state wave function and the ground state energy of the particle.

Now suppose two of these particles are placed in the box. The particles have spin 1/2, and are identical, so they obey the Pauli exclusion principle. The particles do not interact with each other; they see only the potential V, so the Hamiltonian is

$$H = \frac{p_1^2}{2m} + V(\mathbf{r}_1) + \frac{p_2^2}{2m} + V(\mathbf{r}_2). \tag{VI.2}$$

b) Write down the wave function for the lowest energy singlet state for this two-particle system. Use as a basis the simultaneous eigenstates of the positions and z components of the spins of the two particles, so the wave function is of the form

$$\langle \mathbf{r}_1, \mathbf{r}_2, m_1, m_2 | \psi \rangle = \psi(\mathbf{r}_1, \mathbf{r}_2, m_1, m_2), \tag{VI.3}$$

as in equation (41.4).

c) Write down the wave functions for the three degenerate lowest energy triplet states.

Suppose now the two particles in the box interact by a potential energy term that can be approximated as a Dirac delta function in relative position,

$$V = K\delta(\mathbf{r}_1 - \mathbf{r}_2), \tag{VI.4}$$

where K is a real constant.

d) Find in first-order perturbation theory the energy of the lowest energy triplet state for the two particles.

e) Find in first-order perturbation theory the energy of the lowest energy singlet state.

Suppose now three of these identical particles are placed in the box. Ignoring the interaction energy in equation (VI.4), the Hamiltonian is

$$H = \frac{p_1^2}{2m} + V(\mathbf{r}_1) + \frac{p_2^2}{2m} + V(\mathbf{r}_2) + \frac{p_3^2}{2m} + V(\mathbf{r}_3). \tag{VI.5}$$

f) Write down the wave function for the lowest energy state of the Hamiltonian with total spin 1/2.

g) Write down the spin part of the wave function for each of the four lowest energy states with total spin 3/2 and z components $m = 3/2,\ 1/2,\ -1/2$, and $-3/2$.

h) The interaction energy in equation (VI.4) is generalized to

$$V = K[\delta(\mathbf{r}_1 - \mathbf{r}_2) + \delta(\mathbf{r}_2 - \mathbf{r}_3) + \delta(\mathbf{r}_3 - \mathbf{r}_1)], \tag{VI.6}$$

for three particles. Find, in first-order perturbation theory, the perturbation to the energy due to this interaction for the lowest energy spin 3/2 state in part (g).

i) Find in first-order perturbation theory the perturbation to the energy of the lowest energy spin 1/2 state considered in part (f) due to the interaction in equation (VI.6).

VI.2) Two particles of mass m are in the spherical potential well

$$\begin{aligned} V(r) &= 0, \qquad \text{for} \quad r \le r_o, \\ &\to \infty, \qquad \text{for} \quad r > r_o, \end{aligned} \tag{VI.7}$$

where r_o is a positive constant. The single-particle wave functions and energies for this case were derived in problem (I.12). For two particles that do not interact, the Hamiltonian is

$$H = \frac{p_a^2}{2m} + \frac{p_b^2}{2m} + V(r_a) + V(r_b). \tag{VI.8}$$

Suppose the two particles are identical spin 1 bosons. The symmetry condition in this case is that the wave function is symmetric under exchange of all the quantum numbers belonging to the two particles (rather than antisymmetric, as in eq. [41.5], for fermions such as electrons).

a) Find the allowed value(s) of the total spin of the two spin 1 bosons in the ground state, and write down the ground state wave function(s).

b) Find the allowed value(s) of the total spin in the first excited state(s).

VI.3) A relation between the mean kinetic and potential energies of a system of electrons and nucleii in a stationary state (energy eigenstate) is obtained as follows.

The Hamiltonian, ignoring the spin interactions terms, is of the form

$$H = T + V, \tag{VI.9}$$

where the kinetic energy part is

$$T = \sum_i \frac{p_i^2}{2m_i} = -\sum_i \frac{\hbar^2}{2m_i}\nabla_i^2, \tag{VI.10}$$

and the potential energy part is

$$V = \sum_{i<j} \frac{q_i q_j}{r_{ij}}. \tag{VI.11}$$

The charge and mass of the ith particle are q_i and m_i, and r_{ij} is the distance between the ith and j^{th} particles. The sums are over the n particles in the system.

Consider a trial wave function ψ of the form

$$\psi(\mathbf{r}_1, \mathbf{r}_2, \ldots, \mathbf{r}_n) = \phi(\alpha\mathbf{r}_1, \alpha\mathbf{r}_2, \ldots, \alpha\mathbf{r}_n). \tag{VI.12}$$

Here α is a parameter, that multiplies every position variable in the function ϕ.

a) By introducing a suitable change of variables in the integrals, show that the expectation value of the kinetic energy in the state ψ,

$$\langle E \rangle = \frac{\int d^3r_1\, d^3r_2 \ldots d^3r_n\, \psi^* T \psi}{\int d^3r_1\, d^3r_2 \ldots d^3r_n\, \psi^* \psi}, \tag{VI.13}$$

is proportional to a power of α, and find the power.

b) Do the same for the expectation value $\langle V \rangle$ of the potential energy, and use this with the result from part (a) to find how the expectation value of the energy, $\langle H \rangle$, in the state ψ varies with the parameter α.

c) Suppose that, when $\alpha = 1$, the function ψ is the ground state wave function for the system. Use the above results, with the energy variational principle, to show that the expectation value of the kinetic energy in the ground state is equal to half the negative of the expectation value of the potential energy, and that the expectation value of the potential energy in the ground state therefore is twice the energy of the ground state. The result is called the virial theorem.

VI.4) Consider two well-separated hydrogen atoms at rest, each in the ground state. Ignoring the hyperfine structure, this system is degenerate, corresponding to the two choices $\pm 1/2$ for the spin components for the two electrons and the two protons. Suppose the two protons are gently moved together to a separation ~ 0.7 Å, without any external perturbation to the electron spins, after which the atoms are released at low velocity. The experiment is repeated many times, each time with a random choice among the four degenerate states for the electrons in the originally well-separated hydrogen atoms. Sometimes the system flies apart when the protons are released, sometimes it is observed to stay together as a hydrogen molecule. In what fraction of the trials does the system stay together as a hydrogen molecule?

VI.5) The lowest spin 1 energy levels in the bound state of a b quark and a $\bar{b}$ (anti-b) quark are

$$\begin{aligned} \Upsilon(1^3S) &= 9.46\,\text{GeV}, \quad \chi(1^3P) = 9.89\,\text{GeV}, \\ \Upsilon(2^3S) &= 10.02\,\text{GeV}, \quad \chi(2^3P) = 10.27\,\text{GeV}. \end{aligned} \tag{VI.14}$$

These energies include the rest mass, $2Mc^2$, where M is the b and $\bar{b}$ quark mass, and the potential and internal kinetic energy of motion of the quarks. The ^3S states have been given the name Υ and the ^3P states the name χ. The notation in parentheses follows atomic spectroscopy: the superscript 3 means these are triplet states, with the spins of the two spin 1/2 quarks aligned to spin 1. The S states have zero orbital angular momentum, and the P states have orbital angular momentum quantum number $l = 1$. A spin-orbit interaction splits ^3P into three slightly different energy levels with total angular momentum $j = 0$, 1, and 2, but we can ignore this relatively small effect. Colorful names are traditional in this subject. The $b\bar{b}$ bound states are called the bottomonium system.

The motion of the quarks is mildly relativistic, but it is a reasonable approximation to describe the system in nonrelativistic quantum mechanics, with the Hamiltonian

$$H = 2Mc^2 + \frac{p_1^2}{2M} + \frac{p_2^2}{2M} + \kappa r, \tag{VI.15}$$

where κ is a constant and r is the separation of the b and $\bar{b}$ quarks. The potential energy term is that of a string with tension κ that is independent of the length of the string.

a) Write down the single-particle Hamiltonian for the relative motion of the two quarks, as in section 12, with suitable reduced mass.

b) Since the Hamiltonian for the relative motion of the quarks is spherically symmetric, its eigenstates can be assigned definite orbital angular momentum. The wave function with orbital angular momentum quantum numbers l and m is

$$\psi = \frac{u(r)}{r} Y_l^m(\theta, \phi), \tag{VI.16}$$

where $u(r)$ is the radial wave function. Write down the one-dimensional Schrödinger energy eigenvalue equation for the radial wave function, as in section 18.

c) Sketch a graph showing the potential energy $V(r) = \kappa r$ as a function of separation r, and the radial wave functions $u(r)$ for the two lowest energy s-wave ($l = 0$) states as functions of r. Be sure to indicate reasonable behavior of $u(r)$ at $r \to 0$ and $r \to \infty$, and with the right number of zeros.

d) Do the same for the effective potential (taking account of the angular momentum term) and the radial wave functions for the two lowest energy p-wave states. Show that in the p-wave states $u(r)$ has to approach zero at $r \to 0$ as $u \propto r^2$.

e) The energy variational principle in section 38 says the best choice among a set of trial wave functions ψ for the ground state is the one that minimizes the expectation value of the Hamiltonian,

$$\langle H \rangle = \frac{(\psi, H\psi)}{(\psi, \psi)} = \frac{\int d^3r\, \psi^* H \psi}{\int d^3r\, \psi^* \psi}. \tag{VI.17}$$

Find an expression for this condition in terms of a one-dimensional integral over the radial wave function.

f) Use the trial function

$$u(r) = r\, e^{-\alpha r}, \tag{VI.18}$$

with α an adjustable parameter, to estimate the ground state energy in terms of κ and the quark mass.

g) The effective potential $V_{\text{eff}}(r)$ for the p-wave states has a minimum at radius $r_o > 0$. One can get a quick approximation to the p-wave energy levels by approximating $V_{\text{eff}}(r)$ as a parabola, so the problem reduces to a simple harmonic oscillator with the familiar energy levels $(n + 1/2)\hbar\omega$. One way to fit to the parabola is to expand $V_{\text{eff}}(r)$ in a Taylor series around its minimum at separation r_o:

$$V_{\text{eff}}(r) = V_{\text{eff}}(r_o) + \frac{1}{2}K(r - r_o)^2 + \ldots. \tag{VI.19}$$

Find K for the p-wave states, and use it to find approximate expressions for the energies of the 1^3P and 2^3P states. Use the measured energies from equation (VI.14) to estimate the quark mass M (in GeV) and the string tension κ (in GeV fm^{-1}, where 1 fm $= 10^{-13}$ cm).

h) A better approximation uses the energy variational principle. Adapt equation (VI.18) to take account of the behavior of the radial wave function at small radius in a p-wave state, as discussed in part (d), and use it to estimate the 1^3P energy in terms of κ and M.

i) Use the results from parts (f) and (h), along with the measured energies, to find another (and maybe better) set of estimates of M and κ.

VI.6) As discussed in section 44, the relative motion of the two protons in a hydrogen molecule is well approximated as a one-body problem with potential energy $V(r)$ that is a function of the separation of the protons. Equation (44.13) approximates the shape of $V(r)$ near its minimum as a quadratic form. If the orbital angular momentum of the protons is not zero you can still approximate the effective potential near its minimum as a simple harmonic oscillator potential by expanding the effective potential as a power series in r and keeping only the terms through quadratic.

Find numerical estimates of the following quantities, giving energies in electron Volts and lengths in Ångstroms. Useful constants and conversion factors are listed at the beginning of the problem set in chapter 1.

a) The zero point energy of the protons in H_2.

b) The binding energy in the ground state of H_2.

c) The difference of binding energies of H_2 and HD. (In the latter molecule, one of the protons is replaced with a deuteron with the same charge and twice the mass of a proton.)

d) The energy difference between the first vibrationally excited state ($v = 1$ and $l = 0$) and the ground state ($v = 0$ and $l = 0$) of H_2 (where $E_v = (v + 1/2)\hbar\omega$).

e) The mean separation of the protons in H_2 in the first rotationally excited state, with angular momentum quantum number $l = 1$ and vibrational quantum number $v = 0$.

f) The energy difference between the first rotationally excited state ($v = 0$ and $l = 1$) and the ground state of H_2.

VI.7) Suppose a molecular hydrogen gas is at thermal equilibrium at a temperature such that the mean energy in rotation of the molecules is 1 percent of classical energy equipartion. Use the methods of section 1 to calculate the temperature. (Note that for a complete computation one should evaluate an infinite series, but you can make a sensible approximation.)

CHAPTER 7

SCATTERING THEORY

A powerful and commonly used way to explore the interaction between particles is to study the way they scatter off each other. In the scattering problems to be considered here motions are nonrelativistic and particles are conserved: two particles move together, interact, and then move apart again. It will also be assumed that the range of the interaction is finite, so when the particles are well separated they move freely.

In a scattering experiment, one imagines that the particles approach each other as wave packets with fairly definite momenta and positions (within the limits allowed by the uncertainty principle). The motion initially is free, because the particles are separated by great distances compared to the range of their interaction. As the wave packets move together the particles interact through a potential V that is some function of the particle separation. The wave packets then move apart in a scattering pattern that is determined by the interaction potential. We will be considering the wave function for a scattering state with definite energy. This time-independent function is not a good wave function for a single pair of particles, because it is not normalizable. That is to be expected, because the system has positive energy so there is nothing to confine the wave function: it spreads over all space. But we can form a wave packet as a linear combination of scattering state wave functions with slightly different energies. The wave packet is normalizable and describes the time-dependent scattering process.

45 Scattering Amplitude and Scattering Cross Section

Wave Packets in a Scattering State

The purpose of this section is to develop the mathematical description of the scattering process in terms of wave packets. This involves a boundary condition for the solutions of Schrödinger's equation for a scattering solution with definite energy (in the form of eq. [45.8] below); the boundary condition will be seen to contain a scattering amplitude whose square is the cross section that is measured in scattering experiments.

We can describe the two particles in relative and center of mass coordinates, as discussed in section 12. The wave function for the center of mass motion is just a plane wave (eq. [12.21]). The wave function for the relative motion of the particles satisfies the equation

$$i\hbar\frac{\partial\psi}{\partial t} = -\frac{\hbar^2}{2m}\nabla^2\psi + V(\mathbf{r})\psi, \tag{45.1}$$

where m is the reduced mass, and $\mathbf{r} = \mathbf{r}_1 - \mathbf{r}_2$ is the relative position of the two particles. This is Schrödinger's equation for the motion of a single particle with position observable $\mathbf{r}$ in the potential $V(\mathbf{r})$. That is, we can think of the two-body problem as a single-body problem, in which a particle with mass m and the wave function $\psi(\mathbf{r}, t)$ is scattered by a fixed potential, $V(\mathbf{r})$.

The boundary condition for equation (45.1) differs from that of a negative energy bound state solution, where the wave function approaches zero at $r \to \infty$. In a scattering state solution, the boundary condition is that at sufficiently early times the wave function ψ represents a free wave packet, not yet disturbed by the interaction, moving toward the interaction region in the neighborhood of $r \sim 0$. The problem to be discussed here is the way to characterize the behavior of ψ after the particles have interacted and have moved well apart again.

As a first step, let us recall from section 5 how one constructs a wave packet for the motion of a particle (or for the relative motion of a pair of particles, as in eq. [45.1]). A free particle with definite momentum $\mathbf{p} = \hbar\mathbf{k}$ is described by the plane wave

$$\psi_{\mathbf{k}}(\mathbf{r}, t) = e^{i(\mathbf{k}\cdot\mathbf{r} - \omega t)}, \tag{45.2}$$

where the energy is

$$E = \hbar\omega = \frac{\hbar^2 k^2}{2m}. \tag{45.3}$$

Now we want to write down a wave packet, such that the wave function at a given time is negligibly small outside a relatively small region where we are sure to find the particle. This is done by writing the wave function as a linear superposition of these plane waves,

$$\psi(\mathbf{r}, t) = \int d^3k \; a(\mathbf{k})\psi_{\mathbf{k}}(\mathbf{r}, t), \tag{45.4}$$

where the function $a(\mathbf{k})$ is negligibly small outside an isolated peak in the neighborhood of $\mathbf{k} \sim \mathbf{k}_o$. As discussed in section 5, the oscillation of the exponential function $\psi_{\mathbf{k}}$ makes the integral $\psi(\mathbf{r}, t)$ negligibly small except for those values of $\mathbf{r}$ and t for which the phase $\phi = \mathbf{k} \cdot \mathbf{r} - \omega t$ of $\psi_{\mathbf{k}}$ is stationary under small variations of $\mathbf{k}$ around the value $\mathbf{k}_o$ at the peak of $a(\mathbf{k})$. This condition is

$$\frac{\partial \phi}{\partial \mathbf{k}} = \frac{\partial}{\partial \mathbf{k}}(\mathbf{k} \cdot \mathbf{r} - \omega t) = 0 \quad \text{at} \quad \mathbf{k} = \mathbf{k}_o. \tag{45.5}$$

This says the peak of the wave packet defined in equation (45.4) is at position

$$\mathbf{r} = \mathbf{v}t, \tag{45.6}$$

where the group velocity is

$$\mathbf{v} = \frac{\partial \omega}{\partial \mathbf{k}} = \frac{\hbar \mathbf{k}_o}{m}, \tag{45.7}$$

with $\omega(k)$ given by equation (45.3). Equation (45.6) says that the wave packet for this free particle moves at the group velocity $\mathbf{v}$, which is the classical velocity of a particle with momentum $\mathbf{p} = \hbar \mathbf{k}_o$.

As we will see, to produce the wave packet in a scattering experiment when the particles are well separated, either when they are approaching or after they have interacted and moved well apart again, one replaces the plane wave in equation (45.2) with the function

$$\psi_{\mathbf{k}}(\mathbf{r}, t) = \left[e^{i\mathbf{k}\cdot\mathbf{r}} + f_{\mathbf{k}}(\theta, \phi)\frac{e^{ikr}}{r} \right] e^{-i\omega t}. \tag{45.8}$$

This is the form of a scattering state solution to Schrödinger's equation in the limit of large r, well removed from the interaction region. The scattering amplitude $f_{\mathbf{k}}$ in the second term is a function of the direction of $\mathbf{r}$, where θ and ϕ are polar angles for $\mathbf{r}$ relative to the line defined by

$\mathbf{k}$. This function of θ and ϕ is multiplied by the spherical wave e^{ikr}/r that is a function of radius $r = |\mathbf{r}|$. The radius is computed from the origin of coordinates placed within the interaction region, so we can be sure the interaction potential vanishes at sufficiently large r.

The first term of equation (45.8) in equation (45.4) gives a freely moving wave packet, as before. The second term, with the integral over $\mathbf{k}$ in equation (45.4) written in polar coordinates, is

$$\psi_s = \int d\Omega_k k^2 dk\, a(\mathbf{k}) f_{\mathbf{k}} e^{i(kr-\omega t)}/r. \tag{45.9}$$

The volume element for the integral over $\mathbf{k}$ is $d^3k = k^2 dk d\Omega_{\mathbf{k}}$, with $d\Omega_{\mathbf{k}} = \sin\theta_{\mathbf{k}}\, d\theta_{\mathbf{k}} d\phi_{\mathbf{k}}$ the element of solid angle for the integral over the direction of $\mathbf{k}$. As for the plane wave case discussed above, we note that in the limit of very large r and t the exponential factor $e^{i(kr-\omega t)}$ in the integrand is a rapidly oscillating function of k. This makes the integral, ψ_s, negligibly small unless r and t are such that the phase

$$\phi_s = kr - \omega(k)t \tag{45.10}$$

is at an extremum when k is near the value k_o at the peak of $a(\mathbf{k})$. Thus the condition for ψ_s to be appreciable is

$$0 = \frac{d\phi_s}{dk} = r - \hbar\frac{d\omega}{dt}. \tag{45.11}$$

This says the radius r at the wave packet is

$$r \sim \frac{\hbar k_o}{m} t = vt. \tag{45.12}$$

As before, the frequency ω is given by equation (45.3), so v is the magnitude of the group velocity in equation (45.7).

At times $t < 0$ equation (45.12) cannot be satisfied. (Since r is the distance from the interaction region, r is positive.) Thus ψ_s is negligibly small at $t \ll 0$, so the second term in equation (45.8), that contains the scattering amplitude $f_{\mathbf{k}}$, has no effect: the wave packet at $t \ll 0$ is just what we get from the plane wave from the first term, as if the scattered wave were not present. This means we begin with a pure incoming wave packet, moving as in equation (45.6), as required.

At large and positive times the wave function is the sum of two terms. The first is the wave packet from the plane wave in equation (45.8), now moving away from the interaction region toward increasing r in the direction of $\mathbf{k}_o$. This represents the unscattered part of the wave function. The second term in the wave function is the scattered wave packet ψ_s (eq. [45.9]) that is moving radially away from the interaction region, as indicated in equation (45.12).

Equation (45.9) says the amplitude of the scattered wave ψ_s varies inversely as the distance r from the interaction region. This is required to conserve probability. The probability per unit volume for finding the particle at a position off the axis of the incident motion of the wave packet is $|\psi|^2 = |\psi_s|^2$. (We can ignore the plane wave term in eq. [45.8], because the unscattered wave packet is small off the axis.) As the scattered wave moves to larger r, its area spreads as r^2, so the probability density has to vary as $1/r^2$, in agreement with the factor $1/r$ in ψ_s.

To see this another way, recall from section 8 that the probability flux density for the wave function $\psi(\mathbf{r}, t)$ is (eq. [8.13])

$$\mathbf{j} = -\frac{i\hbar}{2m}(\psi^* \nabla \psi - \psi \nabla \psi^*). \tag{45.13}$$

For a plane wave, $\psi = ce^{i\mathbf{k}\cdot\mathbf{r}}$, with c a constant, the flux density is

$$\mathbf{j} = |c|^2 \hbar \mathbf{k}/m = |c|^2 \mathbf{v}. \tag{45.14}$$

This just says the probability density $|\psi|^2 = |c|^2$ is moving at the group velocity $\mathbf{v}$.

Now consider the probability flux density off the axis $\mathbf{k}_o$ of the incoming beam. We can ignore the cross term between the plane and scattered waves in equation (45.8), because after making a wave packet the former is appreciable only in the undisturbed wave packet that runs along the direction $\mathbf{k}_o$. The scattered part of equation (45.8) is

$$w = f_{\mathbf{k}}(\theta, \phi)e^{ikr}/r. \tag{45.15}$$

The radial component of the flux density, in the direction pointed away from the interaction region, is determined by the radial derivative of w:

$$\frac{\partial w}{\partial r} = f_{\mathbf{k}} e^{ikr} \left(\frac{ik}{r} - \frac{1}{r^2} \right). \tag{45.16}$$

At large r, the second term is negligible compared to the first. The probability flux density normal to the radial direction is fixed by the gradient of w in directions perpendicular to the radial direction. The transverse gradient is proportional to $1/r^2$, which again is negligible compared to the first term in the radial derivative of w when r is large. That is, at large r the scattered flux is directed radially away from the scattering region. This is reasonable: the scattered wave is moving away from the interaction region.

The radial derivative of w from equation (45.16) with equation (45.13) gives the flux density in the scattered wave at large r,

$$j_s = \frac{\hbar k}{m} \frac{1}{r^2} |f_{\mathbf{k}}(\theta, \phi)|^2. \tag{45.17}$$

As expected, this varies with distance from the interaction region as $j_s \propto 1/r^2$, consistent with the fact that the area of the scattered beam spreads as r^2.

To summarize, when solutions to Schrödinger's equation with definite energy and the boundary condition of equation (45.8) at large r are used to make a wave packet, as in equation (45.4), it produces the required situation: an incoming wave packet at large negative times that at large and positive times becomes an unscattered outgoing wave packet plus a radially moving outgoing scattered wave. The latter is determined by the scattering amplitude $f_{\mathbf{k}}$.

The flux density in the scattered wave determines the differential scattering cross section, as follows.

Differential Scattering Cross Section

Consider a scattering experiment with an incident beam that contains n particles per unit volume moving toward a fixed target at speed v. The incident particle flux density is $j_{\text{in}} = nv$ particles passing unit area normal to the beam in unit time. As indicated in Figure 45.1, a detector with area A normal to the scattered wave is placed at distance r from the interaction region in a direction θ, ϕ away from the direction of the incident beam. The number of particles detected in A per unit time, dN/dt, is proportional to the product of the incident flux density and the detector area A, and is inversely proportional to the square of the distance r of the detector from the interaction region, because the area of the scattered wave spreads as r^2. The constant of proportionality is

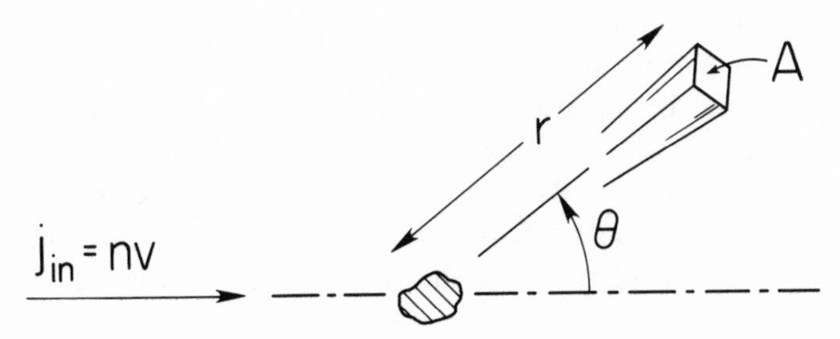

Fig. 45.1 A scattering experiment with a fixed target. The interaction region, where the potential energy is nonzero, is indicated by the shaded area. The beam is incident from the left with n particles per unit volume moving at speed v. The detector is placed at angle θ from the incident direction, and at distance r from the interaction rgegion. The distance r is very large compared to the de Broglie wavelength and the size of the interaction region. The detector has area A normal to the direction to the interaction region; it subtends solid angle $\delta\Omega = A/r^2$ at the interaction region.

the differential scattering cross section, $d\sigma/d\Omega$:

$$\frac{dN}{dt} = j_{\text{in}} \frac{d\sigma}{d\Omega} \frac{A}{r^2}. \tag{45.18}$$

This equation with $j_{\text{in}} = nv$ defines $d\sigma/d\Omega$. The notation $d\Omega$ comes from the fact that the detector subtends solid angle $\delta\Omega = A/r^2$ at the interaction region. The total scattering cross section, σ, is found by summing equation (45.18) over elements of area that cover all directions except forward (as in eq. [45.21] below). In classical physics, σ is the cross section area of the incident beam within which the projectile hits the target.

The incident beam in the wave function in equation (45.8) is represented by the plane wave part. It has a probability density equivalent to one particle per unit volume in the incident beam, so the incident flux density is the incident speed, $j_{\text{in}} = \hbar k/m$, as in equation (45.14). The probability per unit time for detection of the scattered particle is the product of the detector area A and the probability flux j_s per unit area in the scattered beam (eq. [45.17]):

$$\frac{dN}{dt} = j_s A = \frac{\hbar k}{m} \frac{A}{r^2} |f_{\mathbf{k}}(\theta, \phi)|^2. \tag{45.19}$$

Since $j_{\text{in}} = \hbar k/m$ in this wave function, we see from equation (45.18) that the differential scattering cross section is

$$\frac{d\sigma}{d\Omega} = |f_{\mathbf{k}}(\theta, \phi)|^2. \tag{45.20}$$

This is the wanted relation between the scattering cross section, $d\sigma/d\Omega$, which can be measured, and the scattering amplitude, $f_{\mathbf{k}}$. The latter is defined by equation (45.8), which is the asymptotic form of the scattering solution to Schrödinger's equation with definite energy. The scattering amplitude is a function of the polar angles θ and ϕ of the scattered direction relative to the direction of the incident beam. Methods of computing the scattering amplitude are discussed beginning in section 47.

The total scattering cross section, σ, is the integral of the differential cross section over all directions,

$$\sigma = \int d\Omega \frac{d\sigma}{d\Omega}. \tag{45.21}$$

The element of solid angle is $d\Omega = dA/r^2$, as in equation (45.18). The integral has to exclude the forward direction, where the scattered wave overlaps the unscattered wave packet, but by taking r large enough we can make the excluded solid angle negligibly small. The interference between the forward part of the scattered wave and the unscattered wave packet nevertheless is interesting, as is discussed next.

46 Optical Theorem*

The plane wave part of equation (45.8) is $e^{i\mathbf{k}\cdot\mathbf{r}}$. This represents an incident flux density $j_{\rm in} = \hbar k/m = v$. The net flux of probability in the scattered wave is $v\sigma$, where σ is the total scattering cross section defined in equations (45.18) and (45.21). Therefore, the net flux in the outgoing unscattered wave must be reduced from the incoming flux by the amount $v\sigma$. This comes about as follows.

The undeflected outgoing flux comes from the sum of the incident plane wave and the forward part of the scattered wave (at zero scattering angle). Let us take the plane wave to be moving in the direction of the positive z axis, so the wave function (45.8) well away from the interaction region and in directions very near forward is

$$\psi_f = e^{ikz} + f(0)e^{ikr}/r. \tag{46.1}$$

The second term in this expression has to reduce the net flux in the forward direction by the amount σv. To compute this flux, write the radius r in equation (46.1), for directions close to forward, as

$$r = (z^2 + x^2)^{1/2} = z + \frac{1}{2}\frac{x^2}{z}, \tag{46.2}$$

to lowest nontrivial order in perpendicular distance x from the beam axis. This brings the wave function in equation (46.1) in directions very near forward to

$$\psi_f = e^{ikz}[1 + f(0)e^{ikx^2/(2z)}/z]. \tag{46.3}$$

We are taking the limit $z \gg x$, so we need to keep x only in the rapidly varying exponential part of the second term.

Now consider the probability flux in the forward direction. The rapidly varying part of equation (46.3) is the prefactor e^{ikz}, so we can write $\partial\psi_f/\partial z = ik\psi_f$. (The other parts of the derivative have higher powers of $1/z$, so we can ignore them in the limit of large z, as in eq. [45.16]). Then to lowest nontrivial order in $1/z$ we have

$$\begin{aligned}\psi_f^* \frac{\partial\psi_f}{\partial z} &= ik[1 + f(0)^* e^{-ikx^2/(2z)}/z][1 + f(0)e^{ikx^2/(2z)}/z] \\ &= ik[1 + 2\mathrm{Re} f(0)e^{ikx^2/(2z)}/z].\end{aligned} \tag{46.4}$$

The sum of the cross terms gives the real part of $f(0)e^{ikx^2/(2z)}$. The last term in the product can be dropped because it has a higher power of $1/z$. The probability flux density (eq. [45.13]) in the forward direction is then

$$j_f = \frac{\hbar k}{m}\left[1 + 2\frac{\mathrm{Re} f(0)e^{ikx^2/(2z)}}{z}\right]. \tag{46.5}$$

This is the flux of probability per unit area, as a function of the distance x off the beam axis passing through the interaction region. The net flux is the integral of the flux density across the area of the beam, $\int 2\pi x\, dx\, j_f$. The value of this integral over the first term in equation (46.5) is undefined; that depends on the construction of a wave packet, as in the last section. But we are interested in the second term, which gives the effect of the scattering amplitude. This has a finite integral even before the construction of a wave packet because the factor $e^{ikx^2/(2z)}$ makes the integral small off the scattering axis at $x = 0$. The contribution to the net probability flux in the forward direction due to this second term is

$$F_f = \int_0^\infty 2\pi x\, dx \frac{\hbar k}{m} \frac{2\,\mathrm{Re} f(0) e^{ikx^2/(2z)}}{z}. \tag{46.6}$$

We need the integral

$$I = \int_0^\infty 2x dx e^{ikx^2/(2z)}. \tag{46.7}$$

On changing the variable of integration to $w = ikx^2/(2z)$, this is

$$I = \frac{2z}{ik} \int_0^{i\infty} dw\, e^w = -\frac{2iz}{k} \left[e^w\right]_0^{i\infty}. \tag{46.8}$$

The upper limit oscillates to zero mean value, leaving[1]

$$I = 2iz/k. \tag{46.9}$$

Since I in equation (46.9) is imaginary, the real part of the product in equation (46.6) selects the imaginary part of the forward scattering amplitude, $f(0)$, giving

$$F_f = -4\pi\hbar \mathrm{Im} f(0)/m. \tag{46.10}$$

The flux from the first term in equation (46.5) is identical to what it would have been in the absence of the interaction region. Therefore $-F_f$ must be the net amount of probability flux removed by the scattering. Since the net flux removed is $v\sigma = \hbar k\sigma/m$, we have

$$\sigma = 4\pi \mathrm{Im} f(0)/k. \tag{46.11}$$

This relation between the total scattering cross section and the imaginary part of the scattering amplitude is called the optical theorem (perhaps because it was known well before wave mechanics, from the theories of scattering of sound and electromagnetic waves).

---

[1] To be more precise, the exponential $e^w = e^{ikx^2/(2z)}$ is a rapidly oscillating function of perpendicular distance x from the incident axis. The wave packet in equation (45.4) is an average over slightly different incident directions. This smooths the rapidly oscillating exponential to its zero mean value.

47 Born Approximation

We want to solve the Schrödinger equation

$$-\frac{\hbar^2}{2m}\nabla^2\psi + V(\mathbf{r})\psi = E\psi = \frac{\hbar^2 k^2}{2m}\psi, \tag{47.1}$$

with definite energy $E = \hbar^2 k^2/(2m)$ for incident velocity $v = \hbar k/m$. The boundary condition is that at large r the wave function has the scattering wave form of equation (45.8),

$$\psi \to e^{i\mathbf{k}\cdot\mathbf{r}} + fe^{ikr}/r. \tag{47.2}$$

Let us rewrite equation (47.1) as

$$\nabla^2\psi + k^2\psi = \frac{2mV}{\hbar^2}\psi \equiv U\psi. \tag{47.3}$$

The wanted solution is of the form

$$\psi = e^{i\mathbf{k}\cdot\mathbf{r}} + u, \tag{47.4}$$

where u represents the scattered wave. This expression in equation (47.3) is

$$\nabla^2 u + k^2 u = U(\mathbf{r})(e^{i\mathbf{k}\cdot\mathbf{r}} + u). \tag{47.5}$$

If $U = 0$, the solution to equation (47.5) consistent with the boundary condition in equation (47.2) is $u = 0$. If U is relatively small (we will not pause to explore the conditions under which this is so), u is small, so it makes sense to drop the product uU in the right-hand side of equation (47.5), to get the approximate equation

$$\nabla^2 u + k^2 u = U(\mathbf{r})e^{i\mathbf{k}\cdot\mathbf{r}}. \tag{47.6}$$

This is the first equation in the Born series for the scattered wave, u. The next equation in the series is obtained by substituting the solution u to equation (47.6) back into the right-hand side of equation (47.5), and then solving for the new u that appears on the left side of the equation, just as one does in perturbation theory.

The formal solution to equation (47.6) is obtained by the following Fourier transform method. It will be recalled from section 10 that the Fourier transform of the function $u(\mathbf{r})$ may be written as

$$u_{\mathbf{k}'} = \int u(\mathbf{r}')e^{-i\mathbf{k}'\cdot\mathbf{r}'}d^3r', \tag{47.7}$$

and that the reverse relation is (by eq. [10.21])

$$\int u_{\mathbf{k}'}e^{i\mathbf{k}'\cdot\mathbf{r}}d^3k' = (2\pi)^3u(\mathbf{r}). \tag{47.8}$$

The Fourier transform of $\nabla^2 u$ is

$$\int(\nabla^2u)e^{-i\mathbf{k}'\cdot\mathbf{r}'}d^3r' = -(k')^2u_{\mathbf{k}'}, \tag{47.9}$$

as one finds by integration by parts twice. Thus the Fourier transform of equation (47.6) is

$$(k^2-(k')^2)u_{\mathbf{k}'} = \int U(\mathbf{r}')e^{i(\mathbf{k}-\mathbf{k}')\cdot\mathbf{r}'}d^3r'. \tag{47.10}$$

On using this expression for $u_{\mathbf{k}'}$ in the reverse relation (47.8) for $u(\mathbf{r})$, we get

$$u(\mathbf{r}) = \frac{1}{(2\pi)^3}\int d^3k'\frac{e^{i\mathbf{k}'\cdot\mathbf{r}}}{(k^2-(k')^2)}\int d^3r'U(\mathbf{r}')e^{i(\mathbf{k}-\mathbf{k}')\cdot\mathbf{r}'}. \tag{47.11}$$

This is the wanted formal expression for the solution $u(\mathbf{r})$ to equation (47.6).

We can rewrite the solution as

$$u(\mathbf{r}) = \frac{1}{(2\pi)^3}\int d^3r'U(\mathbf{r}')e^{i\mathbf{k}\cdot\mathbf{r}'}I(\mathbf{r}-\mathbf{r}'), \tag{47.12}$$

where

$$I(\mathbf{y}) = \int d^3k'e^{i\mathbf{k}'\cdot\mathbf{y}}/(k^2-(k')^2), \tag{47.13}$$

with $\mathbf{y} = \mathbf{r}-\mathbf{r}'$.

To evaluate the integral I, write $\mathbf{k}'$ in polar coordinates, with polar axis (z axis) along the vector $\mathbf{y}$. Then

$$\mathbf{k}'\cdot\mathbf{y} = k'y\cos\theta, \tag{47.14}$$

where θ is the angle between $\mathbf{k}'$ and $\mathbf{y}$. The volume element in polar coordinates is

$$d^3k' = (k')^2 dk' \sin\theta d\theta d\phi. \tag{47.15}$$

A handy integral is

$$\int_0^\pi \sin\theta\, d\theta e^{ik'y\cos\theta} = \int_{-1}^{+1} e^{ik'y\cos\theta} d\cos\theta = 2\frac{\sin k'y}{k'y}. \tag{47.16}$$

On using this to work the angular integrals in equation (47.13), we get

$$I = 4\pi \int_0^\infty \frac{(k')^2 dk'}{k^2 - (k')^2} \frac{\sin k'y}{k'y}. \tag{47.17}$$

The integrand is an even function of k', so we can write this as an integral over negative as well as positive k':

$$I = -2\pi i \int_{-\infty}^\infty \frac{(k')^2 dk'}{k^2 - (k')^2} \frac{e^{ik'y}}{k'y}. \tag{47.18}$$

The factor $e^{ik'y}$ contains the term $\sin k'y$ in equation (47.17). The term $\cos k'y$ from the real part of $e^{ik'y}$ vanishes because it makes the integrand odd.

Now we come to an apparent problem. Equation (47.18) for I in equation (47.12) seems to give a definite solution to the differential equation (47.6). But we know that there is no definite solution unless boundary conditions are specified. In fact, I is not defined by equation (47.18), because the integral diverges at the singularities at $k' = \pm k$. We will see that the prescription for the treatment of the integral over the singularities is equivalent to the assignment of boundary conditions for the differential equation: different prescriptions give different linear combinations of incoming ($\propto e^{-ikr}$) and outgoing waves.

The wanted pure outgoing wave in equation (47.2) is obtained by taking the path of integration in equation (47.18) to go just above the pole at $k' = -k$ and just below the pole at $k' = k$. As indicated in figure 47.1, we can close the contour integral by going back along the upper complex plane at great distance from the origin. This adds nothing to I because at great distance above the real k' axis $e^{ik'y}$ is negligibly small. (Recall that y is positive because it is the magnitude of $\mathbf{y}$; thus the real part of $ik'y$ is negative above the real axis.) Then we can shrink

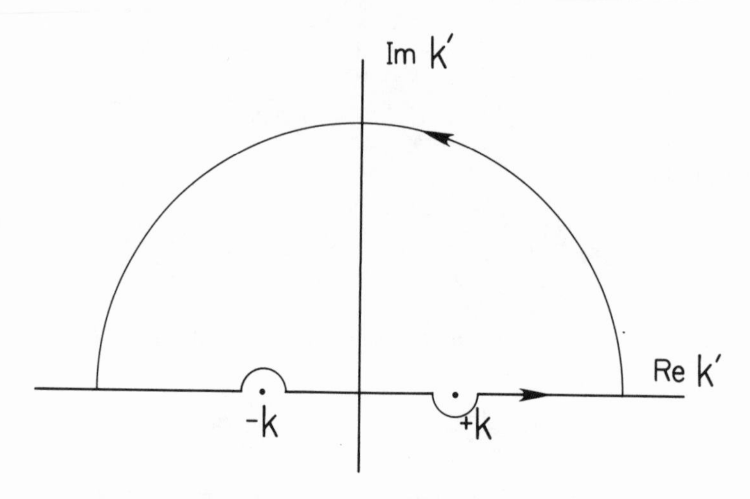

Fig. 47.1 Contour integral for equation 47.18. The prescription for the wanted outgoing wave solution for $u(\mathbf{r})$ is that the integral along the real k' axis passes over the singularity at $-k$ and under the singularity at $+k$. The contour can be closed along the upper half of the complex k' plane at great distance from the origin, where $e^{ik'r}$ is negligible. The contour then can be shrunk to an integral around the pole at $+k$.

the contour to a tight loop going counterclockwise around the pole at $k' = k$. After that, we can replace k' with k everywhere in the integral save of course in the pole. This gives

$$k^2 - (k')^2 = (k - k')(k + k') \rightarrow 2k(k - k'), \tag{47.19}$$

and leaves

$$I = -2\pi i \frac{e^{iky}}{y} \times \frac{1}{2} \oint \frac{dk'}{k - k'}. \tag{47.20}$$

The integral around the pole is

$$\oint \frac{dk'}{k - k'} = -\oint \frac{dz}{z} = -2\pi i, \tag{47.21}$$

so

$$I = -2\pi^2 e^{iky}/y. \tag{47.22}$$

This result in equation (47.12), with $y = |\mathbf{r} - \mathbf{r}'|$, is

$$u(\mathbf{r}) = -\frac{1}{4\pi} \int d^3r' U(\mathbf{r}') e^{i\mathbf{k}\cdot\mathbf{r}'} \frac{e^{ik|\mathbf{r}-\mathbf{r}'|}}{|\mathbf{r} - \mathbf{r}'|}. \tag{47.23}$$

The value of r' in this integral is limited to the finite range of the potential $U(\mathbf{r}')$. When the separation r is much larger than this range of the interaction, we can replace $|\mathbf{r}-\mathbf{r}'|$ with r in the denominator of the last factor. We have to be more careful in the exponential term in the last factor because what matters here is the absolute change in $|\mathbf{r}-\mathbf{r}'|$ as $\mathbf{r}'$ varies over the region of nonzero U. Here we can write

$$|\mathbf{r}-\mathbf{r}'| = r - \mathbf{r}'\cdot\mathbf{r}/r, \tag{47.24}$$

correct to order r'/r. This brings the last factor in equation (47.23) to the form

$$\frac{e^{ik|\mathbf{r}-\mathbf{r}'|}}{|\mathbf{r}-\mathbf{r}'|} = \frac{e^{ikr}}{r}e^{-ik\mathbf{r}'\cdot\mathbf{r}/r}. \tag{47.25}$$

The conclusion is that the wave function (47.4) in the first Born approximation of equation (47.6), and at large distance from the interaction region, is

$$\psi = e^{i\mathbf{k}\cdot\mathbf{r}} + u(\mathbf{r}) = e^{i\mathbf{k}\cdot\mathbf{r}} - \frac{1}{4\pi}\int d^3r' U(\mathbf{r}')e^{i(\mathbf{k}\cdot\mathbf{r}'-k\mathbf{r}'\cdot\mathbf{r}/r)}\frac{e^{ikr}}{r}. \tag{47.26}$$

This expression agrees with the scattering boundary condition in equation (47.2): it is a sum of a plane wave and an outgoing spherical wave $\propto e^{ikr}/r$. It will be recalled that we satisfied this boundary condition by the choice of how to deal with the poles in the integrand in equation (47.18): go over the pole at $k' = -k$, and below the one at $k' = k$. Other prescriptions would add to u terms proportional to $u \propto e^{-ikr}/r$, which are allowed by the differential equation (47.6), but are not allowed by the scattering boundary condition (47.2).

The scattering amplitude in equation (47.26) is

$$f(\theta,\phi) = -\frac{1}{4\pi}\int d^3r' U(\mathbf{r}')e^{i\mathbf{r}'\cdot(\mathbf{k}-k\mathbf{r}/r)}. \tag{47.27}$$

The unit vector $\mathbf{r}/r$ in the exponential points along the scattering direction θ, ϕ in the argument of f.

The unit vector $\mathbf{r}/r$ multiplied by k defines a propagation vector directed along the scattering direction:

$$\mathbf{k}_s \equiv k\mathbf{r}/r. \tag{47.28}$$

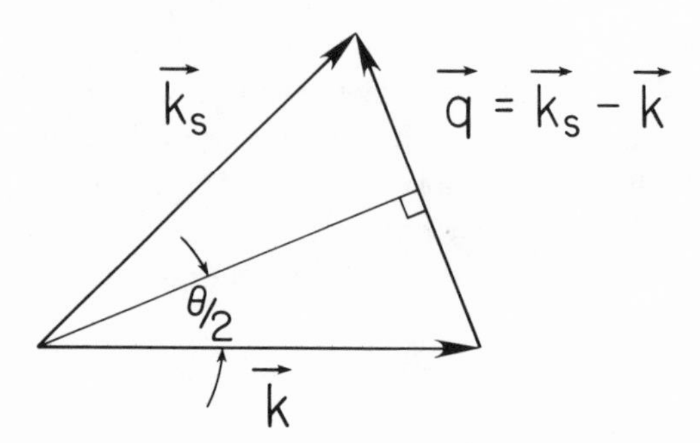

Fig. 47.2 Scattering angle and momentum transfer. The incident momentum is $\hbar\mathbf{k}$, the momentum of the scattered particle is $\hbar\mathbf{k}_s$, and the angle between these vectors is θ. The vectors define an isosceles triangle, with base equal to the momentum transfer, $\hbar(\mathbf{k}_s - \mathbf{k})$.

With this definition, and replacing U with $2mV/\hbar^2$ (eq. [47.3]), we arrive at the final expression for the scattering amplitude in the Born approximation,

$$f(\theta, \phi) = f(\mathbf{k} - \mathbf{k}_s) = -\frac{m}{2\pi\hbar^2} \int d^3r' V(\mathbf{r}')\, e^{i(\mathbf{k}-\mathbf{k}_s)\cdot\mathbf{r}'}. \qquad (47.29)$$

In this expression, $\mathbf{k}$ is the propagation vector of the incident beam in equation (47.2), and $\mathbf{k}_s$ (eq. [47.28]) is the propagation vector in the scattered direction. The lengths of these vectors are the same, $|\mathbf{k}| = |\mathbf{k}_s| = k$, and the energy is $E = \hbar^2 k^2/2m$.

We see that in the Born approximation the scattering amplitude is proportional to the Fourier transform of the potential, with argument

$$\mathbf{q} = \mathbf{k}_s - \mathbf{k}. \qquad (47.30)$$

Since the incident momentum is $\mathbf{p} = \hbar\mathbf{k}$, and the momentum in the scattered direction is $\mathbf{p}_s = \hbar\mathbf{k}_s$, the momentum transfer to the particle is

$$\Delta\mathbf{p} = \mathbf{p}_s - \mathbf{p} = \hbar\mathbf{q}, \qquad (47.31)$$

if the particle is detected in the direction of $\mathbf{k}_s$. In terms of the momentum transfer, the Born approximation to the scattering amplitude is

$$f(\theta, \phi) = -\frac{m}{2\pi\hbar^2} \int d^3r' V(\mathbf{r}')\, e^{-i\Delta\mathbf{p}\cdot\mathbf{r}'/\hbar}. \qquad (47.32)$$

The relation between the momentum transfer and the scattering angle, θ, is indicated in figure 47.2.

As an example, suppose the scattering potential vanishes, $V = 0$, at radius $r > r_o$, and consider low energy scattering, where $E = \hbar^2 k^2/(2m)$ is so small that $kr_o \ll 1$. Then the exponential in equation (47.29) is nearly unity where V is nonzero, and the scattering amplitude is

$$f = -\frac{m}{2\pi\hbar^2}\int d^3r V(\mathbf{r}). \tag{47.33}$$

The differential scattering cross section is

$$\frac{d\sigma}{d\Omega} = |f|^2 = \frac{m^2}{4\pi^2\hbar^4}\left(\int d^3r V(\mathbf{r})\right)^2. \tag{47.34}$$

This scattering cross section is isotropic, that is, independent of the scattering direction $\mathbf{k}_s$ relative to the incident beam along $\mathbf{k}$.

48 Rutherford Scattering Cross Section

The differential cross section for the scattering of particles of charge Q_1 by particles of charge Q_2 through the coulomb potential $V = Q_1Q_2/r$ can be obtained analytically, but the calculation is long and not inspiring, so we will consider only the Born approximation. It is one of the charming coincidences that were so helpful to the discovery of quantum mechanics that one gets the same (experimentally successful) answer from the analytic solution to the Schrödinger coulomb scattering problem, the simple Born approximation, and the classical mechanics Rutherford used to deduce that an atom has to contain a massive point-like charge, the nucleus.

The charge on the massive nucleus of an atom is screened by the electron cloud around it; we will take account of the screening by writing the potential energy as

$$V = Q_1Q_2e^{-\alpha r}/r, \tag{48.1}$$

so the charge is screened at distance $r \gg 1/\alpha$. It will be seen that the factor $e^{-\alpha r}$ drops out of the expression for the scattering cross section at high momentum, where $k \gg \alpha$, because the scattering is dominated by the part of the potential at $r \ll 1/\alpha$. In this limit, the form of the factor multiplying the coulomb potential is irrelevant as long as it smoothly approaches unity at $r \to 0$ and zero at large r; the factor is only needed to make the integral in the Born approximation (47.29) well behaved.

In Born approximation, the differential scattering cross section is

$$\frac{d\sigma}{d\Omega} = \frac{m^2 Q_1^2 Q_2^2}{4\pi^2 \hbar^4} |I|^2, \tag{48.2}$$

where the integral is

$$I = \int \frac{d^3 r}{r} e^{-i\mathbf{q}\cdot\mathbf{r} - \alpha r}. \tag{48.3}$$

Here $\mathbf{q} = \mathbf{k}_s - \mathbf{k}$, where the propagation vector $\mathbf{k}_s$ is along the scattering direction, at angle θ to the incident beam direction along $\mathbf{k}$.

To evaluate the integral in equation (48.3), use polar coordinates, with pole along $\mathbf{q}$. In polar coordinates, the volume element is $d^3 r = r^2 dr\, d\Omega$, where the element of solid angle is $d\Omega = \sin\theta\, d\theta\, d\phi$. The integral over angles is given by equation (47.16), which is worth remembering:

$$\int d\Omega\, e^{-i\mathbf{q}\cdot\mathbf{r}} = 4\pi \frac{\sin qr}{qr}. \tag{48.4}$$

This brings the integral to

$$\begin{aligned} I &= \frac{4\pi}{q} \int_0^\infty dr \sin(qr) e^{-\alpha r} = \frac{4\pi}{q} \mathrm{Im} \int_0^\infty dr\, e^{(iq-\alpha)r} \\ &= \frac{4\pi}{q} \mathrm{Im} \frac{1}{\alpha - iq} = \frac{4\pi}{\alpha^2 + q^2}. \end{aligned} \tag{48.5}$$

At large enough momentum transfer, where the de Broglie wavelength $1/q$ associated with the momentum transfer $\hbar q$ is much smaller than the screening radius $1/\alpha$, we can drop α from the denominator of the last equation. Then equation (48.2) becomes

$$\frac{d\sigma}{d\Omega} = \frac{4m^2 Q_1^2 Q_2^2}{q^4 \hbar^4}. \tag{48.6}$$

In this limit, the screening of the coulomb potential is unimportant because the scattering is dominated by the potential at radii small compared to the screening length $\sim 1/\alpha$.

The scattering angle θ between the scattered direction $\mathbf{k}_s$ and the incident direction $\mathbf{k}$ is indicated in figure 47.2. One sees from the figure that the momentum transfer is given by the equation

$$q = 2k \sin(\theta/2). \tag{48.7}$$

The classical (group) velocity of the beam is $v = \hbar k/m$. On using these equations to eliminate q and k from equation (48.6), we get the Rutherford equation,

$$\frac{d\sigma}{d\Omega} = \frac{Q_1^2 Q_2^2}{4m^2 v^4 \sin^4(\theta/2)}. \tag{48.8}$$

It will be recalled that v is the relative velocity of the pair of particles, and m is the reduced mass.

Planck's constant does not appear in equation (48.8), so it is no surprise that this equation agrees with the result from classical mechanics that must be valid when $\hbar$ is unimportant. This is a fortunate coincidence, because Rutherford used this cross section in the analysis of the scattering of α particles (helium nuclei) by atoms that led him to deduce that an atom contains a massive compact charge, the nucleus. The puzzle of how the electrons could move in stable orbits around the nucleus was one of the elements leading to the discovery of the quantum mechanics by which one more correctly describes Rutherford's scattering experiments.

49 Partial Wave Expansion

Another method for computing scattering cross sections is based on the expansion of the wave function for a scattering state with definite energy E as a linear combination of states with definite orbital angular momentum. It will be recalled that the wave function for a single particle with orbital angular momentum quantum numbers l and m is proportional to the spherical harmonic $Y_l^m(\theta, \phi)$ (section 17). Since the angular momentum eigenstates are a complete set, we can always expand the wave function of a scattering state with definite energy E as a linear combination of angular momentum eigenstates,

$$\psi(\mathbf{r}) = \sum_{l,m} Y_l^m(\theta, \phi) g_{lm}(r). \tag{49.1}$$

This is a general expression for a function $\psi(\mathbf{r})$ of position as a linear combination of spherical harmonics, with expansion coefficients g_{lm} that are functions of radius r. In a scattering problem, $\psi(\mathbf{r})$ is a solution to Schrödinger's equation with definite energy E, with a Hamiltonian that is the Hamiltonian of a free particle outside the interaction region. Therefore, as discussed in section 18, the radial wave function $u(r) = r g_{lm}(r)$

outside the interaction region obeys a one-dimensional Schrödinger equation with an effective potential term due to the angular momentum (eq. 18.18):

$$-\frac{\hbar^2}{2m}\frac{d^2(rg_{lm})}{dr^2}+\frac{\hbar^2 l(l+1)}{2mr^2}(rg_{lm})=Erg_{lm}. \tag{49.2}$$

On setting $E=\hbar^2k^2/(2m)$, $y_l=g_{lm}$ (we can drop the index m because it does not appear in the differential equation), and changing the radius variable to $\rho=kr$, the radial wave equation (49.2) becomes

$$\frac{d^2\rho y_l(\rho)}{d\rho^2}+\left[1-\frac{l(l+1)}{\rho^2}\right]\rho y_l(\rho)=0. \tag{49.3}$$

This equation was studied well before quantum mechanics, in the theories of sound and electromagnetic radiation. The real solution finite at the origin is called the spherical Bessel function $j_l(\rho)$. There also is a real solution singular at the origin; it is the spherical Neumann function $n_l(\rho)$. Only a few properties of these functions will be needed here, as follows.

For $l=0$, the solutions to equation (49.3) for $\rho y_0(\rho)$ are sines and cosines. In the standard normalization, the $l=0$ spherical Bessel and Neumann functions are

$$j_0=\frac{\sin\rho}{\rho},\qquad n_0=-\frac{\cos\rho}{\rho}. \tag{49.4}$$

If $l\geq 1$, it is easy to find the asymptotic behavior of the solutions at large and small radii. At $\rho\ll l$, the second term in square brackets in equation (49.3) dominates. On dropping the first term in the brackets, we get a homogeneous differential equation,

$$\frac{d^2\rho y_l(\rho)}{d\rho^2}-\frac{l(l+1)}{\rho^2}\rho y_l=0. \tag{49.5}$$

This has the power law solutions

$$j_l\propto\rho^l,\quad n_l\propto\rho^{-(l+1)},\qquad \text{for}\quad \rho\ll l. \tag{49.6}$$

As in equation (49.4), the solution regular at the origin is the spherical Bessel function, and the solution singular at the origin is the spherical Neumann function.

At $\rho \gg l$, the constant term in the square brackets in equation (49.3) dominates. In this limit, the solutions $\rho y_l(\rho)$ are linear combinations of $e^{i\rho}$ and $e^{-i\rho}$. Since $\rho = kr$, the solution $y_l \propto e^{i\rho}/\rho \propto e^{ikr}/r$ is an expanding spherical wave, in the standard form of equation (45.8), and $e^{-i\rho}/\rho$ is an incoming spherical wave. That is, the functions g_{lm} in equation (49.1) at large r are sums of incoming and outgoing spherical waves, each with definite orbital angular momentum. The expansion of the wave function at great distance from the interaction region thus looks like

$$\psi \rightarrow \sum C_l^m Y_l^m(\theta,\phi)\frac{e^{ikr}}{r} + \sum D_l^m Y_l^m(\theta,\phi)\frac{e^{-ikr}}{r}, \tag{49.7}$$

where C_l^m and D_l^m are the constant expansion coefficients. This sum over outgoing and incoming spherical waves is called the partial wave expansion of the scattering state ψ.

A plane wave, $e^{i\mathbf{k}\cdot\mathbf{r}}$, satisfies the free Schrödinger equation, so it can be expanded as a sum over partial waves. The result for a wave propagating toward the positive z axis is

$$e^{ikz} = 2\pi^{1/2} \sum_{l=0,\infty} i^l (2l+1)^{1/2} Y_l^0(\theta) j_l(kr), \tag{49.8}$$

where $z = r\cos\theta$. This expansion at large r is a special case of the expression in equation (49.7), as is shown in equation (50.21) below.

Now we can outline the partial wave program. The scattering state boundary condition is that the wave function approach the form in equation (45.8),

$$\psi \rightarrow e^{i\mathbf{k}\cdot\mathbf{r}} + f e^{ikr}/r, \tag{49.9}$$

at great distance from the interaction region. Since ψ outside the interaction region satisfies the free Schrödinger equation, we can expand it as a sum over partial (spherical) incoming and outgoing waves, as in equation (49.7). In particular, the plane wave $e^{i\mathbf{k}\cdot\mathbf{r}}$ is a sum over partial waves, each having a definite amplitude and phase. The interaction potential has a limited range, so it cannot affect the incoming partial waves in a scattering wave solution ψ; they have to be the same as for a plane wave. That is, in a scattering solution the coefficients D_l^m multiplying the incoming waves in equation (49.7) have to be the same as for a plane wave. The effect of the interaction is to alter the coefficients C_l^m of the outgoing waves from what they would be for a pure plane wave. The solution to the scattering problem thus is equivalent to a tabulation of the factors by which the amplitude and phase of each outgoing partial wave has been altered from those of a pure plane wave. These factors can be a useful way to summarize the results of scattering measurements, and they can be computed from Schrödinger's equation for a given interaction potential, as will be seen in the examples in the following sections.

In principle, a scattering solution is specified by an infinite number of amplitudes and phases for all values of l and m for the outgoing partial waves, but in practice only a few may be interestingly different from the plane wave values. To see why this is, note that the partial wave expansion of a plane wave can only contain the spherical Bessel functions $j_l(kr)$, as shown in equation (49.8), because the spherical Neumann functions are singular at the origin (eq. [49.6]). As one might suspect from equation (49.6), the spherical Bessel function $j_l(kr)$ is negligibly small at $\rho \ll l$. Since $\rho = kr$, this means

$$j_l \sim 0 \quad \text{at} \quad kr \ll l. \tag{49.10}$$

In classical mechanics, a particle moving toward the origin with orbital angular momentum $L = l\hbar$ and linear momentum $p = \hbar k$ has impact parameter $r_{\min} = L/p = l/k$. For a free particle, this is the distance of closest approach to the origin. We see from equation (49.10) that quantum mechanics says the particle is not likely to be found much closer to the origin than this classical distance of closest approach. The relevance for the present discussion is that the incoming partial wave with angular momentum l is suppressed at $r < l/k$. If the interaction region is bounded by radius r_o, then incoming partial waves with $l \gg kr_o$ enter the interaction region with strongly suppressed amplitudes, so it is not likely that the interaction can affect the amplitudes and phases of the outgoing parts of the waves. At low energies, where k is small, only the small l partial waves are appreciably affected by the interaction. The following sections for the most part deal with the case where only the $l = 0$ wave (s-wave) is appreciably affected.

50 Phase Shifts and Cross Sections

The s-Wave Case

It is particularly easy to extract the s-wave ($l = 0$) part of the partial wave expansion in equations (49.1) and (49.7). The $l = 0$ spherical harmonic is independent of direction (corresponding to the fact that it is an eigenstate of the generators of rotations with zero eigenvalue). Therefore, if we integrate equation (49.7) over all directions of $\mathbf{r}$ at fixed radius r, and then divide by the total solid angle 4π (so as to average the expression over directions), we eliminate all the terms in this sum over angular momentum eigenstates with $l > 0$ (because they are orthogonal to the spherically symmetric s-wave term). We are left with the s-wave, $l = 0$ part. The result of averaging a plane wave over directions at fixed r has been used several times (eq. 48.4). It is

$$\int \frac{d\Omega}{4\pi} e^{i\mathbf{k}\cdot\mathbf{r}} = \frac{\sin kr}{kr} = \frac{1}{2i}\left(-\frac{e^{-ikr}}{kr} + \frac{e^{ikr}}{kr}\right), \tag{50.1}$$

where as usual $d\Omega = \sin\theta d\theta d\phi$, with θ and ϕ the polar coordinates of $\mathbf{r}$. It follows that the partial wave expansion of a plane wave is of the form

$$e^{i\mathbf{k}\cdot\mathbf{r}} = \frac{1}{2i}\left(-\frac{e^{-ikr}}{kr} + \frac{e^{ikr}}{kr}\right) + \sum_{l>0} g_{lm}(r) Y_l^m. \tag{50.2}$$

Now suppose we introduce a scattering potential with a finite range. This changes equation (50.2) for the wave function outside the interaction region to

$$\psi = \frac{1}{2i}\left(-\frac{e^{-ikr}}{kr} + \eta_0 \frac{e^{ikr}}{kr}\right) + \sum_{l>0} g'_{lm}(r) Y_l^m. \tag{50.3}$$

The incoming s-wave term $\propto e^{-ikr}/r$ is unchanged, because this incoming wave comes in from great distance and cannot know about the potential. The outgoing s-wave is affected; this is represented by the factor η_0. The same applies to all the other terms in the sum over l: the incoming parts are the same as for a plane wave, and the outgoing parts may have new amplitudes and phases.

If the interaction is elastic, so no particles are lost, and if the scattering potential conserves angular momentum, then all the probability flux entering in the s-wave must leave by the s-wave. This means the amplitude of the outgoing s-wave must be the same as for a plane wave; the interaction can only affect the phase. Thus η_0 has to be a pure phase factor,

$$\eta_0 = e^{2i\delta_0}. \tag{50.4}$$

The factor of two is a convenient convention; δ_0 is called the s-wave phase shift.

To get the scattering amplitude, we must write equation (50.3) as the sum of a plane wave and an outgoing spherical wave, as in equation (49.9). This gives

$$\psi = e^{i\mathbf{k}\cdot\mathbf{r}} + \frac{\eta_0 - 1}{2ik}\frac{e^{ikr}}{r} + \ldots. \tag{50.5}$$

The second term on the right-hand side is just the difference between equations (50.3) and (50.2): the scattered wave is the difference between the actual outgoing s-wave and what it would have been in the absence of the interaction. The dots indicate the higher angular momentum outgoing scattered spherical waves.

Equation (50.5) says the s-wave part of the scattering amplitude is

$$f_0 = \frac{\eta_0 - 1}{2ik} = \frac{e^{2i\delta_0} - 1}{2ik}. \tag{50.6}$$

The second equation assumes elastic scattering. If the phase shifts of the higher l partial waves are negligible, then the scattering cross section is

$$\frac{d\sigma}{d\Omega} = |f_0|^2 = \frac{\sin^2\delta_0}{k^2}. \tag{50.7}$$

It is left as an exercise to check that this is consistent with the optical theorem in equation (46.11).

Hard Sphere Scattering Cross Section

By a hard sphere one means that the interaction potential is $V(r) = 0$ at separations $r > r_o$, and that V is positive (repulsive) and very large at smaller r. The s-wave radial wave function $u = rg_0(r)$ for the scattering solution with energy E satisfies the usual one-dimensional Schrödinger equation (with no effective potential term because $l = 0$),

$$-\frac{\hbar^2}{2m}\frac{d^2u}{dr^2} + V(r)u = Eu = \frac{\hbar^2k^2}{2m}u, \tag{50.8}$$

as in equation (18.18). The last expression defines the wave number k in terms of the energy E.

The usual boundary condition is that $u \to 0$ at $r \to 0$ (so that $\psi = u(r)/r$ is well behaved). The effect of the large repulsive potential in equation (50.8) is to keep u small at $r < r_o$, so for a hard sphere we have the condition $u = 0$ at $r = r_o$. Referring to the first expression on the right-hand side of equation (50.3), we see that this condition requires

$$\eta_0 = e^{-2ikr_o}. \tag{50.9}$$

(Formally, this makes the s-wave radial wave function in eq. [50.3] vanish only at $r = r_o$, but that is all we require, because eq. [50.3] only applies outside the interaction region.) Equation (50.9) for η_0 says the phase shift is $\delta_0 = -kr_o$ (eq. [50.4]), so the cross section (eq. [50.7]) is

$$\frac{d\sigma}{d\Omega} = \frac{\sin^2 kr_o}{k^2} = r_o^2. \tag{50.10}$$

The last step follows because we are assuming that $kr_o \ll 1$, so the $l \geq 1$ partial waves are not appreciably affected by the hard sphere (as in the discussion of eq. [49.10]). The total scattering cross section is then $\sigma = 4\pi r_o^2$. This is four times the geometrical cross section of the sphere, πr_o^2.

The general approach to finding the phase shift δ_0 for a given spherical potential $V(r)$ commences with the observation that the solution to equation (50.8) has to approach the form

$$u(r) \propto e^{2i\delta_0} e^{ikr} - e^{-ikr} \propto \sin(kr + \delta_0), \tag{50.11}$$

at r greater than the range of the interaction. This is because the general solution when $V = 0$ is a linear combination of exponentials $e^{\pm ikr}$, or equivalently a sine wave with some constant phase factor, $u = \sin(kr + \delta_0)$. This phase factor is just the s-wave phase shift defined in equations (50.3) and (50.4). To find δ_0, one computes $u(r)$ by integrating equation (50.8) (perhaps numerically), with the boundary condition $u(0) = 0$, through the interaction region where $V(r)$ is nonzero, and then matching the solution outside the interaction region to equation (50.11).

s-*Wave Absorption*

It can happen that incident particles are absorbed, as in the capture of neutrons by an atomic nucleus (which might be followed by the emission of a photon). Outside the nucleus, the neutron wave function satisfies the free Schrödinger wave equation, so it still must be possible to write the wave function outside the nucleus as a linear combination of incoming and outgoing spherical waves, as in equation (50.3). However, here we only have the condition that the outgoing flux cannot exceed the incoming flux. This means η_0 must be a complex number with $|\eta_0| \leq 1$.

The outgoing s-wave flux still is determined by the square of the amplitude of the outgoing scattered wave, so the differential scattering cross section still is $d\sigma/d\Omega = |f|^2$ with $f_0 = (\eta_0 - 1)/2ik$, as in equation (50.6). Thus, if s-wave scattering dominates, the total scattering cross section is

$$\sigma_s = 4\pi|f_0|^2 = \frac{\pi}{k^2}|\eta_0 - 1|^2. \tag{50.12}$$

This scattering cross section is defined by the rate of detection of particles scattered out of the beam, as discussed in section 45. One can also define an absorption cross section, σ_a, by the probability per unit time and per target particle, dN_a/dt, for capture of particles from a beam with incident flux density nv:

$$\frac{dN_a}{dt} = \sigma_a nv. \tag{50.13}$$

We can calculate σ_a in terms of η_0 without a lot of work as follows.

The fluxes of probability entering and leaving by the s-wave are proportional to the squares of the amplitudes of the incoming and outgoing waves. We see from equation (50.3) that the ratio of the s-wave fluxes leaving and entering is $|\eta_o|^2$. The absorption cross section is proportional to the difference of fluxes entering and leaving, so if the absorption is from s-waves, σ_a is proportional to $1 - |\eta_0|^2$. To get the constant of proportionality, note that the maximum value for the total s-wave scattering cross section is $\sigma_s = 4\pi/k^2$, for $\eta_0 = -1$ (eq. [50.12]) We see from equations (50.3) and (50.5) that when $\eta_0 = -1$ the scattered wave has twice the amplitude of the s-wave part of a plane wave, so in this case the scattered s-wave flux is four times the entering flux. The maximum possible value of σ_a for complete absorption of the incident s-wave flux thus must be one quarter of the maximum possible value of σ_s, or π/k^2. Thus the s-wave absorption cross section is

$$\sigma_a = \frac{\pi}{k^2}(1 - |\eta_0|^2). \tag{50.14}$$

The maximum absorption cross section is obtained when $\eta_0 = 0$. In this case, the absorption and scattering cross sections are equal, $\sigma_s = \sigma_a = \pi/k^2$ (eqs. [50.12] and [50.14] with $\eta_0 = 0$). That is, if the target is "black" in the sense that it absorbs all the flux incident on it, then it necessarily scatters as many particles as it absorbs (even though all particles that enter the absorber are lost). This "shadow scattering" or diffraction effect is familiar in optics: when a black object in a beam of light removes a section of the light wave, the remaining part of the wave moves to fill in the hole. This bends the wave away from the original direction. That is, if part of the incident plane wave is removed by absorption, it disturbs the remaining part of the wave that is not absorbed, producing a scattered wave.

Extension to Higher Angular Momentum

We can understand the asymptotic form of the expansion of a plane wave $e^{i\mathbf{k}\cdot\mathbf{r}}$ as a sum over incoming and outgoing spherical waves with definite angular momentum, as in equation (49.7) and (49.8), as follows. If the propagation vector $\mathbf{k}$ points along the z axis the plane wave is symmetric under rotation around the z axis, so the sum over angular momentum eigenstates in equation (49.1) can only contain the axially symmetric $m = 0$ terms. Thus the expansion is of the form

$$e^{ikz} = e^{ikr\cos\theta} = \sum_l g_l(kr) Y_l^0(\theta). \tag{50.15}$$

The spherical harmonics Y_l^m are orthogonal, with the normalization (eq. [17.54])

$$\int Y_l^{-m} Y_{l'}^{m'}\, d\Omega = \delta_{ll'}\delta_{mm'}, \tag{50.16}$$

so we have from equation (50.15)

$$g_l(kr) = 2\pi \int_{-1}^{1} d\cos\theta e^{ikr\cos\theta} Y_l^0(\theta). \tag{50.17}$$

The result of integrating this expression once by parts is

$$\begin{aligned} g_l(kr) &= \frac{2\pi}{ikr} \int Y_l^0(\theta)\, de^{ikr\cos\theta} \\ &= \frac{2\pi}{ikr}\left(\left[Y_l^0(\theta) e^{ikr\cos\theta} \right]_{\cos\theta=-1}^{\cos\theta=1} - \int_0^\pi e^{ikr\cos\theta} \frac{dY_l^0}{d\theta} d\theta \right). \end{aligned} \tag{50.18}$$

Well away from the interaction region, at $kr \gg 1$, the exponential function $e^{ikr\cos\theta}$ oscillates rapidly with θ, making the integral in the second term negligibly small, leaving us with

$$g_l(kr) \rightarrow \frac{2\pi}{ikr}\left(Y_l^0(0)e^{ikr} - Y_l^0(\pi)e^{-ikr}\right), \tag{50.19}$$

at $kr \gg 1$. It will be recalled that Y_l^m has parity $(-1)^l$, so $Y_l^0(\pi) = (-1)^l Y_l^0(0)$. Finally, the value of the $l = 0$ spherical harmonic in the forward direction is

$$Y_l^0(0) = \left(\frac{2l+1}{4\pi}\right)^{1/2}. \tag{50.20}$$

Collecting, we get the wanted asymptotic form of the partial wave expansion of a plane wave,

$$e^{ikr\cos\theta} \rightarrow -i\pi^{1/2}\sum(2l+1)^{1/2}Y_l^0(\theta)\left((-1)^{l+1}\frac{e^{-ikr}}{kr} + \frac{e^{ikr}}{kr}\right). \tag{50.21}$$

This is the asymptotic form of the plane wave expansion in equation (49.8).

If an interaction region is present it can affect only the outgoing part of the scattering solution in equation (50.3). If the interaction is axially symmetric along the incident beam, it can only produce $m = 0$ waves, so the scattering changes the wave function from equation (50.21) to

$$\psi \rightarrow -i\pi^{1/2}\sum(2l+1)^{1/2}Y_l^0(\theta)\left((-1)^{l+1}\frac{e^{-ikr}}{kr} + \eta_l\frac{e^{ikr}}{kr}\right). \tag{50.22}$$

The factors η_l represent the effect of the interaction on the amplitudes and phases of the outgoing partial waves. Since the plane wave expansion in equation (50.21) has $\eta_l = 1$, we can write this expression as

$$\psi \rightarrow e^{ikz} - i\pi^{1/2}\sum(2l+1)^{1/2}Y_l^0(\theta)(\eta_l - 1)\frac{e^{ikr}}{kr}. \tag{50.23}$$

This is the standard form of equation (45.8) for a scattering state, with scattering amplitude

$$f(\theta) = -i\pi^{1/2}\sum(2l+1)^{1/2}Y_l^0(\theta)(\eta_l - 1)/k. \tag{50.24}$$

The differential scattering cross section is $d\sigma/d\Omega = |f|^2$. The total scattering cross section is the integral of this over all solid angles. On using the orthonormality of the spherical harmonics (eq. [50.16]), we get

$$\sigma_s = \int |f|^2 d\Omega = \frac{\pi}{k^2} \sum (2l+1)|\eta_l - 1|^2. \qquad (50.25)$$

For a spherically symmetric scattering potential, with no absorption, the flux of probability in each angular momentum channel is conserved. In this case, the amplitude of each outgoing wave in the scattering solution in equation (50.22) has to be the same as the amplitude of the incoming wave, $|\eta_l|^2 = 1$, so η_l is a pure phase shift:

$$\eta_l = e^{2i\delta_l}. \qquad (50.26)$$

In this case the total scattering cross section becomes

$$\sigma_s = \frac{4\pi}{k^2} \sum (2l+1) \sin^2 \delta_l. \qquad (50.27)$$

This generalizes equation (50.7) for s-wave scattering to a sum over partial waves with phase shifts δ_l.

51 Resonant *s*-Wave Scattering

The scattering cross section as a function of energy may show prominent peaks, or resonances. Here are some ways to deal with these resonances. To keep the discussion relatively simple, we will consider s-wave resonances.

Effect of a Bound State Near Zero Energy

Suppose two particles interact by the potential $V(r)$ that has a bound state of the two-particle system with $l = 0$ and binding energy $B = -E$ that is much less than the depth of the well. It will be supposed also that $V(r)$ is negligibly small beyond some radius r_o, and that r_o satisfies

$$r_o \ll \hbar/(mB)^{1/2}. \qquad (51.1)$$

As you can check, this means that when the scattering energy is comparable to the binding energy B the de Broglie wavelength of the incident particle is large compared to the range r_o of the interaction. Under these two inequalities, we can estimate the low energy scattering cross section as a function of energy as follows.

The radial wave function $u = r\psi$ for the s-wave satisfies the one-dimensional Schrödinger equation

$$-\frac{\hbar^2}{2m}\frac{d^2u}{dr^2} + Vu = Eu. \tag{51.2}$$

For the bound state near zero energy, with $E = -B$, the behavior of the radial wave function $u_b(r)$ may be very complicated within the potential at $r < r_o$, but at $r > r_o$, where $V = 0$, the solution is just

$$u_b \propto \exp -(2mB)^{1/2} r/\hbar. \tag{51.3}$$

Now suppose we increase the energy from $-B$ to a small positive value, E. This changes the exterior solution from an exponential to a sine wave,

$$u_s \propto \sin(kr + \delta_0), \tag{51.4}$$

where the energy is

$$E = \frac{\hbar^2 k^2}{2m}. \tag{51.5}$$

The constant δ_0 in this general solution to equation (51.2) with $V = 0$ and positive energy is the scattering phase shift defined in equations (50.3) and (50.4), as in equation (50.11).

Because the depth of the potential well is supposed to be large compared to the binding energy B, the change in energy in going from $E = -B$ for the bound state to a small positive energy for a low energy scattering state does not have much effect on the value of $E - V$ within the interaction region at $r < r_o$. This means the change of energy does not have much effect on the shape of the solution $u(r)$ to equation (51.2) at $r < r_o$. Therefore, the radial wave functions u_b and u_s for the bound state and the low energy scattering state must have nearly the same shape at $r \sim r_o$. This means we can find the phase shift δ_0 in u_s (eq. [51.4]) by matching the value and derivative of u_s at $r = r_o$ to the value and derivative of u_b at $r = r_o$.

The matching condition on u_b and u_s at $r = r_o$ is conveniently written as

$$\left(\frac{d}{dr}\log u_b\right)_{r_o} = \left(\frac{d}{dr}\log u_s\right)_{r_o}, \tag{51.6}$$

because this automatically eliminates the normalizing factors multiplying the functions. (The same method is used in the cold fusion calculation in eq. [9.5]). Equations (51.3) and (51.4) in equation (51.6) give

$$\tan(kr_o + \delta_0) = -\left(\frac{\hbar^2 k^2}{2mB}\right)^{1/2} = -\left(\frac{E}{B}\right)^{1/2}. \tag{51.7}$$

This is the wanted equation for the s-wave phase shift δ_0. The result is approximate because it ignores the difference between $E - V$ within the interaction region for the bound and scattering states, and because the radius r_o is not sharply defined. The latter does not matter under the assumption of equation (51.1), however, because if the energy E in the scattering solution is on the order of the binding energy B then equation (51.1) says $kr_o \ll 1$, so equation (51.7) further simplifies to

$$\tan^2 \delta_0 = E/B, \tag{51.8}$$

independent of r_o. This is equivalent to

$$\sin^2 \delta_0 = \frac{E}{E + B}. \tag{51.9}$$

The total s-wave scattering cross section (eq. [50.7]) is then

$$\sigma_s = \frac{4\pi}{k^2} \sin^2 \delta_0 = \frac{2\pi\hbar^2}{m} \frac{1}{E + B}. \tag{51.10}$$

Equation (51.5) for the energy E has been used to eliminate k^2.

Equation (51.10) says the scattering cross section is nearly constant at $E < B$, and is on the order of the square of the de Broglie wavelength defined by the binding energy B, $\lambda^2 \sim \hbar^2/(mB)$. By equation (51.1), this is much larger than the geometrical area $\sim r_o^2$ of the scattering potential. This seems paradoxical if one thinks of a classical particle, but of course the particle is moving as a wave that can have a scattering cross section as large as $4\pi/k^2$ in the s-wave channel (eq. [50.7]).

Resonances at $E > 0^*$

The scattering cross section σ_s of neutrons on an atomic nucleus can exhibit sharp peaks as a function of the scattering energy E. Here is a way to interpret these peaks, or resonances.

We seek positive energy scattering solutions to the Schrödinger equation

$$H\psi = E\psi = \frac{\hbar^2 k^2}{2m}\psi. \tag{51.11}$$

Outside the range of the interaction, the s-wave part of the solution is of the form

$$\psi = \alpha(k)\frac{e^{-ikr}}{r} + \beta(k)\frac{e^{ikr}}{r}. \tag{51.12}$$

This agrees with the s-wave part of equation (50.3), up to a multiplicative factor. Since equation (51.11) is real, there must be a real solution, for which

$$\beta(k) = \alpha^*(k), \quad \text{for real } k. \tag{51.13}$$

This allows us to define the usual s-wave phase shift,

$$\eta_0 = \beta/\alpha = \alpha^*/\alpha = e^{2i\delta_0}, \tag{51.14}$$

as in equations (50.3), (50.4), and (50.11).

Now suppose we let k be a complex number, and consider the analytic continuation of the solution (51.12) off the real k axis. This may seem absurd at first glance, but we readily find some interesting physics. Suppose $\alpha(k)$ has a zero on the upper half of the imaginary k axis, at $k = i\kappa$, with κ real and positive. At this value of k, the energy E in equation (51.11) is real and negative, because $k^2 = -\kappa^2$. Also, at $k = i\kappa$ we have $\alpha(i\kappa) = 0$, so the wave function in equation (51.12) is $\psi \propto e^{ikr} \propto e^{-\kappa r}$ (for it is unlikely that the zeros of $\alpha(k)$ and $\beta(k)$ off the real axis coincide). This is the wave function outside the potential well for an $l = 0$ bound state, as in equation (51.3). That is, the zeros of $\alpha(k)$ along the upper half of the imaginary k axis are the negative energy s-wave bound states.

Next suppose $\alpha(k)$ has a zero just below the positive real k axis. It will be convenient here to consider α as a function of energy $E = \hbar^2 k^2/2m$, which is what enters the equation (51.12) that fixes α. The zero is at $E = E_o - i\Gamma/2$, where E_o and Γ are real and positive. It will be supposed that $\Gamma \ll E_o$, so the zero is close to the real axis. The first term in the power series expansion of α around this zero is

$$\alpha = K(E - E_o + i\Gamma/2), \tag{51.15}$$

with K a constant.

If Γ is small, equation (51.15) is a good approximation to α along the real axis at energies near $E \sim E_o$. In this case, we have from equation (51.14), for E real and near E_o,

$$e^{2i\delta_0} = \frac{K^*}{K}\frac{E - E_o - i\Gamma/2}{E - E_o + i\Gamma/2}. \tag{51.16}$$

The first factor is a constant with unit modulus, which we can write as $K^*/K = e^{2i\phi_o}$. We can write the second factor as $e^{i\phi(E)}$, where a little algebra gives

$$\cos\phi(E) = \frac{(E - E_o)^2 - \Gamma^2/4}{(E - E_o)^2 + \Gamma^2/4}, \qquad \sin\phi(E) = -\frac{\Gamma(E - E_o)}{(E - E_o)^2 + \Gamma^2/4}. \tag{51.17}$$

This gives the phase shift

$$\delta_0 = \phi_o + \phi(E)/2. \tag{51.18}$$

At $E \ll E_o - \Gamma$, equation (51.17) says $\cos\phi \sim 1$ and $\sin\phi \sim 0$, so the phase $\phi(E)$ is close to $\phi \sim 0$. The phase increases to $\phi = \pi/2$ at $E = E_o - \Gamma/2$, to $\phi = \pi$ at $E = E_o$, and on to $\phi = 2\pi$ at $E \gg E + \Gamma$. This carries the s-wave phase shift δ_0 through $\pi/2$ or $3\pi/2$, where the scattering cross section (eq. [50.7]) reaches its maximum possible value, $\sigma_s = 4\pi/k^2$. For example, if $|\phi_o| \ll 1$, then $\delta_0 = \phi(E)/2$, and the scattering cross section is

$$\sigma_s = 4\pi\frac{\sin^2\delta_0}{k^2} = \frac{4\pi}{k^2}\frac{\Gamma^2/4}{(E - E_o)^2 + \Gamma^2/4}. \tag{51.19}$$

This is the Breit-Wigner resonance form.

The cross section on resonance, where $E = E_o$, is $\sigma_s = 4\pi/k^2 = \lambda^2/\pi$, where $\lambda = 2\pi/k$ is the de Broglie wavelength of the incident particle. For thermal neutrons, with energy

$$E = \frac{\hbar^2 k^2}{2m} \sim kT, \tag{51.20}$$

with $T \sim 300\,\mathrm{K}$, this works out to $\sigma_s \sim 10^{-16}\,\mathrm{cm}^2$ at the resonance, much larger than the geometrical area of the nucleus, $\sim 10^{-26}\,\mathrm{cm}^2$. This seems paradoxical if one thinks of the neutron as a particle, but of course the neutron moves as a wave.

One can understand the size of the cross section on resonance by the argument in equation (49.10). In classical mechanics, the impact parameter at angular momentum $L \sim \hbar$ and incident momentum $p \sim \hbar k$ is $r_0 = L/p \sim 1/k$. This defines a cross section area $\sim r_0^2 \sim 1/k^2$ in the incident beam within which the angular momentum is less than $\sim \hbar$. In quantum mechanics, the lowest nonzero value of the orbital angular momentum is $\sim \hbar$, so the probability flux within this cross section has to belong to the s-wave part. On resonance the s-wave flux is entirely scattered, so the cross section on resonance is $\sim 1/k^2$.

At the resonance, the particle is trapped in the interaction potential for a time on the order of $\hbar/\Gamma$. To see this, construct a wave packet, as in section 45, as a linear superposition of scattering solutions with energies in the neighborhood of the resonance energy E_o. The s-wave part of the wave packet is

$$\psi(r,t) = \int d^3k\ a(\mathbf{k}) \frac{1}{2i} \left(-\frac{e^{-ikr}}{kr} + \frac{e^{i(kr+2\delta_0)}}{kr} \right) e^{-it\hbar k^2/(2m)}. \tag{51.21}$$

The factor in parentheses is the s-wave part of equation (50.3), with the phase shift defined in equation (50.4). The last factor is the usual time dependence, with $E = \hbar^2 k^2/2m$. The function $a(\mathbf{k})$ in the integral is supposed to be appreciably different from zero only in the neighborhood of $\mathbf{k} = \mathbf{k}_o$, where $k_o = (2mE_o)^{1/2}/\hbar$. This can be compared to the wave packet constructions in equations (5.9), (7.16), and (45.4).

Now the argument proceeds in the usual way. At large r, the exponential phase factors in equation (51.21) oscillate rapidly as functions of k, making the integral negligibly small, unless the phase is stationary where the weight function $a(\mathbf{k})$ peaks up, at $k \sim k_o$. If the radius and time are such that the phase of the first exponential in equation (51.21), for the incoming wave, is stationary at $k = k_o$, we have

$$\frac{\partial}{\partial k} \left(kr + \frac{t\hbar k^2}{2m} \right) = 0, \tag{51.22}$$

or

$$r = -vt, \tag{51.23}$$

where $v = \hbar k_o/m$ is the classical velocity. This gives the position of the incoming wave; it arrives at the interaction region at time $t \sim 0$. The value of the outgoing wave is large where

$$\frac{\partial}{\partial k} \left(kr + 2\delta_0 - \frac{t\hbar k^2}{2m} \right) = 0, \tag{51.24}$$

or

$$r = vt - 2\frac{d\delta_0}{dk}. \tag{51.25}$$

Comparing equations (51.23) and (51.25), we see that the outgoing wave is delayed by the time interval

$$\delta_t = \frac{2m}{\hbar k}\frac{d\delta_o}{dk}, \tag{51.26}$$

where $v = \hbar k/m$. With the resonance equations (51.18) and (51.19) for δ_0, we see that, at $E = E_o$, the time delay is

$$\delta_t = 4\hbar/\Gamma. \tag{51.27}$$

Equation (51.27) indicates that on resonance, where $E \sim E_o$, the scattering particles form an unstable state that has a lifetime $\sim \hbar/\Gamma$. Equation (51.19) says the probability of forming this state is relatively high if the energy differs from resonance by $|E - E_o| \lesssim \Gamma$. The relation between the resonance width Γ and the lifetime $\sim \hbar/\Gamma$ was seen also in the form for the transition probability in equation (35.15).

Problems

VII.1) In a one-dimensional "scattering" or barrier penetration problem, a particle moves toward the positive x direction, initially as a free wave packet. The particle encounters a potential that is a function of x. The potential is nonzero in some finite range of x; this range of x plays the role of the interaction region. The potential transmits part of the wave packet, and reflects part. For a scattering state with definite energy, $E = \hbar^2 k^2/(2m)$, the spatial part of the wave function has the form

$$\psi_- = e^{ikx} + \alpha e^{-ikx}, \tag{VII.1}$$

to the left of the region where the potential is nonzero, and

$$\psi_+ = \beta e^{ikx}, \tag{VII.2}$$

to the right of the region of nonzero potential. Equation (VII.1) is the sum of the incident and reflected waves, and equation (VII.2) is the transmitted wave. The boundary conditions in equations (VII.1) and (VII.2) are the analog of the boundary condition in equation (45.8) for a three-dimensional scattering state.

a) Compute the probability flux densities for ψ_- and ψ_+. Use the condition that the net flux densities have to be the same to the left and right of the barrier (assuming particles are not absorbed) to find a relation between α and β.

b) Suppose the potential is

$$V(x) = A\delta(x), \tag{VII.3}$$

where A is a real constant. Here equations (VII.1) and (VII.2) apply everywhere except in an infinitesimal region around $x = 0$. Since the wave function has to be continuous, we have in this case the condition that $\psi_-(0) = \psi_+(0)$, from which it follows that $1 + \alpha = \beta$. You can get another relation between α and β by solving Schrödinger's equation for $d^2\psi/dx^2$, and then integrating this expression across an infinitesimal range of x that includes $x = 0$. Solve the two equations for α and β, show the results agree with the relation from part (a), and find the probability that a particle incident on this potential is transmitted.

VII.2) The Born approximation can be derived from first-order time-dependent perturbation theory, as follows. Consider a single particle that is scattered by a fixed potential, $V(\mathbf{r})$. Let the unperturbed part of the Hamiltonian be

$$H_o = \frac{p^2}{2m}, \tag{VII.4}$$

and adopt periodic boundary conditions in a cube of volume V_u, so the eigenstates of H_o and momentum are

$$\psi = \frac{1}{V_u^{1/2}} e^{i\mathbf{k}\cdot\mathbf{r}}, \tag{VII.5}$$

with the usual conditions $\mathbf{k} = 2\pi(n_x, n_y, n_z)/V_u^{1/3}$, as in section 1 (eq. [1.55]). The scattering potential $V(\mathbf{r})$ is treated as a perturbation to H_o. This potential does not depend on time, but of course one can still use the method of time-dependent perturbation theory.

a) Given that the initial value of the wave function at time $t = 0$ is one of the plane waves in equation (VII.5), with propagation vector $\mathbf{k} = \mathbf{k}_i$, write down the expression for the probability that at a fixed time $t > 0$ the system is observed to have momentum $\hbar\mathbf{k}_f$, that is, that the state vector is the one of the plane waves in equation (VII.5) with $\mathbf{k} = \mathbf{k}_f \neq \mathbf{k}_i$.

b) The probability that the particle is scattered from the initial momentum $\hbar\mathbf{k}_i$ to a new momentum pointing in a given direction within a small solid angle $\delta\Omega$ is the sum of the transition probabilities from part (a) over all the $\mathbf{k}_f$ contained in the solid angle $\delta\Omega$. One can simplify this sum by approximating it as an integral, as in equations (1.58) and (37.34). The integral expressed in polar coordinates is over a small range $\delta\Omega$ in the direction of $\mathbf{k}_f$ and over all values for the magnitude of $\mathbf{k}_f$. You can evaluate the integral over k_f by a suitably clever approximation and change of variables, as in section 37 (eqs. [37.35] to [37.37]). The resulting probability is equal to the product of the differential scattering cross section $d\sigma/d\Omega$ in the direction of $\mathbf{k}_f$, the element of solid angle $\delta\Omega$, the incident velocity $\hbar k_i/m$, the time interval t, and the effective number density of particles in the initial beam. Show that the result is consistent with the Born approximation in equation (47.29).

VII.3) A particle of mass m scatters off the fixed spherical potential well

$$V(r) = Ae^{-\alpha r}, \tag{VII.6}$$

where A and α are real positive constants. Find the differential scattering cross section in Born approximation.

VII.4) Consider a spherically symmetric flat potential well,

$$\begin{aligned} V(r) &= -V_o, \; r \leq r_o, \\ &= 0, \; r > r_o, \end{aligned} \tag{VII.7}$$

where V_o and r_o are real positive constants.

a) Find the differential scattering cross section for this potential in Born approximation and in the limit of low energy $E = \hbar^2 k^2/(2m)$, such that $kr_o \ll 1$.

b) Find the s-wave phase shift δ_0 produced by this potential. To simplify the calculation, consider the low energy limit, and assume the scattering potential is weak, so

$$(mV_o)^{1/2} r_o \ll \hbar. \tag{VII.8}$$

The first step is to find the solution to the one-dimensional Schrödinger equation for the s-wave radial wave function at $r < r_o$, with the right boundary condition at $r = 0$. The matching condition at $r = r_o$ gives an equation for the phase shift, δ_0. Equation (VII.8) brings this equation to a form in which it is easy to solve for δ_0 at low energy.

c) Find the low energy scattering cross section from the phase shift from part (b). The result ought to agree with what you found in part (a).

VII.5) Consider a spin 1/2 particle that is moving in the direction of the positive z axis with momentum $p = \hbar k$. The particle has positive helicity, which means the component of the spin along the direction of the momentum is $+1/2$. The wave function, $\psi(\mathbf{r}, m)$, where $m\hbar = \pm\hbar/2$ is the eigenvalue of s_z, can be written as

$$\psi(\mathbf{r}, +) \propto e^{ikz}, \qquad \psi(\mathbf{r}, -) = 0, \tag{VII.9}$$

or equivalently as

$$\psi \propto \begin{bmatrix} 1 \\ 0 \end{bmatrix} e^{ikz}. \tag{VII.10}$$

a) Use the methods of section 24 to find the wave function, in the notation of equation (VII.10), for a particle with positive helicity and momentum $\mathbf{k}$ that is in the xz plane and tilted away from the z axis by the angle θ.

b) Suppose this spin 1/2 particle scatters off a fixed pointlike potential,

$$V = A\delta(\mathbf{r}), \tag{VII.11}$$

where A is a constant. Find the Born approximation to the differential cross section $d\sigma/d\Omega$ for scattering from the initial positive helicity state in equation (VII.10) to a state moving at angle θ to the initial direction with positive helicity.

This requires a generalization of the derivation of the Born approximation in section 47. The easiest way is to note that the

Born approximation is equivalent to first-order time-dependent perturbation theory, as discussed in problem (VII.2). Thus the inner product $(\psi_f, V\psi_i)$ in equation (47.29) generalizes to $\langle f|V|i\rangle$ when the particles have spin.

c) Find in Born approximation the differential scattering cross section $d\sigma/d\Omega$ for scattering by angle θ from a positive helicity state to a negative helicity state.

VII.6) The potential energy of interaction of a neutron and proton is spin dependent, but to a reasonable approximation the potential energy is a function only of the total spin of the two particles and of the distance between the particles.

a) The triplet neutron-proton s-wave radial wave function $u_t(r)$ at zero energy is sketched as the lower solid curve in Figure VII.1. The potential energy vanishes at $r > r_o$, and is deep enough at $r < r_o$ that this radial wave function hooks over from the required value $u_t = 0$ at $r = 0$ and slopes back toward the axis at $r = r_o$. At $r > r_o$ and $E = 0$, Schrödinger's equation says $u_t(r)$ is a straight line; the line passes through $u_t = 0$ at $r = a_t$, with

$$a_t = 5.4 \times 10^{-13}\,\text{cm} = 5.4\,\text{fm}. \tag{VII.12}$$

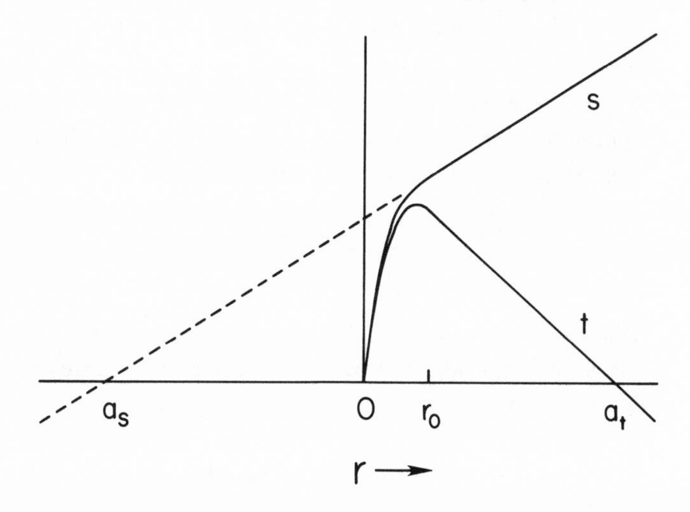

Fig. VII.1 Radial wave functions for the neutron-proton system at zero energy. The lower solid curve is the triplet s-wave function, and the upper solid curve is the singlet s-wave function.

Find the zero energy triplet s-wave phase shift (as defined in eqs. [50.3] and [50.4], or [50.11]), and use the phase shift to find the total low energy triplet scattering cross section, σ_t.

b) The singlet s-wave radial wave function $u_s(r)$ for a neutron and proton at zero energy is the upper solid curve in the figure. The interaction in the singlet state is slightly weaker than in the triplet state; the result is that the singlet radial wave function leaves the interaction region pointing away from the axis. (That is why the deuteron has spin 1: the bound state wave function is a triplet state.) The straight line part of $u_s(r)$ at $r > r_o$ extrapolates back to $u_s = 0$ at $r = a_s$, with

$$a_s = -24\,\text{fm}. \tag{VII.13}$$

Find the low energy singlet scattering cross section, σ_s.

c) Suppose a beam of neutrons with spin up scatters off target protons also with spin up. Find the low energy scattering cross section in terms of σ_t and σ_s.

d) If the beam of neutrons has spin up and the target protons have spin down, what is the low energy scattering cross section in terms of σ_t and σ_s?

e) If the beam and target are unpolarized, so the spins are randomly oriented, what is the low energy scattering cross section in terms of σ_t and σ_s?

VII.7) Particle a with mass m and zero spin is bound in a spherically symmetric three-dimensional simple harmonic oscillator potential well, so its ground state wave function is

$$\psi(\mathbf{r}_a) = \frac{1}{\pi^{3/4} r_o^{3/2}} e^{-r_a^2/(2r_o^2)}, \tag{VII.14}$$

where r_o is a constant. A second particle b with zero spin and mass m' is not affected by this potential, but it weakly interacts with particle a by the short-range potential

$$V_1 = A\delta(\mathbf{r}_a - \mathbf{r}_b), \tag{VII.15}$$

where A is a real constant, and $\mathbf{r}_a$ and $\mathbf{r}_b$ are the position variables for the two particles.

The unconfined particle b is incident with momentum $\mathbf{p} = \hbar\mathbf{k}$ along the z axis, and scatters off the confined particle a. Given that the confined particle initially is in the ground state, and ends up in the ground state, find in Born approximation how the differential scattering cross section varies with scattering angle. You need not bother to compute constants of proportionality.

CHAPTER 8

DIRAC THEORY OF THE ELECTRON

In the treatment of electrons we pulled three rabbits from the hat: an electron has spin 1/2, its magnetic dipole moment is very nearly twice that of the orbital model in which charge and mass move together, and the spin-orbit interaction is a factor of two off the value we arrived at by the heuristic argument in section 42. The factor of two in the last effect is recovered if one does the Lorentz transformations in a more careful (and correct) way, but it is easier to get it from the relativistic Dirac equation to be presented here. This equation applied to an electron also says the particle has spin 1/2, as observed, and it says the gyromagnetic ratio in equation (23.11) is $g = 2$. The small difference from the observed value, $g = 2.002\ldots$ for an electron, is accounted for by the quantum treatment of the electromagnetic field, but that goes beyond the bounds of this book.

52 Electron Spin, Magnetic Dipole Moment, and Spin-Orbit Coupling*

Relativistic de Broglie Relations

It will be recalled that in special relativity theory an event is labeled by its coordinates in space and time: x, y, z, and t. Since these quantities depend on a specific choice for the origin and orientation of the coordinate axes, their values have no fundamental significance: the same event in spacetime can be labeled by another set of quantities, x', y', z', and t', simply by using another coordinate system.

A Lorentz transformation is the set of relations between the coordinates of a given event in spacetime as measured in two coordinate systems that are moving relative to each other at some fixed velocity. If the coordinate axes are perpendicular (a cartesian coordinate system) and the relative velocity of the two coordinate systems is along the x axes, the relations are

$$\begin{aligned} t' &= \frac{t - vx}{(1 - v^2)^{1/2}}, \\ x' &= \frac{x - vt}{(1 - v^2)^{1/2}}, \\ y' &= y, \\ z' &= z. \end{aligned} \tag{52.1}$$

The two coordinate systems have parallel axes and the same origin: the event at $x = y = z = t = 0$ also has coordinate label $x' = y' = z' = t' = 0$. The primed coordinate axes are moving toward the positive x axis at speed v. To check this, note that the second equation says the spatial origin of the primed coordinates, at $x' = 0$, has position $x = vt$ in the unprimed coordinates. Finally, equation (52.1) is written in units of length and time such that the velocity of light is unity:

$$c = 1. \tag{52.2}$$

This standard convention simplifies the equations.

If two events in spacetime are labeled by the coordinates x, y, z, t, and $x + \Delta x$, $y + \Delta y$, $z + \Delta z$, $t + \Delta t$ in a given coordinate system, the invariant "distance" s between the events is given by the equation

$$s^2 = \Delta t^2 - \Delta x^2 - \Delta y^2 - \Delta z^2. \tag{52.3}$$

As you can check, the same value of s^2 is obtained if it is computed in the new coordinate system given by the Lorentz transformation (52.1). The value of s for the two given events also is left unchanged by a rotation of the spatial part of the coordinate system (which of course does not change the spatial separation $[\Delta x^2 + \Delta y^2 + \Delta z^2]^{1/2}$) or by a shift of the origin of the coordinates (which cancels out of the coordinate differences Δx).

We can attach physical significance to s because its value for the given two events is independent of the choice of coordinate system in which it is computed. If s^2 is negative, the magnitude of s is the spatial distance between the events measured by an observer moving so the events are observed to happen at the same time. If $s^2 > 0$, s is the time between the events measured by an observer who moves freely from one event to the other. If the two events are connected by a pulse of light, the pulse moving from spatial position x, y, z at time t to spatial position $x + \Delta x$, $y + \Delta y$, $z + \Delta z$ at time $t + \Delta t$, then in the time interval Δt the pulse has moved through space by the distance

$$r = (\Delta x^2 + \Delta y^2 + \Delta z^2)^{1/2} = \Delta t, \tag{52.4}$$

so $s = 0$. (Recall that we have chosen units of length and time so the velocity of light is unity.)

Another invariant is constructed from the energy and momentum of a particle,

$$\begin{aligned} E &= \frac{m}{(1 - v_1^2)^{1/2}}, \\ \mathbf{p} &= \frac{m\mathbf{v}_1}{(1 - v_1^2)^{1/2}}. \end{aligned} \tag{52.5}$$

The particle is moving with velocity

$$\mathbf{v}_1 = \frac{d\mathbf{r}_1}{dt}, \tag{52.6}$$

in the coordinate system to which these relativistic expressions for energy and momentum refer. The energy and momentum referred to a new coordinate system moving with respect to the first one are given by the Lorentz transformations (52.1), with E replacing t, p_x replacing x, and so on. One says that t, x, y, z and E, p_x, p_y, p_z both are four-vectors, transforming in the same way under a coordinate transformation. An easy way to arrive at the energy and momentum relations in equations (52.5) is to suppose the particle is at rest in the unprimed coordinates of equation (52.1). In this rest coordinate system, the energy is the annihilation energy, $E = m$, and the momentum is $p = 0$ because the particle is not moving. Then on replacing t with $E = m$ and x with $p = 0$ in the first two lines of equation (52.1), we get $E' = m/(1 - v^2)^{1/2}$ and $p' = -mv/(1 - v^2)^{1/2}$. The minus sign appears because the particle is at rest in the unprimed coordinates that are moving toward the negative x' axis at speed v.

The same algebra that showed s^2 in equation (52.3) is an invariant, that is, unaffected by a coordinate transformation, shows that another Lorentz invariant is

$$E^2 - p_x^2 - p_y^2 - p_z^2 = E^2 - p^2 = m^2. \tag{52.7}$$

The invariant here is the square of the particle rest mass, m.

We arrive at yet another invariant in quantum mechanics. In section 5 we wrote down the wave function for a free particle with definite energy and momentum E and $\mathbf{p}$ (eqs. [5.1] and [5.4]):

$$\psi \propto e^{i\phi}, \qquad \phi = \mathbf{k} \cdot \mathbf{r} - \omega t. \tag{52.8}$$

The discussion in section 5 assumes the momentum p is small (energy E, including annihilation energy m, $\gg p$), so we can do mechanics in the nonrelativistic limit. However, we can always make the momentum of the particle large (p comparable to E), for the same physical situation, simply by transforming to a coordinate frame that is moving at high speed relative to the original one. Since the phases, $\phi(\mathbf{r}, t)$, of waves determine how they interfere, and constructive or destructive interference exists independent of the choice of coordinate system in which we choose to compute it, phases have to be invariants. We assure that is the case by postulating that the quantities ω, k_x, k_y, and k_z in equation (52.8) transform as a four-vector under a change of velocity of the coordinate system, following the same Lorentz transformation law we applied to the four-vectors t, x, y, and z, and to E, p_x, p_y, and p_z. Then the algebra that showed s (eq. [52.3]) and m (eq. [52.7]) are invariants shows ϕ in equation (52.8) is invariant.

The key point of this discussion is that the quantities E, p_x, p_y, and p_z transform as a four-vector, as do the quantities ω, k_x, k_y, and k_z. In the nonrelativistic limit, where p is small compared to E, the de Broglie relations (eq. [5.4]) say the two four-vectors are proportional to each other:

$$E = \hbar\omega, \qquad \mathbf{p} = \hbar\mathbf{k}. \tag{52.9}$$

Since they transform in the same way under a change of velocity of the coordinate system, these four-vectors have to be proportional to each other (with constant of proportionality $\hbar$) whatever the particle velocity. That is, equations (52.9) are the general de Broglie relations, where E

and $\mathbf{p}$ are the relativistic energy and momentum defined in equation (52.5). This relativistic relation is used for photons in equation (5.4).

In section 6 the de Broglie relations are used to write down the nonrelativistic Schrödinger equation satisfied by a free particle. The analogous approach here is to note that, by equations (52.7) and (52.9), the de Broglie frequency and wave number satisfy

$$\hbar^2(\omega^2 - k^2) = m^2, \tag{52.10}$$

so the wave function in equation (52.8) satisfies the differential equation

$$-\hbar^2 \frac{\partial^2 \psi}{\partial t^2} + \hbar^2 \nabla^2 \psi = m^2 \psi. \tag{52.11}$$

If we wanted to use this as a relativistic Schrödinger equation, we would have to come up with expressions for the probability distribution in position, $\rho(\mathbf{r}, t)$, and the probability flux, $\mathbf{j}(\mathbf{r}, t)$, that satisfy the conservation law (eq. [8.11])

$$\frac{\partial \rho}{\partial t} + \nabla \cdot \mathbf{j} = 0. \tag{52.12}$$

Candidates are

$$\begin{aligned} \rho &= +\frac{i\hbar}{2m}\left(\psi^* \frac{\partial \psi}{\partial t} - \psi \frac{\partial \psi^*}{\partial t}\right), \\ \mathbf{j} &= -\frac{i\hbar}{2m}\left(\psi^* \nabla \psi - \psi \nabla \psi^*\right). \end{aligned} \tag{52.13}$$

One readily checks that if ψ satisfies the wave equation (52.11) then ρ and $\mathbf{j}$ satisfy the conservation law (52.12). The probability flux density $\mathbf{j}$ in equation (52.13) is the same as the nonrelativistic expression in equation (8.13), so this looks like the right relativistic generalization. However, the expression for ρ is unacceptable because it can be negative, and a probability density has to be greater than or equal to zero.

It is easy to see why we have a problem with probability density. The discussion in equations (21.12) to (21.15) indicates that Schrödinger's equation has to be first order in the time derivative, while equation (52.11) is second order. To get a first-order equation we want to write equation (52.7) as

$$E = (p^2 + m^2)^{1/2}, \tag{52.14}$$

and then replace E and $\mathbf{p}$ with time and space derivative operators. But how does one deal with the square root of an operator? Dirac's beautiful trick for taking the "square root" in equation (52.14) is discussed next.

The Dirac Equation

Dirac introduced the Hamiltonian

$$H = \vec{\alpha} \cdot \mathbf{p} + \beta m. \tag{52.15}$$

The rest mass of the particle is m, β is a constant, $\vec{\alpha}$ is a constant space vector, and the momentum operator is

$$\mathbf{p} = -i\hbar\nabla. \tag{52.16}$$

This familiar expression for the momentum operator is consistent with the second of the relativistic de Broglie relations in equation (52.9), that is, it is the proper relativistic expression for the momentum operator. Consistent with the first de Broglie relation, Schrödinger's equation is the usual form

$$i\hbar\frac{\partial\psi}{\partial t} = H\psi, \tag{52.17}$$

with

$$H\psi = E\psi, \tag{52.18}$$

for a state with definite relativistic energy E.

Now we want the energy operator H to agree with equation (52.7), so it must satisfy

$$H^2 = p^2 + m^2. \tag{52.19}$$

The square of the Dirac form for H in equation (52.15) is

$$H^2 = \alpha_a\alpha_b p_a p_b + m(\alpha_a\beta + \beta\alpha_a)p_a + m^2\beta^2. \tag{52.20}$$

The order of $\vec{\alpha}$ and β has been preserved, because to get this expression to agree with equation (52.19) we have to assume $\vec{\alpha}$ and β do not commute. But since $\vec{\alpha}$ and β are constants (independent of position) we can put them on either side of the derivative operator $\mathbf{p}$. The summation convention is used here: repeated indices are to be summed from 1 to 3, representing the x, y, and z components. Thus the dot product between vectors $\mathbf{A}$ and $\mathbf{B}$ is

$$\mathbf{A} \cdot \mathbf{B} = A_a B_a. \tag{52.21}$$

Previously we used this index notation with Greek letters $\alpha, \beta, \ldots$; Latin letters are used here to avoid confusion with Dirac's vector $\vec{\alpha}$.

The two expressions for H in equations (52.19) and (52.20) agree if

$$\beta^2 = 1,$$
$$\alpha_a\beta + \beta\alpha_a = [\alpha_a, \beta]_+ = 0, \tag{52.22}$$
$$\alpha_a\alpha_b + \alpha_b\alpha_a = [\alpha_a, \alpha_b]_+ = 2\delta_{ab}.$$

The plus sign in the subscript means an anticommutator, rather than the commutators we have been dealing with up to now. Thus

$$[A, B]_+ = AB + BA. \tag{52.23}$$

To see that the first term in the right-hand side of equation (52.20) is reduced to the wanted form, write it as

$$\frac{1}{2}(\alpha_a\alpha_b + \alpha_b\alpha_a)p_a p_b = p_a p_b \delta_{ab} = p^2. \tag{52.24}$$

The conclusion is that if Dirac's constants $\vec{\alpha}$ and β satisfy the anti-commutation relations in equation (52.22) then the square of Dirac's Hamiltonian (eq. [52.15]) satisfies the wanted relation between energy, momentum, and rest mass.

Dirac had a good model for what $\vec{\alpha}$ and β might be from the Pauli spin matrices (eq. 24.20]),

$$\sigma_x = \begin{bmatrix} 0 & 1 \\ 1 & 0 \end{bmatrix}, \qquad \sigma_y = \begin{bmatrix} 0 & -i \\ i & 0 \end{bmatrix}, \qquad \sigma_z = \begin{bmatrix} 1 & 0 \\ 0 & -1 \end{bmatrix}. \tag{52.25}$$

It will be recalled that the square of any one of these matrices is the identity matrix,

$$\sigma_a^2 = 1, \tag{52.26}$$

and that the product of two different matrices satisfies

$$\sigma_x\sigma_y = i\sigma_z, \qquad \sigma_y\sigma_x = -i\sigma_z, \tag{52.27}$$

and so on. Thus these matrices satisfy the anticommutation relations

$$[\sigma_a\sigma_b]_+ = 2\delta_{ab}. \tag{52.28}$$

This is just what is wanted for the Dirac constants (eq. [52.22]). However, there are only three Pauli matrices and we need four for β and the three components of $\vec{\alpha}$. This means we have to go to matrices of larger size; four-by-four does it.

Dirac's choices for the four-by-four matrices are

$$\alpha_a = \begin{bmatrix} 0 & \sigma_a \\ \sigma_a & 0 \end{bmatrix}, \qquad \beta = \begin{bmatrix} I & 0 \\ 0 & -I \end{bmatrix}. \tag{52.29}$$

These are two-by-two matrices each of whose elements are two-by-two matrices, where I is the identity,

$$I = \begin{bmatrix} 1 & 0 \\ 0 & 1 \end{bmatrix}. \tag{52.30}$$

Thus for example

$$\alpha_x = \begin{bmatrix} 0 & \sigma_x \\ \sigma_x & o \end{bmatrix} = \begin{bmatrix} 0 & 0 & 0 & 1 \\ 0 & 0 & 1 & 0 \\ 0 & 1 & 0 & 0 \\ 1 & 0 & 0 & 0 \end{bmatrix}. \tag{52.31}$$

Here is a notation that helps keep track of multiplications of these matrices. If A and B are matrices,

$$A = \begin{bmatrix} A_{11} & A_{12} \\ A_{21} & A_{22} \end{bmatrix}, \qquad B = \begin{bmatrix} B_{11} & B_{12} \\ B_{21} & B_{22} \end{bmatrix}, \tag{52.32}$$

then we can define a four-by-four matrix

$$A \otimes B = \begin{bmatrix} A_{11}B & A_{12}B \\ A_{21}B & A_{22}B \end{bmatrix} = \begin{bmatrix} A_{11}B_{11} & A_{11}B_{12} & A_{12}B_{11} & A_{12}B_{12} \\ A_{11}B_{21} & A_{11}B_{22} & A_{12}B_{21} & A_{12}B_{22} \\ A_{21}B_{11} & A_{21}B_{12} & A_{22}B_{11} & A_{22}B_{12} \\ A_{21}B_{21} & A_{21}B_{22} & A_{22}B_{21} & A_{22}B_{22} \end{bmatrix}. \tag{52.33}$$

That is, each element of $A \otimes B$ is the matrix element of A multiplied by the matrix B. In computing the product of $A \otimes B$ with $C \otimes D$ we can multiply out the product AC in the usual way, but we note that all the matrix elements of A are multiplied by B, and all the elements of C are multiplied by D, so the elements of the product AC all are multiplied by the matrix product BD:

$$(A \otimes B)(C \otimes D) = AC \otimes BD. \tag{52.34}$$

The Dirac matrices in equation (52.29) are

$$\alpha_a = \sigma_1 \otimes \sigma_a \qquad \beta = \sigma_3 \otimes I. \tag{52.35}$$

Here σ_1 and σ_3 are the same as σ_x and σ_z in equation (52.25); the use of numbers may help distinguish them from the matrices σ_a and I that appear in their matrix elements. As an example of the product of two of these matrices, consider

$$\alpha_x\beta = (\sigma_1 \otimes \sigma_x)(\sigma_3 \otimes I) = \sigma_1\sigma_3 \otimes \sigma_x I = -i\sigma_2 \otimes \sigma_x. \tag{52.36}$$

The product $\sigma_1\sigma_3$ is $-i\sigma_2$ (eq. [52.27]). The product of σ_x with the identity is just σ_x. The same product in the opposite order is

$$\beta\alpha_x = \sigma_3\sigma_1 \otimes I\sigma_x = i\sigma_2 \otimes \sigma_x. \tag{52.37}$$

The sum is

$$\alpha_x\beta + \beta\alpha_x = [\alpha_x, \beta]_+ = 0. \tag{52.38}$$

These matrices anticommute, as desired. It is left as an exercise to check that the Dirac matrices in equation (52.35) satisfy the other relations in equation (52.22).

Since $\vec{\alpha}$ and β are four-by-four matrices, the wave function has four components, of the form

$$\psi = \begin{bmatrix} \psi_1(\mathbf{r},t) \\ \psi_2(\mathbf{r},t) \\ \psi_3(\mathbf{r},t) \\ \psi_4(\mathbf{r},t) \end{bmatrix}. \tag{52.39}$$

As for the matrices, we can write the wave function as a two-component vector,

$$\psi = \begin{bmatrix} \psi_+ \\ \psi_- \end{bmatrix}, \tag{52.40}$$

where ψ_+ and ψ_- each are two-element column vectors. Then in the notation in equation (52.29) for $\vec{\alpha}$ and β as two-by-two matrices, each of whose elements is a two-by-two matrix, Dirac's equation for a free particle is (eqs. [52.15], [52.17], and [52.35])

$$i\hbar\frac{\partial}{\partial t}\begin{bmatrix}\psi_+\\ \psi_-\end{bmatrix} = \mathbf{p}\cdot\begin{bmatrix}0 & \vec{\sigma}\\ \vec{\sigma} & 0\end{bmatrix}\begin{bmatrix}\psi_+\\ \psi_-\end{bmatrix} + m\begin{bmatrix}I & 0\\ 0 & -I\end{bmatrix}\begin{bmatrix}\psi_+\\ \psi_-\end{bmatrix}. \qquad (52.41)$$

The two components of this equation are

$$\begin{aligned} i\hbar\frac{\partial}{\partial t}\psi_+ &= \vec{\sigma}\cdot\mathbf{p}\psi_- + m\psi_+,\\ i\hbar\frac{\partial}{\partial t}\psi_- &= \vec{\sigma}\cdot\mathbf{p}\psi_+ - m\psi_-. \end{aligned} \qquad (52.42)$$

Each of these two equations has two components.

Spin

Let us place the Dirac particle in a spherically symmetric potential well that is represented by adding the potential energy $V(r)$ to the Hamiltonian, giving

$$H = \vec{\alpha}\cdot\mathbf{p} + \beta m + V(r). \qquad (52.43)$$

To see that the particle has spin, consider angular momentum conservation. The orbital angular momentum operator for this particle is $\mathbf{L} = \mathbf{r}\times\mathbf{p}$; in the notation of equation (17.8) this is

$$L_a = \epsilon_{abc} r_b p_c. \qquad (52.44)$$

As usual, repeated indices are summed. The permutation symbol is $\epsilon_{123} = 1$, and ϵ_{abc} changes sign if indices are exchanged.

The commutator of the angular momentum component L_a with the potential $V(r)$ vanishes because V is spherically symmetric. Since βm is a constant it commutes with L_a. Thus the commutator of H in equation (52.43) with the $a^{\rm th}$ component of orbital angular momentum is

$$[H, L_a] = \epsilon_{abc}\alpha_d[p_d, r_b p_c]. \qquad (52.45)$$

We have from equation (52.16) the canonical commutation relation $[p_d, r_b] = -i\hbar\delta_{bd}$, so this is

$$[H, L_a] = -i\hbar\epsilon_{abc}\alpha_b p_c. \qquad (52.46)$$

Since this commutator does not identically vanish, orbital angular momentum is not conserved. This means there has to be another contribution to the total angular momentum.

We arrive at the angular momentum conservation law by considering the three matrices

$$\Sigma_a = I \otimes \sigma_a = \begin{bmatrix} \sigma_a & 0 \\ 0 & \sigma_a \end{bmatrix}, \tag{52.47}$$

for $a = 1, 2, 3$. Let us find the commutator of Σ_a with the Dirac Hamiltonian. The commutator of β with Σ_a is

$$[\beta, \Sigma_a] = [\sigma_3 \otimes I, I \otimes \sigma_a] = \sigma_3 \otimes [I, \sigma_a] = 0. \tag{52.48}$$

The multiplication procedure works as in equation (52.36): the identity matrix multiplied by σ_3 gives σ_3; its matrix elements are multiplied by the commutator of the identity matrix with σ_a, which vanishes. In the same way, one finds

$$[\alpha_c, \Sigma_a] = [\sigma_1 \otimes \sigma_c, I \otimes \sigma_a] = \sigma_1 \otimes [\sigma_c, \sigma_a]. \tag{52.49}$$

The commutator here is

$$[\sigma_c, \sigma_a] = 2i\epsilon_{abc}\sigma_b. \tag{52.50}$$

This follows from equation (52.27), or from the general angular momentum commutation relations (eq. [24.28]). Equation (52.49) is then

$$[\alpha_c, \Sigma_a] = 2i\epsilon_{abc}\sigma_1 \otimes \sigma_b = 2i\epsilon_{abc}\alpha_b. \tag{52.51}$$

In the last step we recognize the Dirac matrix (eq. [52.35]).

With equations (52.48) and (52.51), we see that the commutator of Σ_a with the Hamiltonian (52.43) is

$$[H, \Sigma_a] = p_c[\alpha_c, \Sigma_a] = 2i\epsilon_{abc}\alpha_b p_c. \tag{52.52}$$

This differs from the commutator of H with the orbital angular momentum component L_a by the factor $-\hbar/2$ (eq. [52.46]). This suggests we form the vector sum

$$J_a = L_a + \frac{\hbar}{2}\Sigma_a. \tag{52.53}$$

Equations (52.46) and (52.52) say the sum J_a commutes with the Hamiltonian, that is, J_a represents a conserved quantity.

The conclusion is that the orbital angular momentum $\mathbf{L}$ of a Dirac particle in a spherically symmetric potential well is not conserved, but that we get a conservation law by adding to $\mathbf{L}$ the observables called the components of the particle spin,

$$\mathbf{s}_a = \frac{\hbar}{2}\Sigma_a. \tag{52.54}$$

One sees from equations (52.47) and (52.50) that the spin operators obey the usual angular momentum commutation relations,

$$[s_a, s_b] = i\hbar\epsilon_{abc}s_c. \tag{52.55}$$

Since $\Sigma_a^2 = I$ (eq. [52.26]), the square of the spin operator defined in equation (52.54) is

$$s^2 = s_x^2 + s_y^2 + s_z^2 = 3\hbar^2/4 = \hbar \times 1/2 \times 3/2. \tag{52.56}$$

This is the eigenvalue of the square of the angular momentum operator for spin 1/2. That is, the Dirac particle has spin 1/2 (for the Σ_a are spin 1/2 matrices).

It is easy to write down the eigenvectors of the z component of the spin. The z spin matrix is (eqs. [52.47] and [52.54])

$$s_z = \frac{\hbar}{2}\begin{bmatrix} \sigma_z & 0 \\ 0 & \sigma_z \end{bmatrix}. \tag{52.57}$$

Let the two-component eigenvector of the two-by-two Pauli matrix σ_z with eigenvalue +1 be $\psi(\uparrow)$. Then in the wave function notation of equation (52.40) two eigenvectors of s_z both having eigenvalue $+\hbar/2$, are

$$\begin{bmatrix} \psi_+(\uparrow) \\ 0 \end{bmatrix} \quad \text{and} \quad \begin{bmatrix} 0 \\ \psi_-(\uparrow) \end{bmatrix}. \tag{52.58}$$

Since we need two components to represent a spin 1/2 particle, why does the Dirac equation have four components? Having arranged to express the Hamiltonian as the "square root" of the relativistic relation between energy and momentum (eq. [52.14]), we should not be surprised that the mathematics allows either sign for the square root, that is, that it allows negative energy solutions as well as positive energy ones. The two pairs of components in the wave function reflect the possibility of choosing either positive or negative energy. As will be seen, in the non-relativistic limit the positive energy solutions have ψ_- small compared to ψ_+ in equation (52.40).

Magnetic Moment

The prescription for the Hamiltonian of a particle with charge q in a magnetic field $\mathbf{B}$ was discussed in section 19: wherever the momentum operator $\mathbf{p}$ appears in the Hamiltonian in the absence of $\mathbf{B}$, replace $\mathbf{p}$ with

$$\mathbf{p} \rightarrow \mathbf{p} - q\mathbf{A}, \tag{52.59}$$

where the magnetic field is the curl of the vector potential $\mathbf{A}$ (eqs. [19.14] and [19.32]):

$$\mathbf{B} = \nabla \times \mathbf{A}. \tag{52.60}$$

The Hamiltonian for a Dirac particle of charge q in a magnetic field thus becomes (eq. [52.15])

$$H = \vec{\alpha} \cdot (\mathbf{p} - q\mathbf{A}) + \beta m, \tag{52.61}$$

where $\mathbf{p}$ is the derivative operator in equation (52.16).

With this Hamiltonian, equation (52.42) is changed to

$$\begin{aligned} i\hbar\frac{\partial}{\partial t}\psi_+ &= \vec{\sigma} \cdot (\mathbf{p} - q\mathbf{A})\psi_- + m\psi_+, \\ i\hbar\frac{\partial}{\partial t}\psi_- &= \vec{\sigma} \cdot (\mathbf{p} - q\mathbf{A})\psi_+ - m\psi_-. \end{aligned} \tag{52.62}$$

To see why this describes a particle with a magnetic dipole moment, suppose the energy E is positive and the motions are nonrelativistic, so E is close to the annihilation energy, m. Since the wave function varies with time as $\psi \sim e^{-iEt/\hbar}$, and E is close to m, we remove the dominant part of the time dependence by defining the functions $\phi_\pm$ by the equations

$$\psi_+ = \phi_+ e^{-imt/\hbar}, \qquad \psi_- = \phi_- e^{-imt/\hbar}. \tag{52.63}$$

This in equation (52.62) gives

$$
\begin{aligned}
i\hbar\frac{\partial}{\partial t}\phi_+ &= \vec{\sigma}\cdot(\mathbf{p}-q\mathbf{A})\phi_-,\\
i\hbar\frac{\partial}{\partial t}\phi_- &= \vec{\sigma}\cdot(\mathbf{p}-q\mathbf{A})\phi_+ - 2m\phi_-.
\end{aligned}
\tag{52.64}
$$

The substitution eliminates m from the first equation, but not the second, where the mass appears with the opposite sign.

The time rate of change of the $\phi_\pm$ is determined by the difference between the total energy and the annihilation energy, m. In the nonrelativistic limit this difference is small compared to m. Thus in the second of equations (52.64) we can neglect $i\hbar\partial\phi_-/\partial t$ as small compared to $m\phi_-$. This leaves

$$
\phi_- = \frac{1}{2m}\vec{\sigma}\cdot(\mathbf{p}-q\mathbf{A})\phi_+, \tag{52.65}
$$

in the nonrelativistic limit. The large denominator in this expression makes ϕ_- much smaller than ϕ_+ in this limit.

The result of substituting equation (52.65) for ϕ_- back into equation (52.64) for ϕ_+ is

$$
i\hbar\frac{\partial}{\partial t}\phi_+ = \frac{1}{2m}\sigma_a\sigma_b(p_a-qA_a)(p_b-qA_b)\phi_+. \tag{52.66}
$$

This expression is simplified by using the identity

$$
\sigma_a\sigma_b = \delta_{ab} + i\epsilon_{abc}\sigma_c. \tag{52.67}
$$

If $a = b$ this is equation (52.26); if $a = 1$ and $b = 2$ it says $\sigma_x\sigma_y = i\sigma_z$, which is equation (52.27). The second term in this identity in the right-hand side of equation (52.66) gives

$$
\begin{aligned}
\frac{i}{2m}\sigma_c\epsilon_{abc}(p_a-qA_a)(p_b-qA_b)\phi_+ &= -\frac{iq}{2m}\sigma_c\epsilon_{abc}(p_aA_b+A_ap_b)\phi_+\\
&= -\frac{q\hbar}{2m}\sigma_c\epsilon_{abc}(\nabla_aA_b-A_b\nabla_a)\phi_+\\
&= -\frac{q\hbar}{2m}\sigma_c\epsilon_{abc}\frac{\partial A_b}{\partial r_a}\phi_+\\
&= -\frac{q\hbar}{2m}\vec{\sigma}\cdot\mathbf{B}\,\phi_+.
\end{aligned}
\tag{52.68}
$$

The first step eliminates the even terms $p_a p_b$ and $A_a A_b$, because ϵ_{abc} is odd. The second step replaces the momentum with the derivative operator and exchanges the indices of summation a and b in the second term. The latter changes the sign of the term, because ϵ_{abc} is antisymmetric. This yields the commutator that produces the derivative of the vector potential in the next line. The last step is to note that we have arrived at the curl of the vector potential $\mathbf{A}$, which is the magnetic field.

This result with the first term in the identity (52.67) brings equation (52.66) to

$$i\hbar\frac{\partial\phi_+}{\partial t} = \frac{(\mathbf{p} - q\mathbf{A})^2}{2m}\phi_+ - \frac{q}{m}\mathbf{s}\cdot\mathbf{B}\phi_+, \qquad (52.69)$$

with $\mathbf{s} = \hbar\vec{\sigma}/2$, as in equation (52.57). This is a two-component wave equation, representing the two spin components of the particle. The first term on the right-hand side of this equation is the nonrelativistic form for the kinetic energy in a magnetic field (eq. [19.32]). The second term is equations (23.9) and (23.11) for the energy of a particle with gyromagnetic ratio $g = 2$. As we have noted, the slightly larger measured value of g for the electron is accounted for in a quantum treatment of the electromagnetic field.

Spin-Orbit Coupling

Consider again a Dirac particle in a spherically symmetric potential well $V(r)$, with no magnetic field, so the Hamiltonian is given in equation (52.43). The goal here is to write the energy eigenvalue equation at low energy in a form that reveals the spin-orbit coupling discussed in section 42.

Let us write the total energy of the particle as $m + E$, so the term E represents the contribution of the kinetic and potential energies. The energy eigenvalue equation written out in the notation of equation (52.42) is

$$\begin{aligned} \vec{\sigma}\cdot\mathbf{p}\psi_- + m\psi_+ + V\psi_+ &= (E+m)\psi_+, \\ \vec{\sigma}\cdot\mathbf{p}\psi_+ - m\psi_- + V\psi_- &= (E+m)\psi_-. \end{aligned} \qquad (52.70)$$

We can use the second equation to express ψ_- in terms of ψ_+:

$$\psi_- = \frac{1}{2m + E - V}\vec{\sigma}\cdot\mathbf{p}\psi_+. \qquad (52.71)$$

This in the first of equations (52.70) gives

$$\vec{\sigma} \cdot \mathbf{p} \frac{1}{2m + E - V} \vec{\sigma} \cdot \mathbf{p} \psi_+ + V \psi_+ = E \psi_+. \tag{52.72}$$

We have arrived at a two-component energy eigenvalue equation. The first term in the equation contains the product $\sigma_a \sigma_b$, which we can replace with the identity in equation (52.67). The second term in this identity gives an expression of the form

$$i \epsilon_{abc} \sigma_c p_a F p_b \psi_+. \tag{52.73}$$

In the nonrelativistic limit, the function F is

$$F = \frac{1}{2m + E - V} = \frac{1}{2m} \left(1 + \frac{V - E}{2m} \right), \tag{52.74}$$

to lowest nontrivial order in the small number $(V - E)/m$. In equation (52.73) we can write

$$p_a F p_b = [p_a, F] p_b + F p_a p_b = -i\hbar \frac{\partial F}{\partial r_a} p_b + F p_a p_b. \tag{52.75}$$

The second term vanishes in equation (52.73), because $p_a p_b$ is symmetric in a and b and ϵ_{abc} is antisymmetric. The derivative of F in equation (52.74) is

$$\frac{\partial F}{\partial r_a} = \frac{1}{4m^2} \frac{\partial V}{\partial r_a} = \frac{1}{4m^2} \frac{dV}{dr} \frac{r_a}{r}. \tag{52.76}$$

The last step assumes the potential V is a function of radius r, and uses $\partial r / \partial r_a = r_a / r$ (eq. [6.11]). On collecting all these results, equation (52.72) becomes

$$\frac{1}{2m} \left(p^2 + \mathbf{p} \cdot \frac{V - E}{2m} \mathbf{p} \right) \psi_+ + \frac{\hbar}{4m^2 r} \frac{dV}{dr} \epsilon_{abc} \sigma_c r_a p_b \psi_+ = E \psi_+. \tag{52.77}$$

We recognize $\epsilon_{abc} r_a p_b$ as L_c, the cth component of the orbital angular momentum operator. On writing the spin operator as $\mathbf{s} = \hbar \vec{\sigma}/2$, as in equation (52.69), we get finally

$$\frac{1}{2m} \left(p^2 + \mathbf{p} \cdot \frac{V - E}{2m} \mathbf{p} \right) \psi_+ + \frac{1}{2m^2 r} \frac{dV}{dr} \mathbf{L} \cdot \mathbf{s} \psi_+ = E \psi_+. \tag{52.78}$$

The second term is the same as in equation (42.11), with potential $V = q\phi = -e\phi$, where ϕ is the electrostatic potential, and the right numerical factor for the spin-orbit coupling. The first term contains the first-order relativistic correction to the kinetic energy.

Index